Physics
DeMYSTiFieD®

DeMYSTiFieD® Series

Accounting Demystified
Advanced Calculus Demystified
Advanced Physics Demystified
Advanced Statistics Demystified
Algebra Demystified
Alternative Energy Demystified
Anatomy Demystified
asp.net 2.0 Demystified
Astronomy Demystified
Audio Demystified
Biology Demystified
Biotechnology Demystified
Business Calculus Demystified
Business Math Demystified
Business Statistics Demystified
C++ Demystified
Calculus Demystified
Chemistry Demystified
Circuit Analysis Demystified
College Algebra Demystified
Corporate Finance Demystified
Databases Demystified
Data Structures Demystified
Differential Equations Demystified
Digital Electronics Demystified
Earth Science Demystified
Electricity Demystified
Electronics Demystified
Engineering Statistics Demystified
Environmental Science Demystified
Everyday Math Demystified
Fertility Demystified
Financial Planning Demystified
Forensics Demystified
French Demystified
Genetics Demystified
Geometry Demystified
German Demystified
Home Networking Demystified
Investing Demystified
Italian Demystified
Java Demystified
JavaScript Demystified
Lean Six Sigma Demystified

Linear Algebra Demystified
Macroeconomics Demystified
Management Accounting Demystified
Math Proofs Demystified
Math Word Problems Demystified
MATLAB® Demystified
Medical Billing and Coding Demystified
Medical Terminology Demystified
Meteorology Demystified
Microbiology Demystified
Microeconomics Demystified
Nanotechnology Demystified
Nurse Management Demystified
OOP Demystified
Options Demystified
Organic Chemistry Demystified
Personal Computing Demystified
Pharmacology Demystified
Physics Demystified
Physiology Demystified
Pre-Algebra Demystified
Precalculus Demystified
Probability Demystified
Project Management Demystified
Psychology Demystified
Quality Management Demystified
Quantum Mechanics Demystified
Real Estate Math Demystified
Relativity Demystified
Robotics Demystified
Sales Management Demystified
Signals and Systems Demystified
Six Sigma Demystified
Spanish Demystified
sql Demystified
Statics and Dynamics Demystified
Statistics Demystified
Technical Analysis Demystified
Technical Math Demystified
Trigonometry Demystified
uml Demystified
Visual Basic 2005 Demystified
Visual C# 2005 Demystified
XML Demystified

Physics
DeMYSTiFieD®

Stan Gibilisco

Second Edition

McGraw Graw Hill

New York   Chicago   San Francisco   Lisbon   London   Madrid   Mexico City
Milan   New Delhi   San Juan   Seoul   Singapore   Sydney   Toronto

The McGraw·Hill Companies

Cataloging-in-Publication Data is on file with the Library of Congress

Physics DeMYSTiFieD®, Second Edition

5 6 7 8 9 0   DOC/DOC   1 9 8 7 6 5

ISBN 978-0-07-174450-8
MHID 0-07-174450-9

Sponsoring Editor Judy Bass	**Proofreader** Eina Malik, Glyph International
Acquisitions Coordinator Michael Mulcahy	**Production Supervisor** Pamela A. Pelton
Editing Supervisor David E. Fogarty	**Composition** Glyph International
Project Manager Arushi Chawla, Glyph International	**Art Director, Cover** Jeff Weeks
Copy Editor Priyanka Sinha	**Cover Illustration** Lance Lekander

To Samuel, Tony, and Tim
from Uncle Stan

About the Author

Stan Gibilisco, an electronics engineer, researcher, and mathematician, has authored multiple titles for the McGraw-Hill *Demystified* series, along with numerous other technical books and dozens of magazine articles. His work has been published in several languages.

Contents

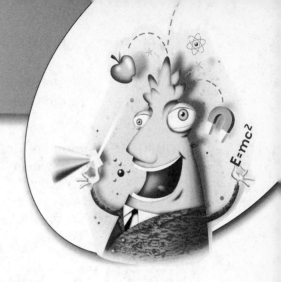

Acknowledgments

I extend thanks to my nephew Tim Boutelle, a graduate of the University of Chicago. He spent many hours helping me proofread the manuscript, and he offered insights and suggestions from the viewpoint of the intended audience.

How to Use This Book

This book can help you learn basic physics without taking a formal course. It can also serve as a supplemental text in a classroom, tutored, or home-schooling environment.

In order to learn physics, you *must* have some mathematical skill. Mathematics constitutes the *prime language* of physics. If I told you otherwise, I'd be cheating you. None of the mathematics in this book goes beyond high-school level. If you need a refresher, you can select from several *Demystified* books dedicated to mathematics topics. If you want to build a "rock-solid" mathematics foundation before you start this course, I recommend that you go through *Algebra Know-It-All* and *Pre-Calculus Know-It-All*.

This book contains abundant multiple-choice questions written in standardized test format. You'll find an "open-book" quiz at the end of every chapter. You may (and should!) refer to the chapter texts when taking these quizzes. Write down your answers, and then give your list of answers to a friend. Have your friend tell you your score, but not which questions you missed. The correct answers appear in the back of the book. Stick with a chapter until you get most of the quiz answers correct.

This book comprises three major sections. A multiple-choice test concludes each section. Take these tests when you're done with the respective sections and have taken all the chapter quizzes. Don't look back at the text when taking the section tests. They're easier than the chapter-ending quizzes, and they don't require you to memorize trivial things. A satisfactory score is at least 35 to 40 correct answers out of 50. Answers appear at the back of the book.

You'll encounter a final exam at the end of this course. The questions involve less mathematics than those in the quizzes. The final exam contains questions drawn equally from all the chapters except Chapter 0. Take this exam when you've finished all the sections, all the section tests, and all of the chapter quizzes. A satisfactory score is at least 75 correct answers out of 100.

With the section tests and the final exam, as with the quizzes, have a friend tell you your score without letting you know which questions you missed. That way, you won't subconsciously memorize the answers. You might want to take each test, and the final exam, two or three times. When you get a score that makes you happy, you can (and should!) check to see where your strengths and weaknesses lie.

Strive to complete one chapter of this book every week or 10 days. An hour or two daily ought to prove sufficient. Don't rush yourself. Give your mind time to absorb the material. But don't go too slowly either. Proceed at a steady pace and keep it up. (As much as we all wish otherwise, nothing can substitute for "good study habits.") You'll complete the course in a few months. When you're done with the course, you can use this book as a permanent reference.

I welcome your suggestions for future editions.

Stan Gibilisco

Physics
DeMYSTiFieD®

Part I

Classical Physics

chapter **0**

Review of Scientific Notation

Physicists and engineers often encounter minuscule or gigantic numbers. How many atoms compose our galaxy? What's the ratio of the volume of a marble to the volume of the known universe? If we try to denote numbers such as these in ordinary form, we must scribble long strings of numerals. There's a better way! We can take advantage of *scientific notation*. Let's review this technique.

CHAPTER OBJECTIVES

In this chapter, you will

- Express quantities as powers of 10.
- Do arithmetic with powers of 10.
- Perform numerical approximations.
- Determine degrees of accuracy.
- Truncate and round off quantities.

Subscripts and Superscripts

We can use *subscripts* to modify the meanings of units, constants, and variables. *Superscripts* usually represent *exponents* (raising to a power). Italicized, lower-case English letters from the second half of the alphabet (n through z) denote variable exponents.

Examples of Subscripts

We never italicize numeric subscripts, but in some cases we'll want to italicize alphabetic subscripts. For example:

Z_0 read "Z sub zero";
stands for characteristic impedance of a transmission line

R_{out} read "R sub out";
stands for output resistance in an electronic circuit

y_n read "y sub n";
represents a variable

Some physical constants include subscripts as a matter of convention. For example, m_e represents the mass of an electron at rest (not moving relative to the observer). Coordinate subscripting can help us to denote points in multi-dimensional space. For example, we might denote a point Q in rectangular 11-space as

$$Q = (q_1, q_2, q_3, ..., q_{11})$$

Examples of Superscripts

As with subscripts, we don't italicize numeric superscripts. However, we'll usually italicize alphabetic superscripts. For example:

2^3 read "two cubed";
represents $2 \times 2 \times 2$

e^x read "e to the xth" power;
represents the base-e exponential function of x

$y^{1/2}$ read "y to the one-half power";
represents the positive square root of y

Still Struggling

You can easily get superscripts and subscripts mixed up. If you have any doubt about the meaning of a particular expression, check the context carefully. A subscript generally tells you how to *think* about a quantity or variable, while a superscript almost always tells you what to *do* with it.

Power-of-10 Notation

Scientists and engineers express extreme numerical values using *power-of-10 notation*. That's usually what "scientific notation" means in practice.

Standard Form

We denote a numeral in *standard power-of-10 notation* as follows:

$$m.n \times 10^z$$

where the dot is a period written on the base line (not a raised dot indicating multiplication). We call this dot the *radix point* or *decimal point*. The variable m (to the left of the radix point) represents a positive integer from the set $\{1, 2, 3, 4, 5, 6, 7, 8, 9\}$. The variable n (to the right of the radix point) represents a nonnegative integer from the set $\{0, 1, 2, 3, 4, 5, 6, 7, 8, 9\}$. The variable z, which represents the power of 10, can be any integer: positive, negative, or zero. For example:

> 2.56×10^6
> represents 2.56 times exactly 1,000,000
> or 2,560,000

> 8.0773×10^{-8}
> represents 8.0773 times exactly 0.00000001
> or 0.000000080773

> 1.000×10^0
> represents 1.000 times exactly 1
> or 1.000

Alternative Form

In some countries, scientists use a variation of the preceding theme. The *alternative power-of-10 notation* requires $m = 0$. When we express the previously described quantities this way, they appear as decimal fractions greater than 0 but lesser than 1, and the value of the exponent increases by 1 compared with the standard form. Therefore, we get the following:

$$0.256 \times 10^7$$
$$0.80773 \times 10^{-7}$$
$$0.1000 \times 10^1$$

These expressions represent the same three numerical values as the previous three, although they appear different at first glance.

The "Times Sign"

You'll encounter several variations on the arithmetic "times sign" (multiplication symbol) in power-of-10 expressions. Most scientists in America use the tilted cross ($\times$), as we have done in the examples preceding. Some writers use a small dot raised above the base line ($\cdot$) to represent multiplication in power-of-10 notation. When written that way, the above numbers look as follows in the standard form:

$$2.56 \cdot 10^6$$
$$8.0773 \cdot 10^{-8}$$
$$1.000 \cdot 10^0$$

We must exercise caution when writing expressions like these, so we don't confuse the raised dot with a radix point! For example, consider

$$m.n \cdot 10^z$$

The radix point between m and n lies along the base line, while the multiplication dot between n and 10^z lies above the base line. Most scientists prefer the elevated dot symbol when expressing the dimensions of physical units. For example, we would normally denote the *kilogram-meter per second squared* as $kg \cdot m/s^2$ or $kg \cdot m \cdot s^{-2}$.

When using an old-fashioned typewriter, or in word processors that lack a good repertoire of symbols, we can use the nonitalicized letter x to indicate multiplication, but we'd better not mistake this x for the variable x (which should appear in italics, a subtle but significant difference)! To escape this confusion, we can use an asterisk (∗). Then the previous three expressions would appear as follows:

$$2.56 * 10^6$$
$$8.0773 * 10^{-8}$$
$$1.000 * 10^0$$

Plain-Text Exponents

Occasionally you'll want to express numbers in power-of-10 notation using plain, unformatted text, such as when you transmit information within the body of an e-mail message. Some calculators and computers use this scheme because their displays can't render superscripts. The uppercase letter E means that the quantity immediately following represents an exponent. In this format, you'd write the above quantities as follows:

$$2.56E6$$
$$8.0773E{-}8$$
$$1.000E0$$

Sometimes the exponent always has two numerals and includes a plus sign or a minus sign, such as the following:

$$2.56E{+}06$$
$$8.0773E{-}08$$
$$1.000E{+}00$$

Some authors use an asterisk to indicate multiplication and the symbol ^ to indicate a superscript, rendering the expressions as follows:

$$2.56 * 10{\wedge}6$$
$$8.0773 * 10{\wedge}{-}8$$
$$1.000 * 10{\wedge}0$$

Orders of Magnitude

As you can see, power-of-10 notation makes it easier to represent numbers with extreme absolute values. Consider the quantities

$$2.55 \times 10^{45,589}$$

and

$$-9.8988 \times 10^{-7,654,321}$$

Imagine the task of writing either of these numbers out in ordinary decimal form! In the first case, you'd have to write the numeral 2, then the numeral 5, then another numeral 5, and finally follow with a string of 45,587 ciphers (numerals 0). In the second case, you'd have to write a minus sign, then a numeral 0, then a radix point, then a string of 7,654,320 ciphers, and then the numerals 9, 8, 9, 8, and 8. Now consider

$$2.55 \times 10^{45,599}$$

and

$$-9.8988 \times 10^{-7,654,311}$$

Still Struggling

The expressions immediately above closely resemble the preceding ones, but don't let the resemblance deceive you! Both of these new numbers represent quantities ten thousand million (10,000,000,000) times larger than the previous two numbers depict. You can tell the difference by scrutinizing the exponents, both of which are larger by 10 in the second pair of expressions than in the first pair. (Remember that numbers grow larger in the mathematical sense as they become more positive or less negative.) The second pair of numbers measures 10 *orders of magnitude* larger than the first pair. "Ratio-wise," that's the radius of the moon's orbit compared to the span of a butterfly's wings.

Rules for Use

Most scientists use power-of-10 notation only for quantities involving integer exponents smaller than −3 or larger than 3. For example, we'll normally want to write out 2.458, 24.58, 245.8, and 2458 as decimal expressions, but we'll find that scientific notation conveniently shortens 20,458,000,000 to 2.458×10^{10}. We'll usually render 0.24, 0.024, and 0.0024 in conventional decimal form, but we'll likely express 0.00000002458 as 2.458×10^{-8}. With this rule in mind, let's see how power-of-10 notation works with arithmetic.

Addition

When we want to add two quantities to each other, we should write them out in ordinary decimal form if possible. Three examples follow:

$$(3.045 \times 10^5) + (6.853 \times 10^6)$$
$$= 304{,}500 + 6{,}853{,}000$$
$$= 7{,}157{,}500$$
$$= 7.1575 \times 10^6$$

$$(3.045 \times 10^{-4}) + (6.853 \times 10^{-7})$$
$$= 0.0003045 + 0.0000006853$$
$$= 0.0003051853$$
$$= 3.051853 \times 10^{-4}$$

$$(3.045 \times 10^5) + (6.853 \times 10^{-7})$$
$$= 304{,}500 + 0.0000006853$$
$$= 304{,}500.0000006853$$
$$= 3.045000000006853 \times 10^5$$

Subtraction

Subtraction follows the same basic rules as addition. Here's what happens when we subtract the preceding three pairs of numbers instead of adding them:

$$(3.045 \times 10^5) - (6.853 \times 10^6)$$
$$= 304{,}500 - 6{,}853{,}000$$
$$= -6{,}548{,}500$$
$$= -6.548500 \times 10^6$$

$$(3.045 \times 10^{-4}) - (6.853 \times 10^{-7})$$
$$= 0.0003045 - 0.0000006853$$
$$= 0.0003038147$$
$$= 3.038147 \times 10^{-4}$$

$$(3.045 \times 10^5) - (6.853 \times 10^{-7})$$
$$= 304{,}500 - 0.0000006853$$
$$= 304{,}499.9999993147$$
$$= 3.044999999993147 \times 10^5$$

If the absolute values of two numbers differ by many orders of magnitude, the quantity having the smaller absolute value (the one closer to zero) often loses practical significance in a sum or difference. We'll examine that phenomenon later in this chapter.

Multiplication

When we want to multiply two quantities using power-of-10 notation, we start by calculating the product of the *coefficients* (the numbers that don't have exponents). Then we add the powers of 10. Finally, we reduce the expression to scientific standard form. Three examples follow:

$$(3.045 \times 10^5) \times (6.853 \times 10^6)$$
$$= (3.045 \times 6.853) \times (10^5 \times 10^6)$$
$$= 20.867385 \times 10^{(5+6)}$$
$$= 20.867385 \times 10^{11}$$
$$= 2.0867385 \times 10^{12}$$

$$(3.045 \times 10^{-4}) \times (6.853 \times 10^{-7})$$
$$= (3.045 \times 6.853) \times (10^{-4} \times 10^{-7})$$
$$= 20.867385 \times 10^{[-4+(-7)]}$$
$$= 20.867385 \times 10^{-11}$$
$$= 2.0867385 \times 10^{-10}$$

$$(3.045 \times 10^5) \times (6.853 \times 10^{-7})$$
$$= (3.045 \times 6.853) \times (10^5 \times 10^{-7})$$
$$= 20.867385 \times 10^{[5+(-7)]}$$
$$= 20.867385 \times 10^{-2}$$
$$= 2.0867385 \times 10^{-1}$$
$$= 0.20867385$$

We can write out the last number in plain decimal form because, in scientific notation, its exponent lies within the range of −3 to 3 inclusive.

Division

When we want to divide one quantity by another in power-of-10 notation, we begin by calculating the quotient of the coefficients. Then we subtract the second power of 10 from the first one. Finally, we put the expression in standard form. Three examples follow:

$$(3.045 \times 10^5) / (6.853 \times 10^6)$$
$$= (3.045/6.853) \times (10^5/10^6)$$
$$\approx 0.444331 \times 10^{(5-6)}$$
$$\approx 0.444331 \times 10^{-1}$$
$$\approx 0.0444331$$

$$(3.045 \times 10^{-4}) / (6.853 \times 10^{-7})$$
$$= (3.045/6.853) \times (10^{-4}/10^{-7})$$
$$\approx 0.444331 \times 10^{[-4-(-7)]}$$
$$\approx 0.444331 \times 10^3$$
$$\approx 4.44331 \times 10^2$$
$$\approx 444.331$$

$$(3.045 \times 10^5) / (6.853 \times 10^{-7})$$
$$= (3.045/6.853) \times (10^5/10^{-7})$$
$$\approx 0.444331 \times 10^{[5-(-7)]}$$
$$\approx 0.444331 \times 10^{12}$$
$$\approx 4.44331 \times 10^{11}$$

Note the wavy equals signs ($\approx$) in the calculated coefficients. We can use that symbol when quotients don't divide neatly, forcing us to approximate the resultant.

Exponentiation

When we want to raise a quantity to a power using scientific notation, we raise both the coefficient and the power of 10 to that power, and then we multiply the result. Consider the following example:

$$(4.33 \times 10^5)^3$$
$$= 4.33^3 \times (10^5)^3$$
$$= 81.182737 \times 10^{(5 \times 3)}$$
$$= 81.182737 \times 10^{15}$$
$$= 8.1182737 \times 10^{16}$$

Now let's look at an example containing a negative power of 10:

$$(5.27 \times 10^{-4})^2$$
$$= 5.27^2 \times (10^{-4})^2$$
$$= 27.7729 \times 10^{(-4 \times 2)}$$
$$= 27.7729 \times 10^{-8}$$
$$= 2.77729 \times 10^{-7}$$

Taking Roots

When we want to find the root of a number expressed in scientific notation, we can think of the root as a fractional exponent. For example, a positive square root becomes the $1/2$ power; a cube root becomes the $1/3$ power. With the help of this tactic, we calculate the positive square root of 5.27×10^{-4} as follows:

$$(5.27 \times 10^{-4})^{1/2}$$
$$= 5.27^{1/2} \times (10^{-4})^{1/2}$$
$$\approx 2.2956 \times 10^{[-4 \times (1/2)]}$$
$$\approx 2.2956 \times 10^{-2}$$
$$\approx 0.022956$$

Still Struggling

When you see or write a fractional exponent, the *numerator* represents the *power* while the *denominator* represents the *root*. In their pure mathematical form, some roots have two values, one positive and the other negative! If the denominator in a fractional exponent equals an even number, always think of the positive value for the root. For example,

$$16^{1/2} = 4$$
$$16^{1/4} = 2$$
$$16^{3/4} = 8$$

If you want to express the negative values for roots in cases of this sort, place a minus sign in front of the entire quantity. For example,

$$-16^{1/2} = -4$$
$$-16^{1/4} = -2$$
$$-16^{3/4} = -8$$

Approximation, Error, and Precedence

In experimental physics, we rarely work with exact values, so we must usually rely on *approximation*. We can approximate the value of a quantity with *truncation* or *rounding*. When we have meaningful numbers to work with, we can calculate the extent of our *error*. As we make calculations, we must perform our operations according to the correct mathematical *precedence*.

Truncation

When we employ truncation to approximate a numerical value, we delete all the numerals to the right of a certain point. For example, we can truncate 3.830175692803 to fewer digits in steps as follows:

$$3.830175692803$$

$$3.83017569280$$

$$3.8301756928$$

$$3.830175692$$

$$3.83017569$$

$$3.8301756$$

$$3.830175$$

$$3.83017$$

$$3.8301$$

$$3.830$$

$$3.83$$

$$3.8$$

$$3$$

Rounding

When rounding, we must follow a more complex procedure, but we get a better approximation. When we delete a particular digit (call it r) at the right-hand extreme of an expression, we don't change the digit q to its left (which becomes the new r after we delete the old r) if $0 \leq r \leq 4$. However, if $5 \leq r \leq 9$, then we increase q by 1 (we "round it up"). We can round 3.830175692803 to fewer digits in steps as follows:

$$3.830175692803$$
$$3.83017569280$$
$$3.8301756928$$
$$3.830175693$$
$$3.83017569$$
$$3.8301757$$
$$3.830176$$
$$3.83018$$
$$3.8302$$
$$3.830$$
$$3.83$$
$$3.8$$
$$4$$

Error

When experimenters measure physical quantities, errors occur because of imperfections in the instruments (and in the people who use them!). Suppose that x_a represents the actual value of a quantity that we measure in an experiment. Let x_m represent the measured value of that quantity, in the same units as x_a. We can calculate the *absolute error*, D_a (in the same units as x_a), as follows:

$$D_a = x_m - x_a$$

We determine the *proportional error*, D_p, when we divide the absolute error by the actual value of the quantity:

$$D_p = (x_m - x_a)/x_a$$

The *percentage error*, $D_\%$, equals 100 times the proportional error:

$$D_\% = 100(x_m - x_a)/x_a$$

Error values and percentages turn out positive if $x_m > x_a$, and negative if $x_m < x_a$. If the measured value exceeds the actual value, we have a positive error. If the actual value exceeds the measured value, we have a negative error.

When you want to express a positive error figure, you should precede it with a plus sign (+). When you want to express a negative error, you should precede it with a minus sign (−). Once in a while, you'll find a quoted specification in which the error spans a range on either side of a certain reference value. Then you'll see a plus-or-minus sign (±) in front of the error figure.

Still Struggling

The denominators of all three of the foregoing equations contain x_a, the actual value of the quantity under scrutiny—a quantity that we don't exactly know! You must wonder, "How can we calculate an error based on formulas involving a quantity that suffers from that very uncertainty?" That's an astute question! We must make an *educated guess* at the value of x_a. To get the most accurate possible result, we must make as many measurements, during the course of our experiments, as we have time to make. Then we can average all of these values to make a good educated guess at x_a.

Precedence

Mathematicians, scientists, and engineers agree on a certain order for the performance of operations that occur together in a single expression. This agreement prevents confusion and ambiguity. When various operations such as addition, subtraction, multiplication, division, and exponentiation appear in an expression, and if you need to simplify that expression, you should do the operations in the following sequence:

- Simplify all expressions within parentheses, brackets, and braces from the inside out.
- Perform all exponential operations going from left to right.
- Perform all products and quotients going from left to right.
- Perform all sums and differences going from left to right.

Let's look at two examples of expressions, simplified according to the above rules of precedence. The order of the numerals and operations is the same in each case, but the groupings differ:

$$[(2+3)(-3-1)^2]^2$$
$$= [5 \times (-4)^2]^2$$
$$= (5 \times 16)^2$$
$$= 80^2$$
$$= 6400$$

$$[(2+3 \times (-3) - 1)^2]^2$$
$$= [(2 + (-9) - 1)^2]^2$$
$$= (-8^2)^2$$
$$= 64^2$$
$$= 4096$$

Suppose that you encounter a complicated expression containing no parentheses, brackets, or braces? You can avoid ambiguity if you adhere to the previously mentioned rules. Consider the equation

$$z = -3x^3 + 4x^2y - 12xy^2 - 5y^3$$

When you rewrite this equation with plenty of grouping symbols (too many, some people might say) to emphasize the rules of precedence, you obtain

$$z = [-3(x^3)] + \{4[(x^2)y]\} - \{12[x(y^2)]\} - [5(y^3)]$$

If you have doubt about a scientific equation, you're better off using some unnecessary grouping symbols than risking a massive mistake in your calculations.

Significant Figures

When we do multiplication or division using power-of-10 notation, we must never let the number of significant figures in our answer exceed the number of significant figures in the least-exact factor. Consider, for example, the numbers

$$x = 2.453 \times 10^4$$

and

$$y = 7.2 \times 10^7$$

In plain arithmetic, we can say that

$$xy = 2.453 \times 10^4 \times 7.2 \times 10^7$$
$$= 2.453 \times 7.2 \times 10^{11}$$
$$= 17.6616 \times 10^{11}$$
$$= 1.76616 \times 10^{12}$$

If x and y represent measured quantities, as they would in experimental physics, the above statement needs qualification. We can't claim more than a certain amount of accuracy.

How Accurate Are We?

When you see a product or quotient containing numbers in scientific notation, count the number of single digits in each coefficient. Then look at the quantity whose coefficient contains the smallest number of digits. That's the number of significant figures you can claim in the final answer or solution.

In the foregoing example, the coefficient of x has four digits, and the coefficient of y has two digits. You must therefore round off (not truncate) the product to two significant figures. When you do that, you'll get

$$xy = 2.453 \times 10^4 \times 7.2 \times 10^7$$
$$= 1.76616 \times 10^{12}$$
$$\approx 1.8 \times 10^{12}$$

Most experimentalists use ordinary equals signs instead of wavy equals signs when rounding off to the appropriate number of significant figures. Henceforth, let's use plain, straight equals signs when we reduce expressions to the correct number of significant figures.

Suppose that you want to find the quotient x/y instead of the product xy. You can proceed as follows:

$$x/y = (2.453 \times 10^4) / (7.2 \times 10^7)$$
$$= (2.453/7.2) \times 10^{-3}$$
$$= 0.3406944444 \ldots \times 10^{-3}$$
$$= 3.406944444 \ldots \times 10^{-4}$$
$$= 3.4 \times 10^{-4}$$

What about "Extra Zeros?"

Sometimes, when you make a calculation, you'll get an answer that lands on a neat, seemingly whole-number value. Consider the quantities

$$x = 1.41421$$

and

$$y = 1.41422$$

Both of these numbers are expressed to six significant figures. When you multiply them and take the significant figures into account, you'll obtain

$$xy = 1.41421 \times 1.41422$$
$$= 2.0000040662$$
$$= 2.00000$$

The five "extra zeros" tell you how close to the exact value of 2 you should imagine the real-world product. In this case, an uncertainty of ± 0.000005 exists. Those five zeros mean that the product equals at least 1.999995 but less than 2.000005. In pure mathematical terms, you can write

$$1.999995 \leq xy < 2.000005$$

If you truncate all the zeros and simply state that $xy = 2$, you technically allow for an uncertainty of up to ± 0.5. That would provide a range

$$1.5 \leq xy < 2.5$$

You're entitled to a lot more accuracy than that in this situation. When you claim a certain number of significant figures, you must give "extra zeros" as much consideration as you give any other digit. Always write down enough zeros (but never too many).

In Addition and Subtraction

When you add or subtract measured quantities, you might need to make subjective judgments when determining the allowable number of significant figures. For best results in most situations, you should expand all the values out to their plain decimal form (if possible), make the calculation, and then, at the end of the process, decide how many significant figures you can reasonably claim.

In some cases, the outcome of determining significant figures in a sum or difference resembles what happens with multiplication or division. Take, for example, the sum $x + y$, where

$$x = 3.778800 \times 10^{-6}$$

and

$$y = 9.22 \times 10^{-7}$$

When you expand the values out to plain decimal form, you obtain

$$x = 0.000003778800$$

and

$$y = 0.000000922$$

Adding these two numbers directly, you get

$$x + y = 0.000003778800 + 0.000000922$$

$$= 0.0000047008$$

$$= 4.7008 \times 10^{-6}$$

$$= 4.70 \times 10^{-6}$$

Here's a Twist!

In some instances, one of the values in a sum or difference has no significance with respect to the other. Suppose that you encounter the values

$$x = 3.778800 \times 10^{4}$$

and

$$y = 9.225678 \times 10^{-7}$$

Expanding to decimal form, you get

$$x = 37,788.00$$

and

$$y = 0.0000009225678$$

The arithmetic sum, including all the digits, works out as

$$x + y = 37,788.00 + 0.0000009225678$$
$$= 37,788.0000009225678$$
$$= 3.77880000009225678 \times 10^4$$

In this scenario, the value of y doesn't "play into" the value of the sum when you round the answer off. You might think of y, in relation to x or to the sum $x + y$, as a snowflake falling onto a mountain. If a snowflake lands on a mountain, the mass of the mountain does not appreciably change. For real-world purposes, the sum equals the larger quantity x, while the smaller quantity y constitutes a mere nuisance (if you notice it at all). Therefore, you're perfectly well justified to state that

$$x + y = 3.778800 \times 10^4$$

Still Struggling

Imagine that you encounter two quantities expressed in scientific notation, with the absolute value of one quantity many orders of magnitude larger than the absolute value of the other quantity (as in the foregoing scenario). Suppose, as before, that the accuracy extends to different numbers of significant figures in each expression. In a case like this, you should claim the number of significant figures in the quantity whose *absolute value* is *larger*. Consider the sum of

$$x = -3.778800 \times 10^4$$

and

$$y = 9.2 \times 10^{-7}$$

In plain decimal form, these quantities are

$$x = -37,788.00$$

and

$$y = 0.00000092$$

Before taking significant digits into account, the arithmetic sum works out as

$$x + y = -37{,}788.00 + 0.00000092$$

$$= -37{,}787.99999908$$

$$= -3.778799999908 \times 10^4$$

This quantity essentially equals x; the much smaller (in absolute-value terms) y vanishes into insignificance. You can claim all seven significant figures that appear in the original expression of x, but no more. You should round off the last expression in the above equation to obtain

$$x + y = -3.778800 \times 10^4$$

QUIZ

Refer to the text in this chapter if necessary. A good score is eight correct. Answers are at the back of the book.

1. What's the arithmetic value of $-4 \times 2^2 + 16$? Don't worry about rounding or significant figures.

   A. 16
   B. 80
   C. 0
   D. You can't define this expression because it's ambiguous.

2. What's the difference $7.776 \times 10^5 - 6.2 \times 10^{-17}$, taking significant figures into account?

   A. 7.776×10^5
   B. 7.78×10^5
   C. 7.8×10^5
   D. You'll need more information to work this out.

3. Suppose that you encounter the quantity 0.045, and you want to express it in a better way. What should you write?

   A. 4.5×10^{-2}
   B. 4.5E–02
   C. 4.5*10^–2
   D. Nothing! The expression is okay as stated.

4. Your professor tells you that two quantities differ in size by precisely four orders of magnitude. Based on this information, you know that one of the quantities is

   A. 10,000 times as large as the other.
   B. larger than the other by 10,000.
   C. four times as large as the other.
   D. 1/4 as large as the other.

5. Imagine that you use your Web browser to measure the download speed of your Internet connection and come up with a value of 14.05 megabits per second (Mbps). Suppose the speed of the connection holds absolutely constant over time. You call in a professional computer engineer who tests your Internet connection and obtains a speed reading of 13.88 Mbps, which you regard as the "actual value." What's the error of the speed measurement that you made using your Web browser, expressed to the nearest percentage point?

   A. 13.88 / 14.05, or +99 %
   B. (14.05 – 13.88) / 13.88, or +1.2 %
   C. (13.88 – 14.05) / 14.05, or –1.2 %
   D. 14.05 / 13.88, or +101%

6. Suppose that you measure a quantity, call it q, and get 1.5×10^4 units. Within what range of whole-number values can you conclude that q actually lies?

    A. $14{,}950 \leq q < 15{,}050$
    B. $14{,}500 \leq q < 15{,}500$
    C. $14{,}000 \leq q < 16{,}000$
    D. $10{,}000 \leq q < 20{,}000$

7. Suppose that you measure a quantity, call it q, and get 1.50×10^4 units. Within what range of whole-number values can you conclude that q actually lies?

    A. $14{,}995 \leq q < 15{,}005$
    B. $14{,}950 \leq q < 15{,}050$
    C. $14{,}500 \leq q < 15{,}500$
    D. $14{,}000 \leq q < 16{,}000$

8. What's the product of 9×10^5 and 1.1494×10^{-6}, taking significant figures into account?

    A. 0.99978
    B. 0.99
    C. 0.9
    D. 1

9. While reading a technical paper, you see a complicated arithmetic expression of sums, differences, and products with no grouping symbols (parentheses, brackets, or braces) whatsoever. If you want to figure out the value of the expression, what should you do first?

    A. Calculate the left-most sum.
    B. Calculate the left-most difference.
    C. Calculate the left-most product.
    D. Execute the left-most operation no matter what it is.

10. Suppose that a certain country contains 12,006,000 people. What's this quantity, rounded and expressed to four significant figures in the power-of-10 notation that a computer might use?

    A. 1.200E+07
    B. 1.201E+07
    C. 1.2006E+07
    D. 1.210E+07

chapter 1

Units and Constants

Units allow us to quantify physical objects and effects. For example, we think of time in *seconds, minutes, hours, days,* or *years*. We define mass in *grams* or *kilograms*, electric current in *amperes*, and visual-light intensity in *candela*.

CHAPTER OBJECTIVES

In this chapter, you will

- Review the common unit systems.
- Combine basic units to get derived units.
- Learn fundamental physical constants.
- Do arithmetic with units and constants.
- Convert different unit systems.

Systems of Units

At least three systems of physical units exist. Most scientists and engineers favor the *meter/kilogram/second (mks) system*, also called the *metric system* or the *International System*. A few people prefer the *centimeter/gram/second (cgs) system*. Physicists rarely talk or write about the *foot/pound/second (fps) system*, also called the *English system*, although lay people in some countries cling to it with fondness bordering on the irrational!

The International System (SI)

We can denote the International System as *SI*, a French abbreviation for *Système International*. The fundamental or *base* units in SI quantify *displacement*, *mass*, *time*, *temperature*, *electric current*, *luminous intensity* (visible-light brightness), and *material quantity* (the number of atoms or molecules in a sample):

- The SI base unit of displacement is the *meter*.
- The SI base unit of mass is the *kilogram*.
- The SI base unit of time is the *second*.
- The SI base unit of temperature is the *kelvin*.
- The SI base unit of electric current is the *ampere*.
- The SI base unit of luminous intensity is the *candela*.
- The SI base unit of material quantity is the *mole*.

The cgs System

In the centimeter/gram/second (cgs) system, the base units are the *centimeter* (exactly 0.01 meter), the *gram* (exactly 0.001 kilogram), the second, the *degree Celsius* (approximately the number of kelvins minus 273), the ampere, the candela, and the mole. The second, the ampere, the candela, and the mole are identical in cgs and SI.

The fps System

In the fps system, the base units are the *foot* (approximately 30.5 centimeters), the *pound* (equivalent to about 0.454 kilograms for an object at the earth's surface), the second, the *degree Fahrenheit* (where water freezes at 32 degrees and boils at 212 degrees at sea-level atmospheric pressure), the ampere, the candela, and the mole. The second, the ampere, the candela, and the mole are identical in fps and SI.

Still Struggling

Does the fps system seem illogical to you? If so, you have good company. Scientists and engineers dislike the fps system because its units break down into components whose sizes vary, and whose definitions defy easy recall. For example, an object that weighs a pound consists of 16 *ounces avoirdupois* (16 oz av), whereas 32 *fluid ounces* (32 fl oz) give us a *quart* (1 qt) and 128 fl oz yield a *gallon* (1 gal). In contrast, the units in the SI and cgs systems nearly always break down or build up neatly as powers of 10.

Base Units

In any system of measurement, we can derive all the units from the base units, which represent the elementary properties or phenomena we see in the natural world. Let's take a close look at how scientists define the SI base units.

The Meter

The SI base unit of distance is the meter, symbolized by the nonitalicized, lowercase English letter m. While walking at a brisk pace, an adult's full stride spans approximately 1 m.

Originally, people defined the meter as the distance between two scratches on a platinum bar displayed in Paris, France. That span, multiplied by 10 million (10,000,000 or 10^7), would go from the earth's north pole through Paris to the equator over the earth's surface (Fig. 1-1).

Nowadays, the meter is defined as the distance over which a ray of light travels through a vacuum in 3.33564095 thousand-millionths (0.0000000033564095 or $3.33564095 \times 10^{-9}$) of a second.

The Kilogram

The SI base unit of mass is the kilogram, symbolized by the lowercase, nonitalicized English letters kg. Originally, scientists defined the kilogram as the mass of 0.001 cubic meter (or 1 *liter*) of pure liquid water (Fig. 1-2). That's still a good definition, but these days, scientists define 1 kg as the mass of a sample of platinum-iridium alloy kept under lock and key at the International Bureau of Weights and Measures.

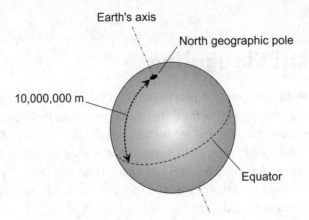

FIGURE 1-1 • The distance from the earth's north geographic pole to the equator, as measured over the surface, is approximately 10,000,000 m.

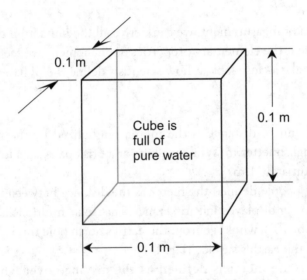

FIGURE 1-2 • A cube of pure water measuring 0.1 m on each edge masses a kilogram (not counting the mass of the cubical container, of course).

Mass does *not* constitute *weight*. An object having a mass of 1 kg always has a mass of 1 kg, no matter where we take it. That standard platinum-iridium ingot would mass 1 kg on the moon, on Mars, on Mercury, or in interplanetary space. Weight expresses the force exerted by gravitation or acceleration on an object having a specific mass.

On the surface of the earth, an object having a mass of 1 kg weighs about 2.2 pounds. But on board a vessel coasting through interplanetary space, that same object would weigh nothing. A 1-kg mass would weigh something less than 2.2 pounds on the moon, on Mars, or on Mercury; it would weigh something more than 2.2 pounds on Jupiter or on the sun (if those bodies had solid surfaces).

The Second

The SI base unit of time is the second, symbolized by the lowercase, nonitalic English letter s (or sometimes abbreviated as sec). Informally, we can define 1 s as 1/60 of a *minute* (min), which is 1/60 of an *hour* (h), which is 1/24 of a *mean solar day* (msd). Figure 1-3 illustrates these relationships. As things work out, 1 s = 1/86,400 msd. Scientists formally define 1 s in a more sophisticated fashion: the amount of time it takes for a certain cesium atom to oscillate through 9,192,631,770 (9.192631770×10^9) complete electromagnetic cycles.

In a time interval of precisely 1 s, a ray of light travels a distance of 2.99792458×10^8 m through outer space. That's roughly three-quarters of the way from the earth to the moon. The moon therefore has an orbital radius of slightly more than 1 *light-second*. Are you old enough to recall the conversations between earth-based personnel and moon-based Apollo astronauts in the 1970s?

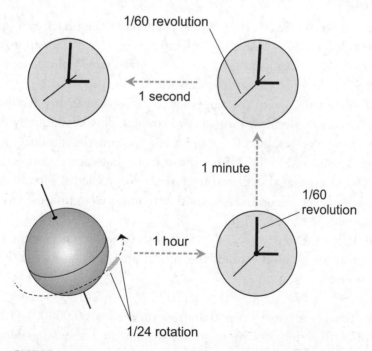

FIGURE 1-3 · One second equals 1/60 of 1/60 of 1/24 (that is, 1/86,400) of a mean solar day.

When "Houston" asked a question, the astronauts' response never came back until more than 2 s had passed. It took that long for the radio signals to make a round trip between the earth and the moon.

The Kelvin

The SI base unit of temperature is the kelvin, symbolized by the uppercase, nonitalic English letter K. A figure in kelvins tells us the temperature relative to *absolute zero*, which represents the complete absence of heat (the coldest possible temperature). Kelvin temperatures therefore cannot drop below zero.

Formally, scientists define the kelvin as a temperature increment (an increase or decrease) of 0.003661 part of the thermodynamic temperature of the *triple point* of pure water. Pure water at sea level freezes (or melts) at 273.15 K, and boils (or condenses) at 373.15 K.

What, you ask, do we mean by the term "triple point"? For water, it's the temperature and pressure at which the compound can exist as vapor, liquid, and ice in equilibrium. For most practical purposes, you can think of the triple point as the freezing or melting point.

The Ampere

The ampere, symbolized by the nonitalic, uppercase English letter A (or abbreviated as amp), is the SI base unit of electric current. A flow of approximately 6.241506×10^{18} electrons per second, past a given fixed point in an electrical conductor, produces an electric current of 1 A.

The formal definition of the ampere involves a bit of theory. It's the rate of constant charge-carrier flow through two straight, parallel, infinitely thin, perfectly conducting wires, placed 1 m apart in a vacuum, that produces a force of 0.0000002 *newton* (2×10^{-7} N) per linear meter between the wires. If you're astute, you'll see that this definition poses two problems! First, we haven't defined *newton* yet. Second, this definition requires us to imagine *ideal objects* that can't exist in the real world.

We'll define the newton shortly. Ideal objects, while not tangible in themselves, constitute an ideal tool for theoretical physicists, so we had better get used to them.

A *milliampere* (mA) equals 0.001 (10^{-3}) A, or 6.241506×10^{15} electrons passing a fixed point in 1 s. A *microampere* (μA) equals 0.000001 (10^{-6}) A, or 6.241506×10^{12} electrons passing a fixed point in 1 s. A *nanoampere* (nA) equals 0.000000001 (10^{-9}) A, or 6.241506×10^9 electrons passing a fixed point in 1 s.

The Candela

The candela, symbolized by the nonitalicized, lowercase English letters cd, constitutes the SI base unit of luminous intensity. It's equivalent to 1/683 of a *watt* of radiant power, emitted at a frequency of 5.4×10^{14} *hertz* (cycles per second), in a solid angle of 1 *steradian*. (We'll define the steradian shortly.) We specify that frequency because it's where the human eye has the greatest sensitivity.

An alternative reference doesn't force us to rely on derived units. According to this definition, 1 cd equals the amount of power emitted by a perfectly radiating surface (or *blackbody*) having an area of 0.000001667 square meter (1.667×10^{-6} m²) at the solidification temperature of platinum.

A third, informal definition tells us that 1 cd represents approximately the amount of light that a common candle flame emits.

The Mole

The mole, symbolized by the nonitalicized, lowercase English letters mol, is the SI base standard unit of material quantity. Chemists sometimes call it *Avogadro's number*, a quantity equal to about 6.022169×10^{23} and representing the number of atoms in 0.012 kg of *carbon 12*, the most common natural form of elemental carbon.

The mole is one of those mysterious fixed numbers that nature, for reasons unknown (and perhaps unknowable), seems to favor. If not for the unique properties of the mole, scientists of earlier times would probably have chosen a more "sensible" standard constant for material quantity such as a thousand, a million, or a "baker's dozen."

Still Struggling

Until now, we've rigorously reiterated the fact that symbols and abbreviations consist of lowercase or uppercase, nonitalicized letters or strings of letters. That's important! If we get careless about case or italics, we risk confusing physical units with the constants and variables that appear in mathematical equations.

When you see a letter in italics, it usually represents a constant or variable. When nonitalicized, a letter typically represents a physical unit. For example, the nonitalic letter s means "second" or "seconds," while the italic *s* can stand for displacement or physical separation. From now on, we won't belabor this issue every time a unit symbol or abbreviation comes up—but let's never forget it!

Derived Units

We can combine the foregoing units to get *derived units*. The technical expressions for derived units sometimes seem strange at first glance, such as seconds cubed (s^3) or kilograms to the minus-one power (kg^{-1}).

The Radian

The standard unit of *plane angular measure* is the *radian* (rad). Suppose that you take a string and run it out from the center of a circular disk to some point on the edge, and then lay that string down around the periphery of the disk. Between the ends of the string, you'll have an angle of 1 rad at the center of the disk.

You might also think of a radian as the angle between the two straight sides of a wedge within a circle of radius r, such that the wedge's straight and curved edges all have length r as shown in Fig. 1-4. In a flat geometric plane, an angle of 1 rad measures $360/(2\pi)$, or about 57.2958 *angular degrees*.

The Angular Degree

The angular degree, symbolized by a small elevated circle (°) or by the three-letter abbreviation deg, equals 1/360 of a complete circle. The history of the degree presents a good subject for debate! According to one popular theory, ancient Egyptian and Greek mathematicians invented this unit to represent a day in the course of a calendar year. (If that story represents the truth, our

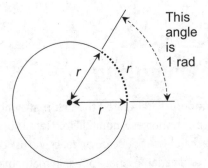

FIGURE 1-4 · One radian comprises the angle at the apex of a wedge within a circle, such that the lengths of the straight and curved edges all equal the circle's radius r.

distant ancestors slightly miscounted the number of days in a year, either accidentally or on purpose!) One angular degree measures 2π divided by 360, or approximately 0.0174533 radians.

The Steradian

The standard unit of *solid angular measure* is the *steradian*, symbolized sr. You can think of a steradian as a "conical angle" encompassing a defined region in space. Imagine a cone whose apex lies at the center of a sphere. Suppose that the cone intersects the sphere's surface in a circle such that, within the circle, the enclosed area on the sphere's surface equals the square of the sphere's radius, as shown in Fig. 1-5. In this situation, the *solid angle* at the cone's apex measures 1 sr. A complete sphere contains 4π, or approximately 12.56636 steradians.

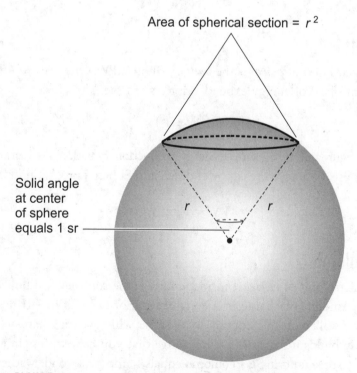

Area of spherical section = r^2

Solid angle
at center
of sphere
equals 1 sr

r r

FIGURE 1-5 • One steradian comprises a solid (or "conical") angle defining the area on a sphere that equals the square of the sphere's radius r^2.

The Newton

The standard unit of *mechanical force* is the *newton*, which we can symbolize as N. In the absence of friction, an applied force of 1 N causes a 1-kg object to accelerate through space at 1 meter per second per second (1 m/s^2). Engineers express jet and rocket engine propulsion in newtons. When we break newtons down into SI base units, we get *kilogram-meters per second squared*, so that

$$1 \text{ N} = 1 \text{ kg} \cdot \text{m/s}^2$$

The Joule

The standard unit of *energy* is the *joule*, symbolized J. In real-world terms, 1 J doesn't represent much energy. It's the equivalent of a newton-meter (n · m). When we reduce energy to base units in SI, we get mass times distance squared divided by time squared. Mathematically, we have

$$1 \text{ J} = 1 \text{ kg} \cdot \text{m}^2/\text{s}^2$$

The Watt

The standard unit of *power* is the *watt*, symbolized W. One watt represents the equivalent of 1 J of energy expended for 1 s of time. Therefore

$$1 \text{ W} = 1 \text{ J/s}$$

A power figure tells us the rate of energy production, radiation, dissipation, expenditure, or use. When we break power down into SI base units, we find that

$$1 \text{ W} = 1 \text{ kg} \cdot \text{m}^2/\text{s}^3$$

The Coulomb

The standard unit of *electric charge quantity* is the *coulomb*, symbolized C. It's the total amount of charge in a congregation of 6.241506×10^{18} electrons. When you walk along a carpet in dry weather while wearing hard-soled shoes, your body builds up a *static electric charge* that you can express in coulombs. Reduced to base units in SI, 1 coulomb equals 1 ampere-second. Mathematically, we have

$$1 \text{ C} = 1 \text{ A} \cdot \text{s}$$

The Volt

The standard unit of *electric potential* or *potential difference*, also called *electromotive force* (EMF), is the *volt*, symbolized V. Volts represent joules per coulomb, so we can say that

$$1 \text{ V} = 1 \text{ J/C}$$

In real-world terms, 1 V constitutes a small electric potential. A standard *dry cell* of the sort you find in a flashlight (often erroneously called a "battery") produces about 1.5 V. Most automotive batteries in the United States produce around 13.5 V.

The Ohm

The standard unit of *electrical resistance* is the *ohm*, symbolized by the uppercase Greek letter omega (Ω) or written out in full as ohm or ohms. If we apply 1 V of EMF across a component having a resistance of 1 ohm, then a current of 1 A flows through that component. Ohms represent volts per ampere, therefore

$$1 \text{ ohm} = 1 \text{ V/A}$$

The Siemens

The standard unit of *electrical conductance* is the *siemens*, symbolized S. In the olden days (that is, before about 1960), engineers called it the *mho*. (You'll still come across that term in some papers and texts.) The conductance of a component in siemens equals the reciprocal of that component's resistance in ohms. Siemens represent amperes per volt. Mathematically, we have

$$1 \text{ S} = 1 \text{ A/V}$$

The Hertz

The standard unit of *frequency* is the *hertz*, symbolized Hz and formerly called the *cycle per second* or simply the *cycle*. The hertz constitutes a tiny unit in everyday experience; 1 Hz represents an extremely low frequency. In most real-world situations, sounds and signals attain frequencies of thousands, millions, billions (thousand-millions), or trillions (million-millions) of hertz, giving us larger frequency units as follows:

- 1000 (10^3) Hz represents a *kilohertz* (1 kHz).
- 1,000,000 (10^6) Hz represents a *megahertz* (1 MHz).

- 1,000,000,000 (10^9) Hz represents a *gigahertz* (1 GHz).
- 1,000,000,000,000 (10^{12}) Hz represents a *terahertz* (1 THz).

In terms of SI units, the hertz breaks down so simply that it confuses some people when they first hear or read the expression. A hertz is an "inverse second," therefore

$$1 \text{ Hz} = 1 \text{ s}^{-1}$$

The Farad

The standard unit of *capacitance* is the *farad*, which we symbolize by writing F. Capacitance expresses the ability of an object or electrical component to store energy in the form of an *electric field*. The farad represents a coulomb per volt. Mathematically, we have

$$1 \text{ F} = 1 \text{ C/V}$$

In practice, 1 F constitutes a lot of capacitance. Most capacitance values in real-world systems equal tiny fractions of 1 F. In engineering, you'll see smaller capacitance units as follows:

- 0.000001 (10^{-6}) F represents a *microfarad* (1 μF).
- 0.000000001 (10^{-9}) F represents a *nanofarad* (1 nF).
- 0.000000000001 (10^{-12}) F represents a *picofarad* (1 pF).

The Henry

The standard unit of *inductance* is the *henry*, symbolized H. Inductance expresses the ability of an object or electrical component to store energy as a *magnetic field*. The henry represents a volt-second per ampere, therefore

$$1 \text{ H} = 1 \text{ V} \cdot \text{s} \cdot \text{A}^{-1}$$

The henry constitutes a large unit (although not as huge in practical terms as the farad). In electrical and electronic circuits, most inductance values constitute small fractions of 1 H. In electrical engineering, you'll commonly come across the following units:

- 0.001 (10^{-3}) H represents a *millihenry* (1 mH).
- 0.000001 (10^{-6}) H represents a *microhenry* (1 μH).
- 0.000000001 (10^{-9}) H represents a *nanohenry* (1 nH).

The Weber

The standard unit of *magnetic flux quantity* is the *weber*, symbolized Wb. It's a large unit. One weber equals 1 ampere-henry, so we can state the equation

$$1 \text{ Wb} = 1 \text{ A} \cdot \text{H}$$

A magnetic flux quantity of 1 Wb represents the amount magnetism produced by a constant, direct current of 1 A flowing through an electrical component (such as a large wire coil) having an inductance of 1 H.

The Tesla

The standard unit of *magnetic flux density* is the *tesla*, symbolized T. One tesla represents 1 weber per meter squared. Mathematically, we get

$$1 \text{ T} = 1 \text{ Wb} \cdot \text{m}^{-2}$$

at a surface through which the *magnetic lines of flux* pass at right angles. Lines of flux manifest themselves mathematically as *vectors* that tell us the intensity and orientation of a magnetic field at various points in space.

 Still Struggling

In practical physics, you'll often find it unwieldy to speak or write directly in terms of standard units, because a particular unit doesn't fit the magnitude of the phenomenon you observe. Scientists use *prefix multipliers* to express power-of-10 fractions or multiples of standard units. A common range goes from 10^{-24} (septillionths) up to 10^{24} (septillions). Table 1-1 outlines the prefix multipliers and what they represent. You don't have to memorize all of these (but you'll suffer no harm if you do). Try to remember at least the ones from 10^{-12} to 10^{12}.

TABLE 1-1 Power-of-10 prefix multipliers and their abbreviations.

Designator	Symbol	Multiplier
yocto–	y	10^{-24}
zepto–	z	10^{-21}
atto–	a	10^{-18}
femto–	f	10^{-15}
pico–	p	10^{-12}
nano–	n	10^{-9}
micro–	μ or mm	10^{-6}
milli–	m	10^{-3}
centi–	c	10^{-2}
deci–	d	10^{-1}
(none)	—	10^{0}
deka–	da or D	10^{1}
hecto–	h	10^{2}
kilo–	K or k	10^{3}
mega–	M	10^{6}
giga–	G	10^{9}
tera–	T	10^{12}
peta–	P	10^{15}
exa–	E	10^{18}
zetta–	Z	10^{21}
yotta–	Y	10^{24}

PROBLEM 1-1

Suppose that someone says a computer's microprocessor has a clock frequency of 5.3 GHz. What's this frequency in hertz? In kilohertz? In megahertz?

SOLUTION

From Table 1-1, we observe that a gigahertz (GHz) represents 10^9 Hz. Therefore

$$5.3 \text{ GHz} = 5.3 \times 10^9 \text{ Hz}$$
$$= 5,300,000,000 \text{ Hz}$$

A kilohertz (kHz) represents 10^3 Hz, so our microprocessor frequency equals 5,300,000 kHz. A megahertz represents 10^6 Hz, so the same microprocessor frequency equals 5,300 MHz.

PROBLEM 1-2

Suppose that an electronic component exhibits 0.5 μF of capacitance. What's this value in farads? In nanofarads? In picofarads?

SOLUTION

From Table 1-1, we note that the symbol μ represents "micro-," indicating a multiplier of 10^{-6}. Therefore, 0.5 μF stands for 0.5 microfarad, which is 0.5×10^{-6} F. Mathematically, we can calculate that

$$0.5 \times 10^{-6} \text{ F} = 500 \times 10^{-9} \text{ F}$$

$$= 500 \text{ nF}$$

and

$$0.5 \times 10^{-6} \text{ F} = 500{,}000 \times 10^{-12} \text{ F}$$

$$= 500{,}000 \text{ pF}$$

$$= 5 \times 10^{5} \text{ pF}$$

PROBLEM 1-3

Suppose that an inductor has a value of 33 mH. What's this in henrys? In microhenrys? In nanohenrys?

SOLUTION

From Table 1-1, we see that the prefix multiplier m stands for "milli-," indicating a multiplier of 10^{-3}. Therefore, 33 mH equals 33 millihenrys. That's 33×10^{-3} H, or (more simply) 0.033 H. To get the value in microhenrys, we calculate

$$0.033 \text{ H} = 0.033 \times 10^{6} \text{ μH}$$

$$= 3.3 \times 10^{4} \text{ μH}$$

and to get it in nanohenrys, we calculate

$$0.033 \text{ H} = 0.033 \times 10^{9} \text{ nH} = 3.3 \times 10^{7} \text{ nH}$$

Constants

Constants define quantities that we "take for granted." We should never expect any physical constant to change unless other factors intervene with dramatic effect.

Mathematics versus Physics

In pure mathematics, constants appear as plain numbers without any units associated. We call such quantities *dimensionless constants*. Examples include π, the circumference-to-diameter ratio of a circle, and e, the natural logarithm base. In physics and engineering, we'll usually find a unit associated with a constant. For example, most scientists express the speed of light in free space, c, in meters per second (m/s) or kilometers per second (km/s). Lay people often express the speed of light in miles per second (mi/s).

Table 1-2 lists some constants that you'll encounter in physics. Do you find them unfamiliar or arcane? Don't worry about that right now. As your knowledge of physics expands, you'll learn about them. You can use Table 1-2 as a reference after you've completed this course.

Following are some practical examples of how fundamental physical constants present themselves in scientific work.

Mass of the Sun

How can we express the mass of our sun in comprehensible terms? Using scientific notation, we come up with 1.989×10^{30} kg if we go to four significant figures. That amount of mass constitutes nearly 2 *nonillion kilograms*, or 2 *octillion metric tons*.

We can represent 2 octillion numerically as a 2 followed by 27 zeros. In scientific notation, we write it as 2×10^{27}. Let's split that up into $2 \times 10^9 \times 10^9 \times 10^9$. Imagine a huge box that measures 2000 km tall by 1000 km wide by 1000 km deep. Suppose that some taskmaster tells us to stack that box neatly full of little cubes measuring 1 millimeter (1 mm) on an edge, comparable in size to grains of coarse sand.

Just to humor the taskmaster, we begin stacking these little cubes with the help of tweezers and a magnifying glass. We gaze up at the box towering high above the earth's atmosphere. We suspect that it will take us a long time to finish this job. If we could live long enough to actually do it, we'd have stacked up two octillion little cubes. Two octillion: the number of *metric tons* in the mass of our sun!

TABLE 1-2 Some well-known physical constants and their abbreviations.

Quantity or Phenomenon	Value	Symbol
Mass of the sun	1.989×10^{30} kg	m_{sun}
Mass of the earth	5.974×10^{24} kg	m_{earth}
Avogadro's number	6.022169×10^{23}	N or N_A
Mass of the moon	7.348×10^{22} kg	m_{moon}
Mean radius of the sun	6.970×10^{8} m	r_{sun}
Speed of electromagnetic–field propagation in free space	2.99792×10^{8} m/s	c
Faraday constant	9.64867×10^{4} C/mol	F
Mean radius of the earth	6.371×10^{6} m	r_{earth}
Mean orbital speed of the earth	2.978×10^{4} m/s	
Base of natural logarithms	2.718282	e or ε
Ratio of circle circumference to diameter	3.14159	π
Mean radius of the moon	1.738×10^{6} m	r_{moon}
Characteristic impedance of free space	376.7 ohms	Z_0
Speed of sound in dry air at standard atmospheric temperature and pressure	344 m/s	
Gravitational acceleration at sea level	9.8067 m/s^2	g
Gas constant	8.31434 J/K/mol	R or R_0
Fine structure constant	7.2974×10^{-3}	α
Wien's constant	0.0029 m $\cdot$ K	σ_W
Second radiation constant	0.0143883 m $\cdot$ K	c_2
Permeability of free space	1.257×10^{-6} H/m	μ_0
Stefan–Boltzmann constant	5.66961×10^{-8} W/m^2/K^4	σ
Gravitational constant	6.6732×10^{-11} N $\cdot$ m^2/kg^2	G
Permittivity of free space	8.85×10^{-12} F/m	ε_0
Boltzmann's constant	1.380622×10^{-23} J/K	k
First radiation constant	4.99258×10^{-24} J $\cdot$ m	c_1
Atomic mass unit (AMU)	1.66053×10^{-27} kg	u
Bohr magneton	9.2741×10^{-24} J/T	μ_B
Bohr radius	5.2918×10^{-11} m	α_0

TABLE 1-2 Some well-known physical constants and their abbreviations. (*Continued*)

Quantity or Phenomenon	Value	Symbol
Nuclear magneton	5.0510×10^{-27} J/T	μ_n
Mass of alpha particle	6.64×10^{-27} kg	m_α
Mass of neutron at rest	1.67492×10^{-27} kg	m_n
Mass of proton at rest	1.67261×10^{-27} kg	m_p
Compton wavelength of proton	1.3214×10^{-15} m	λ_{cp}
Mass of electron at rest	9.10956×10^{-31} kg	m_e
Radius of electron	2.81794×10^{-15} m	r_e
Elementary charge	1.60219×10^{-19} C	e
Charge–to–mass ratio of electron	1.7588×10^{11} C/kg	e/m_e
Compton wavelength of electron	2.4263×10^{-12} m	λ_c
Planck's constant	6.6261×10^{-34} J $\cdot$ s	h
Quantum–charge ratio	4.1357×10^{-15} J $\cdot$ s/C	h/e
Rydberg constant	1.0974×10^7 m^{-1}	R_∞
Euler's constant	0.577216	γ

Mass of the Earth

The sun's mass dwarfs the earth's mass. Expressed to four significant figures, the earth masses 5.974×10^{24} kg, or roughly 6 septillion metric tons (where a septillion equals 10^{24}, represented as the numeral 1 followed by 24 zeros).

To get an idea of the magnitude of 6 septillion, let's imagine a cubical box measuring 2.45×10^5 m, or 245 km, on an edge, along with an unlimited supply of little cubes measuring 1 centimeter (1 cm) on an edge, the size of a gambling die. Suppose that we begin stacking up all the little cubes in the huge box. When we've completed the job, we'll have placed approximately 6 septillion little cubes in the box.

Speed of Light

Visible light travels through a vacuum at roughly 3×10^5 km/s. Radio waves, infrared rays, visible light beams, ultraviolet rays, x rays, and gamma rays all propagate through space at this same speed, which we symbolize by writing c.

To get an idea of the magnitude of c, we can calculate the length of time necessary for a ray of light to travel from home plate to the center-field fence in a major-league baseball stadium. Most baseball parks measure about 120 m (or 0.12 km) from the batter to the center-field fence. To calculate the time t for a ray of light to travel that far, we divide 0.12 km by 2.99792×10^5 km/s, getting

$$t = 0.12/(2.99792 \times 10^5)$$
$$= 0.040 \times 10^{-5} \text{ s}$$
$$= 4.0 \times 10^{-7} \text{ s}$$

We must remember the principles of significant figures! We can justify going to only two significant figures here. Also, we must remember to keep the units consistent with each other throughout the calculation process.

Some Arithmetic with Units

If we take the above problem and calculate in terms of units without using any numbers at all, we get

$$t = \text{kilometers}/(\text{kilometers per second}) = \text{seconds}$$

which we might symbolize as

$$t = \text{km}/(\text{km/s}) = \text{km} \times \text{s}/\text{km} = \text{s}$$

In this calculation, kilometers cancel out, leaving only seconds.

Still Struggling

Let's try to make the calculation using feet (ft), rather than meters, as the figure for the distance from home plate to the center-field fence, while leaving the speed of light in meters per second. If we round off the center-field depth to 400 ft, then we get a figure for the visible-light transit time t as

$$t = \text{feet}/(\text{meters per second})$$

or

$$t = \text{ft}/(\text{m/s}) = \text{ft} \times \text{s}/\text{m}$$

Feet don't neatly cancel out meters, so we've created a nonstandard time unit, the "foot-second per meter." While the unit does express time in theory, a physicist would likely call it nonsense.

Gravitational Acceleration

The term "acceleration," used in reference to gravitation, can confuse the uninitiated. Isn't gravity merely a force that pulls things toward the ground? What role does motion play here? Let's delve into these subtle but significant questions.

You can see that "gravity" pulls physical objects, including your own body, downward with constant force. If you were on another planet, you'd notice gravitation there, but it would probably pull on your body with more or less force than it does on the earth. If you weigh 150 pounds (lb) here on earth, for example, you'd weigh only about 56 lb on Mars. Your body mass, 68 kg, would not change (unless you ate too much or too little during your journey to Mars), but the effect of the *gravitational field* on your body's mass would change.

Physicists measure the intensity of a gravitational field according to the rate at which an object accelerates when allowed to fall freely in a vacuum (so that no atmospheric resistance interferes with the measurement). On and near the earth's surface, the *gravitational acceleration* equals approximately 9.8 meters per second per second, or 9.8 m/s^2. If you drop a heavy brick from a great height, it will attain speeds as follows:

$$9.8 \text{ m/s after 1 s}$$
$$9.8 \times 2 \text{ m/s after 2 s}$$
$$9.8 \times 3 \text{ m/s after 3 s}$$
$$\downarrow$$
$$9.8 \times n \text{ m/s after } n \text{ s}$$

The speed increases by 9.8 m/s with every passing second. On and near the surface of Mars, the gravitational acceleration equals approximately 3.7 m/s^2, so you would observe a less dramatic speed increase:

$$3.7 \text{ m/s after 1 s}$$
$$3.7 \times 2 \text{ m/s after 2 s}$$
$$3.7 \times 3 \text{ m/s after 3 s}$$
$$\downarrow$$
$$3.7 \times n \text{ m/s after } n \text{ s}$$

On Jupiter (if Jupiter had a definable surface), you'd observe more gravitational acceleration than you would on the earth. Jupiter's gravitational acceleration equals roughly 25 m/s^2, so your speed-versus-time figures would come out like this:

$$25 \text{ m/s after } 1 \text{ s}$$
$$25 \times 2 \text{ m/s after } 2 \text{ s}$$
$$25 \times 3 \text{ m/s after } 3 \text{ s}$$
$$\downarrow$$
$$25 \times n \text{ m/s after } n \text{ s}$$

The rate of gravitational acceleration *does not* depend on the mass of the object being "pulled on." The rate of gravitational acceleration *does* depend on the mass of the object that "does the pulling," and also on the distance between the "pulling mass" and the "pulled-on mass." The Italian astronomer and experimentalist *Galileo Galilei* surmised this fact in the 17th century, although other scientists had observed the effect before him. Popular legend holds that Galileo demonstrated gravitational acceleration at a "leaning tower" in Pisa, Italy by dropping objects of unequal mass and watching them fall alongside each other all the way to the ground—baffling (and angering) old-school "scientists" who believed that heavy objects fall faster than light ones.

Let's not fault the ancients too much. They probably conducted honest (though flawed) experiments and got results supporting their beliefs, neglecting the role of the atmosphere! If you drop a feather next to a golf ball, the ball falls faster than the feather because the ball overcomes air resistance better than the feather does. If you drop both objects in a vacuum, however, they'll fall right next to each other.

Unit Conversions

At the time of this writing, the National Institute of Standards and Technology's physics page on the Internet offers links to online conversion calculators. Navigate to the Web site www.physics.nist.gov and type "unit conversions" in the search box. Alternatively, you can enter the phrase "unit conversions" in your favorite search engine. Table 1-3 lists some factors for converting base SI units to other common units and vice versa.

TABLE 1-3 Conversions for base units in the international system (SI) to units in other systems. When you see no coefficient, assume that it equals precisely 1.

To Convert	To	Multiply by	Conversely, Multiply by
meters (m)	Angstroms	10^{10}	10^{-10}
meters (m)	nanometers (nm)	10^9	10^{-9}
meters (m)	micrometers (µm)	10^6	10^{-6}
meters (m)	millimeters (mm)	10^3	10^{-3}
meters (m)	centimeters (cm)	10^2	10^{-2}
meters (m)	inches (in)	39.37	0.02540
meters (m)	feet (ft)	3.281	0.3048
meters (m)	yards (yd)	1.094	0.9144
meters (m)	kilometers (km)	10^{-3}	10^3
meters (m)	statute miles (mi)	6.214×10^{-4}	1.609×10^3
meters (m)	nautical miles	5.397×10^{-4}	1.853×10^3
meters (m)	light seconds	3.336×10^{-9}	2.998×10^8
meters (m)	astronomical units (AU)	6.685×10^{-12}	1.496×10^{11}
meters (m)	light years	1.057×10^{-16}	9.461×10^{15}
meters (m)	parsecs (pc)	3.241×10^{-17}	3.085×10^{16}
kilograms (kg)	atomic mass units (amu)	6.022×10^{26}	1.661×10^{-27}
kilograms (kg)	nanograms (ng)	10^{12}	10^{-12}
kilograms (kg)	micrograms (µg)	10^9	10^{-9}
kilograms (kg)	milligrams (mg)	10^6	10^{-6}
kilograms (kg)	grams (g)	10^3	10^{-3}
kilograms (kg) on the earth's surface	ounces (oz)	35.28	0.02834
kilograms (kg) on the earth's surface	pounds (lb)	2.205	0.4535
kilograms (kg) on the earth's surface	English tons	1.103×10^{-3}	907.0
seconds (s)	minutes (min)	0.01667	60.00
seconds (s)	hours (h)	2.778×10^{-4}	3.600×10^3
seconds (s)	days (dy)	1.157×10^{-5}	8.640×10^4

TABLE 1-3 Conversions for base units in the international system (SI) to units in other systems. When you see no coefficient, assume that it equals precisely 1. (*Continued*)

To Convert	To	Multiply by	Conversely, Multiply by
seconds (s)	years (yr)	3.169×10^{-8}	3.156×10^{7}
seconds (s)	centuries	3.169×10^{-10}	3.156×10^{9}
seconds (s)	millenia	3.169×10^{-11}	3.156×10^{10}
kelvins (K)	degrees Celsius (°C)	Subtract 273	Add 273
kelvins (K)	degrees Fahrenheit (°F)	Multiply by 1.80, then subtract 459	Multiply by 0.556, then add 255
kelvins (K)	degrees Rankine (°R)	1.80	0.556
amperes (A)	carriers per second	6.24×10^{18}	1.60×10^{-19}
amperes (A)	statamperes (statA)	2.998×10^{9}	3.336×10^{-10}
amperes (A)	nanoamperes (nA)	10^{9}	10^{-9}
amperes (A)	microamperes (μA)	10^{6}	10^{-6}
amperes (A)	abamperes (abA)	0.10000	10.000
amperes (A)	milliamperes (mA)	10^{3}	10^{-3}
candela (cd)	microwatts per steradian (μW/sr)	1.464×10^{3}	6.831×10^{-4}
candela (cd)	milliwatts per steradian (mW/sr)	1.464	0.6831
candela (cd)	lumens per steradian (lum/sr)	Identical; no conversion	Identical; no conversion
candela (cd)	watts per steradian (W/sr)	1.464×10^{-3}	683.1
moles (mol)	coulombs (C)	9.65×10^{4}	1.04×10^{-5}

Dimensions

Scientists call the "thing" that a unit quantifies the *dimension* of that unit. Meters per second, kilometers per hour, and feet per minute express the speed dimension; seconds, minutes, hours, and days express the time dimension. Physical units always have dimension. So do most constants—although some constants, called *dimensionless constants*, can exist independently of physical units.

PROBLEM 1-4

Table 1-2 on pages 41 and 42 lists a few dimensionless constants along with the unit-bearing constants. Name them.

SOLUTION

Dimensionless constants lack unit associations. When you search through the table for units having no units, you'll find five, as follows:

- Avogadro's number
- Base of natural logarithms
- Ratio of circle circumference to diameter
- Fine structure constant
- Euler's constant

PROBLEM 1-5

Suppose that you step on a scale and it says that you weigh 132 lb. What's your mass in kilograms, assuming you conduct this experiment on earth? Round your answer off to the appropriate number of significant figures.

SOLUTION

Refer to Table 1-3. Multiply 132 lb by 0.4535 kg/lb to get 59.862 kg. Because you know your weight to three significant figures, you should round this result off to 59.9 kg.

PROBLEM **1-6**

Imagine that you're native to France, where nearly everyone expresses distances in meters and kilometers. You're visiting the United States for the first time, where feet and miles are the rule. You see that the a certain highway has a posted speed limit of 55 miles per hour (mi/h). How many kilometers per hour (km/h) does this represent? Round your answer off to the appropriate number of significant figures.

SOLUTION

In the United States, people usually express distances over land in *statute miles*. (Sailors more often use *nautical miles*, which are a little longer than statute miles.) According to Table 1-3, a statute mile measures 1.609×10^3 m. That's 1.609 km. You can multiply 55 mi/h by 1.609 km/mi to obtain a speed equivalent of 88.495 km/h. The posted speed limit only gives you two significant figures, so you should round this result off to 88 km/h.

TIP   *When you convert a physical quantity or phenomenon from one unit system to another, you must refer to that same quantity or phenomenon throughout the process. For example, you can't convert meters squared to meters cubed, or candela to amperes. Keep in mind what you're expressing to ensure that you don't, in effect, attempt to change an apple into an orange.*

QUIZ

Refer to the text in this chapter if necessary (and don't forget the tables!).
A good score is eight correct. Answers are at the back of the book.

1. We can express gravitational acceleration in

    A. meters.
    B. kilograms per meter.
    C. kilograms per pound.
    D. meters per second squared.

2. Suppose that we find a certain quantity expressed in kilowatt-seconds. We can convert this quantity to

    A. amperes.
    B. coulombs.
    C. joules.
    D. ohms per henry.

3. A light source generates 5.0 cd of energy at the peak visible wavelength. How many milliwatts per steradian does this represent?

    A. 7.3
    B. 3.4
    C. 5.0
    D. None, because the unit dimensions don't agree.

4. We can convert an object's weight to its mass if, but only, if, we know the

    A. number of molecules in the object.
    B. mass of the object.
    C. density of the object.
    D. gravitational acceleration.

5. Suppose that a direct current of 500 mA flows through a coil that exhibits an inductance of 500 mH. How much magnetic flux does this current produce?

    A. 0.0625 Wb
    B. 0.250 Wb
    C. 0.500 Wb
    D. 0.707 Wb

6. What's the kelvin equivalent of the lowest possible temperature?

    A. −273.15 K
    B. −459.15 K
    C. 0 K
    D. Nothing, because there's no limit to how cold things can get.

7. **We can express electrical current in terms of charge carriers**
   A. per kilogram.
   B. per square meter.
   C. per cubic meter.
   D. per second.

8. **Which of the following pairs of units *do not* agree in dimension?**
   A. Candela and microwatts
   B. Moles and coulombs
   C. Meters and feet
   D. Meters and light years

9. **When we talk about steradians, we talk about**
   A. volume.
   B. surface area.
   C. solid angular measure.
   D. gravitational acceleration.

10. **When we talk about astronomical units, (AU) we talk about**
    A. mass.
    B. time.
    C. distance.
    D. temperature.

chapter 2

Mass, Force, and Motion

In this chapter, you'll learn the fundamentals of *classical mechanics*, the science of motion. Classical mechanics, also called *Newtonian mechanics* (after *Isaac Newton*, the well-remembered English scientist who lived in the 17th and 18th centuries), constitutes a cornerstone of physics.

CHAPTER OBJECTIVES

In this chapter, you will

- Define scalar and vector quantities.
- Distinguish between mass and weight.
- Measure mass, force, and weight.
- Distinguish between speed and velocity.
- Measure displacement, speed, velocity, and acceleration.

Mass

The term *mass* refers to the extent to which an object resists a change in speed or direction when an external *force* acts on it. We can think of mass as *heft*. A hefty object contains a lot of matter.

Mass Is a Scalar Quantity

We express mass as a *scalar* quantity, meaning that it has *magnitude* (quantity or extent), but not *direction*. We represent known masses in defined units such as kilograms, grams, milligrams, or micrograms. When we want to represent mass as a variable in an equation or formula, we use the lowercase italic letter *m*.

When you stand on a horizontal surface, your body presses downward on that surface. If your friend masses more than you, his body presses downward too, but harder. If you get in a car and accelerate, your body presses backward in the seat as well as downward toward the center of the earth. The "pull" constitutes force, not mass. The force represents the effect of gravitation or acceleration on an object that has mass.

Mass, all by itself, has no direction. If you go into outer space and become "weightless," you'll have the same mass as you do on earth (assuming you do not lose or gain any body heft in between times). No force will exist unless your spaceship changes speed or direction.

Mass on a Spring Scale

We can indirectly measure mass with a simple *spring scale*. Actually, when you place an object on a spring scale, you measure that object's *weight* in the gravitational field of the earth. Weight is the force produced by a certain mass in a certain gravitational field, or when subjected to a certain force caused by acceleration. Remember:

- Mass is the actual quantity of matter in a sample.
- Weight is the force that an object having a given mass experiences as a result of gravitation or acceleration.

When we want to express weight in rigorous theoretical terms, we must use a unit of force, the most common of which is the newton (N), or kilogram-meter per second squared (kg · m/s^2). The pound (lb), used by lay people in the

United States, represents the force produced by approximately 0.454 kg of mass in a gravitational or acceleration field of approximately 9.8 m/s². If you multiply things out, you'll see that

$$1 \text{ lb} \approx 4.4 \text{ N}$$

where the "wavy" pair of lines translates into the words "approximately equals."

How to Measure Mass

Imagine that you drift along in a spaceship on your way to Mars, or in orbit around the earth, so that everything inside the vessel is weightless. How can you measure the mass of a small, dense object such as a fishing sinker (a lump of lead) under these conditions? It floats around in the cabin along with your body, the pencils you write with, and everything else that's not tied down. You realize that the sinker has more mass than, say, a pea, but how can you verify that fact?

You can measure mass, independently of gravity, using a "dual-spring scale" mounted in a rigid frame (Fig. 2-1). You place the object in the middle of the assembly. If you pull the object to one side and then release it, the object oscillates

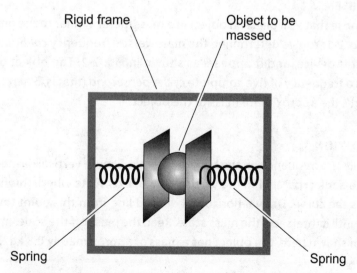

FIGURE 2-1 · We can indirectly measure mass by allowing an object to oscillate between a pair of springs in a weightless environment.

back and forth as the springs alternately compress and expand. Suppose that you try this with a pea (as you pass the time on your journey to Mars), and the springs oscillate rapidly. You try it again with a fishing sinker, and the springs oscillate slowly. Imagine that you've anchored your *mass meter* to a desk in the spaceship's cabin, which is in turn anchored to the floor. Securing the device in this way keeps the assembly from wagging back and forth in midair after you start the pea or lead shot oscillating.

You must *calibrate* a scale of this type before it can render meaningful figures for masses of objects. The calibration tests will give you a graph that shows the oscillation *period* (time required for one complete cycle) or *frequency* (number of cycles per unit time) as a function of the mass. Once you've calibrated your scale in a weightless environment and drawn the graph, you can use the device to measure the mass of anything within reason.

TIP *Every lab instrument has practical limitations! You've designed a mass meter that works okay in the absence of external forces, but if you try to use it on the earth, on the moon, or on Mars, you'll encounter inaccuracy because gravitation exerts a distracting force. You'll run into similar trouble if you try to use your mass meter when the spaceship accelerates, rather than merely coasting through space.*

PROBLEM 2-1

Suppose that you place an object in a mass meter similar to the one shown in Fig. 2-1. You've determined the mass-versus-frequency *calibration curve* for this device, and it appears as shown in Fig. 2-2. The object oscillates with a frequency of five complete *cycles per second* (that is, 5 *hertz* or 5 Hz). What's the approximate mass of the object?

SOLUTION

Locate the frequency on the horizontal scale. Draw a vertical dashed line (or place a ruler) parallel to the vertical (mass) axis. Note where this line intersects the curve. Draw a horizontal dashed line from this point toward the left until it intersects the mass scale. Read the mass off the scale. In the situation shown here, the object has a mass of approximately 0.8 kg (Fig. 2-3).

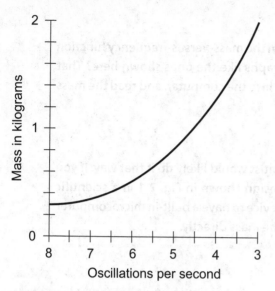

FIGURE 2-2 • Graph of mass versus oscillation frequency for a hypothetical "mass meter" such as the one shown in Fig. 2-1.

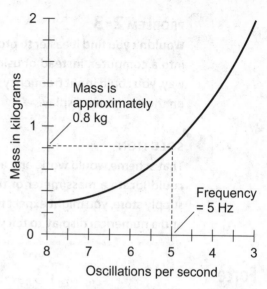

FIGURE 2-3 • Solution to Problem 2-1.

PROBLEM 2-2

What will the mass meter shown in Fig. 2-1, and whose mass-versus-frequency function graph appears in Figs. 2-2 and 2-3, do if you place nothing between the springs and let them oscillate all by themselves?

✔ SOLUTION

The scale will oscillate at a high frequency. In this example, the point corresponding to zero mass lies outside the range of the graph. You might at first suppose that the zero-mass oscillation frequency ought to be infinite. That would be true if the springs and surrounding structure had no mass of their own, but your mass meter has a certain maximum oscillation frequency because the springs and the clamps themselves have mass.

PROBLEM 2-3

Wouldn't you find it easier to program the mass-versus-frequency function into a computer, instead of using graphs like the ones shown here? That way, you could input frequency data into the computer, and read the mass on the computer display.

SOLUTION

That scheme would work, and a scientist would likely do it that way. If you could locate a mass meter of the design shown in Fig. 2-1 in a scientific supply store, you might expect the device to have a built-in microcomputer and a numerical display to tell you the mass directly.

Force

Imagine again that you are in a spacecraft orbiting the earth. Nothing in the cabin has any weight. Two objects float in front of you: a brick and a marble. You know that the brick possesses more mass than the marble. You flick your index finger against the marble. It flies across the cabin and bounces off the wall. Then you flick your finger just as hard (no more, no less) against the brick. The brick moves a little, but it takes several minutes to float across the cabin and bump into the opposite wall. The flicking of your finger imparts a force to the marble or the brick for a moment, but that force has a different effect on the brick than on the marble.

Force as a Vector

Force constitutes a *vector* quantity. That means it has two independent properties: magnitude and direction. We symbolize vectors by writing boldface letters of the alphabet. For example, we can denote a force vector by writing **F**. Sometimes we don't have to worry about the direction in which a force vector occurs; we only have to consider its magnitude. In that case, we can denote force as an italicized (but not bold) letter such as F. Scientists use the newton (N) to express force vector magnitude. As we've learned,

$$1 \text{ N} = 1 \text{ kg} \cdot \text{m/s}^2$$

Suppose the brick in your spacecraft has a mass of 1 kg, and you push against it with a force of 1 N for 1 s, and then let go. The brick will then move at a speed of 1 m/s. It will have gone from stationary (with respect to its surroundings) to a speed of 1 m/s.

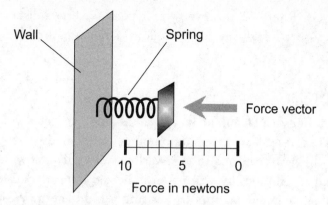

FIGURE 2-4 · A simple device for measuring force.

How We Determine Force

We can measure force by observing its effect on an object having known mass. We can also measure force by noting the amount of deflection or distortion it produces in an elastic object such as a spring. We can modify the mass meter described in the previous section to make a *force meter* if we remove one spring and replace it with a graduated scale as shown in Fig. 2-4. To construct an accurate force meter, we would have to calibrate the scale in a controlled laboratory environment before using it.

Still Struggling

You might ask, "Can forces exist in the real, everyday, humdrum physical world without any direction, but only as phenomena having variable intensity?" The answer: "Absolutely not!" You can talk or write about the magitude of a force, but you'll never know what a force does in the real world unless you know the direction in which it operates.

Displacement

In the simplest situations, lay people define *displacement* as the distance along a straight line between two points in a specific direction. We might say that the distance from Minneapolis, Minnesota to Rochester, Minnesota equals 100 km along a straight line going more or less toward the southeast. If you drive along U.S. Route 52, however, you have to traverse a distance of about 120 km, because that highway doesn't follow a straight path from Minneapolis to Rochester.

Vector or Scalar?

When we define displacement in a straight line, we get a vector having magnitude (expressed in meters, kilometers, or other distance units) and direction (which we can quantify in various ways). For now, let's denote displacement magnitudes by writing an italicized letter q, and let's denote complete displacement vectors by writing a boldface $\mathbf{q}$. We can attach subscripts to identify these quantities in various situations.

We can describe the straight-line displacement vector $\mathbf{q}_{rm}$ from Minneapolis to Rochester as approximately 100 km in a southeasterly direction. If we express the direction as an *azimuth* (or "compass") bearing, it's around 135° measured *clockwise* from *due north*. A mathematician might determine the direction in the *counterclockwise* sense from *due east*, in which case she would think of it as 315°.

Still Struggling

When we drive along U.S. Route 52, we cannot define our positional change from Minneapolis to Rochester as a constant vector, because we must alter our direction of travel as the road bends, passes over hills, and dips into valleys.

How to Determine Displacement

We can ascertain displacement vector magnitude by mechanically measuring distance, or by inferring it with observations and calculations. A car or truck driving along U.S. Route 52 would keep track of displacement with an *odometer* that counts the number of wheel rotations, and then multiplies that tally by the circumference of the wheel. In a laboratory environment, we can measure displacement directly with a *meter stick*, indirectly by *triangulation*, or by measuring the time it takes for a ray of light to travel between two points, given the constancy of the speed of light ($c \approx 2.99792 \times 10^8$ m/s).

We can figure out displacement vector direction by measuring angles relative to a *reference axis*. In the case of a local region on the earth's surface, we can specify the *azimuth*, the angle in degrees around the horizon clockwise from true north. In three-dimensional (3D) space, we would more likely specify *direction angles*. We could begin by defining the reference axis as a vector pointing toward *Polaris*, the North Star. We could then define two angles relative to that axis. The coordinate system preferred by astronomers and space scientists involves angles called *celestial latitude* and *celestial longitude*, or, alternatively, *right ascension* and *declination*.

Speed

Speed defines the rate at which an object moves relative to a specific *reference frame* (point of view). We imagine that reference frame as "stationary." When you stand still on the surface of the earth, you're "stationary." You might think that you're "stationary" as you sit in a train reading a book, but when you look out the window and see the landscape whizzing by, you think that you're "moving."

Speed Is a Scalar

The standard unit of speed is the *meter per second* (m/s). A car driving along U.S. Route 52 might have a "cruise control" that you can set at, say, 25 m/s. Then, assuming the cruise control works properly, you'll travel at a constant speed of 25 m/s relative to the highway pavement. If the "cruise control" works perfectly, you'll move at 25 m/s on a level straightaway, rounding a curve, cresting a hill, or passing the bottom of a valley. You can always express speed as a scalar. You don't have to take direction into account. Let's symbolize speed by v.

Speed can change from moment to moment. At any particular instant in time, you can specify a number that quantifies *instantaneous speed*. If you hit the brakes to avoid a deer crossing the road, your instantaneous speed abruptly decreases. As you pass the deer (relieved to see it bounding off into a forest unharmed), you push on the "gas pedal" and your instantaneous speed increases again.

You can average your speed over a particular period of time. In the foregoing example, suppose that you travel at 25 m/s and then see a deer, put on the brakes, slow down to a minimum of 10 m/s, watch the deer run away, and then speed up to 25 m/s again, all in a time span of 1 minute. Your *average speed* over that minute is somewhere between 10 m/s and 25 m/s—maybe 17 m/s. Your instantaneous speed varies, and attains a value of 17 m/s at only two points in time (once as you slow down, and again as you speed back up).

How We Determine Speed

A motor vehicle determines its instantaneous speed using the same odometer that measures distance. However, instead of simply counting up the number of wheel rotations from a given starting point, a *speedometer* measures the rate at which the wheels turn. Given the wheels' circumference, the device can translate the number of wheel rotations per unit time into an average speed value over that period of time. Most speedometers respond almost immediately to a change in speed. These instruments measure the rotation rate of a car or truck axle by another method, similar to that used by the engine's *tachometer* (a device that measures *revolutions per minute*, or rpm).

If, in a given period of time t, an object travels a distance q at an average speed v_{avg}, then we can determine that average speed in terms of the total distance traveled and the elapsed time using the formula

$$v_{avg} = q/t$$

We can calculate the total distance traveled in terms of the average speed and the elapsed time using the formula

$$q = v_{avg}\, t$$

We can calculate the elapsed time in terms of the total distance traveled and the average speed using the formula

$$t = q/v_{avg}$$

PROBLEM 2-4

Examine the graph in Fig. 2-5. Assume that plot A is a perfectly straight line. What's the instantaneous speed v_{inst} at $t = 5$ seconds?

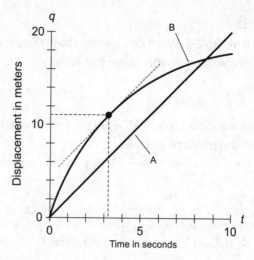

FIGURE 2-5 · Illustration for Problems 2-4 through 2-8.

✔ **SOLUTION**

We know that the speed depicted by plot A remains constant over the entire time interval shown, because the plot constitutes a straight line. Therefore, the instantaneous speed v_{inst} at every point in time equals the average speed v_{avg} over the interval. In 10 seconds, the object travels 20 meters, so we can calculate the instantaneous speed as

$$v_{inst} = v_{avg} = q/t$$
$$= 20\,\text{m}/10\,\text{s}$$
$$= 2.0\,\text{m/s}$$

PROBLEM 2-5

What's the average speed v_{avg} of the object denoted by plot A in Fig. 2-5, during the time span from $t = 3$ s to $t = 7$ s?

✔ **SOLUTION**

Because plot A is a straight line, the speed never varies, and we've calculated it as 2.0 m/s. Therefore, $v_{avg} = 2.0$ m/s during the time span from $t = 3$ s to $t = 7$ s, and also between any two time points within the interval from $t = 0$ s to $t = 10$ s.

PROBLEM 2-6

Examine plot B in Fig. 2-5. What can we say about the instantaneous speed of the object whose motion this curve portrays?

SOLUTION

The object starts out moving relatively fast, and the instantaneous speed decreases with the passage of time.

PROBLEM 2-7

In the situation shown by Fig. 2-5, when does the instantaneous speed v_{inst} of the object described by curve B equal 2.0 m/s?

SOLUTION

Let's use visual approximation. We can take a ruler or other straight-edged object and find a straight line that's *tangent* to curve B, and whose *slope* (rise over run) equals that of curve A. We manipulate the ruler until we find a straight line, parallel to line A, that intersects curve B at a single point. In Fig. 2-5, that's the dashed, sloping line. We draw another dashed line going straight down, parallel to the displacement (q) axis, until it intersects the time (t) axis. We can read the answer to the problem directly from the t axis. In this example, we get roughly $t = 3.2$ s.

PROBLEM 2-8

Consider the object whose motion is described by curve B in Fig. 2-5. At the point in time t where the instantaneous speed v_{inst} is 2.0 m/s, how far has the object traveled since time $t = 0$?

SOLUTION

We locate the same point that we found in Problem 2-7, corresponding to the tangent point of curve B and the line parallel to curve A. From this point we draw a dashed, horizontal line toward the left, parallel to the time (t) axis, until it intersects the displacement (q) axis. Then we can read the value from the graduated scale on the q axis. In this example, we obtain approximately $q = 11$ m.

TIP *A real-life car or truck speedometer measures instantaneous speed, not average speed. In fact, many motor vehicles (especially older ones) lack any system for measuring average speed. If you want to know the average speed at which you've traveled during a certain period of time, you must measure the distance on the odometer, and then divide by the time elapsed.*

Velocity

Velocity consists of two independent components: *speed* and *direction*. We can define direction in a single dimension (either way along a straight line), in two dimensions (within a defined geometric plane), or in three dimensions (in space). Some theoretical physicists get involved with expressions of velocity in more than three spatial dimensions, but we won't worry about that sort of thing in this course!

Velocity Is a Vector

Because velocity possesses two independent components (magnitude and direction), it's a vector quantity. We can't express velocity without defining both of these components. In the earlier example of a car driving along a highway from one town to another, the vehicle might travel at a constant speed, but the velocity will inevitably change from moment to moment unless the highway remains straight and level over the entire distance. If we travel at 25 m/s and then we come to a bend in the road, our velocity *must* change because our direction changes.

We can geometrically illustrate a vector by drawing a line segment with an arrowhead at one end. The length of the line segment tells us the vector magnitude (speed component). The arrowhead indicates the direction component. Figure 2-6A portrays a car as it travels along a curved, level road. Let's call the points P, Q, and R, such that:

- At point P, the car has velocity vector **p**.
- At point Q, the car has velocity vector **q**.
- At point R, the car has velocity vector **r**.

We can see by examining Fig. 2-6 that our car's speed and direction both change from moment to moment as time passes.

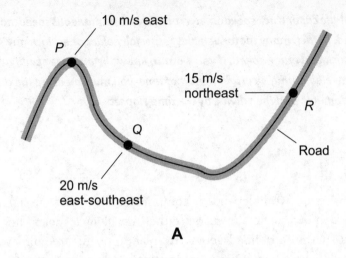

10 m/s east

P

15 m/s
northeast

R

Q

Road

20 m/s
east-southeast

A

Each division represents 5 m/s

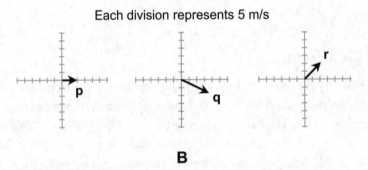

p

q

r

B

FIGURE 2-6 · A car travels along a curved, level road at a constant forward speed of 25 m/s. At A, we see three points *P*, *Q*, and *R*; at B, we see the corresponding velocity vectors **p**, **q**, and **r**.

How to Determine Velocity

As we ride in a car or truck, we can get a good idea of our *instantaneous velocity* by using a speedometer in combination with a device such as a magnetic compass that indicates the instantaneous direction of travel. But these two devices don't tell us the whole story unless we're driving on a level surface.

In certain parts of the American Midwest, a speedometer and compass can completely define the instantaneous velocity of a car, because the land stays as flat as a table-top. In most real-life travel, however, our true motion takes place in three dimensions as we encounter rolling terrain, hills, or mountains. Then we need a *clinometer* (a device for measuring the steepness of the grade we're ascending or descending) along with the speedometer and compass to get a complete description of our instantaneous velocity.

We can define two-dimensional direction components as azimuth bearings, or as angles measured counterclockwise with respect to a reference axis pointing due east. Hikers, mariners, and aviators prefer the former system. Theoretical physicists and mathematicians generally prefer the latter scheme. When we examine Fig. 2-6B, we can see that

- The azimuth of **p** is roughly 90°.
- The azimuth of **q** is roughly 120°.
- The azimuth of **r** is roughly 45°.

In the theoretician's system, we denote the direction of a vector by writing "dir" and then the name of the vector. Therefore, according to illustration B in Fig. 2-6:

- dir **p** ≈ 0°
- dir **q** ≈ 330°
- dir **r** ≈ 45°

Remember that in the theoretician's system, our reference axis points due east, not due north, and we turn counterclockwise to define the direction angles.

Acceleration

Acceleration expresses instantaneous changes in an object's velocity. Acceleration can manifest as a change in speed, a change in direction, or both. Sometimes we can define acceleration in one dimension (along a line). More often, however, we need two dimensions (a plane) or three dimensions (space).

Acceleration Is a Vector

Occasionally, we'll see the magnitude of an *acceleration vector* described as "acceleration" and symbolized by a lowercase italic letter such as *a*. Technically, however, we should think of acceleration as a vector quantity, use vector notation, and symbolize it as a lowercase bold letter such as **a**.

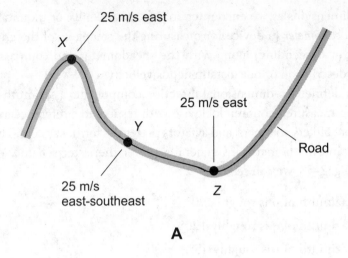

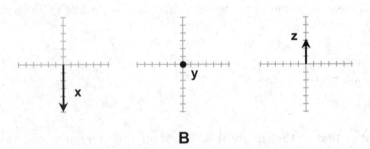

FIGURE 2-7 · A car travels along a curved, level road at a constant forward speed of 25 m/s. At A, we see three points X, Y, and Z; at B, we see the corresponding acceleration vectors **x**, **y**, and **z**. No acceleration occurs at Y, so **y** = 0.

Once again, imagine that a car travels along a highway at a speed of 25 m/s. The velocity changes whenever the car rounds a curve, crests a hill, or bottoms-out in a valley. If the car moves along a straight, level stretch of the road and its forward speed increases, then the acceleration vector points in the same direction as the car moves. If the car's driver puts on the brakes, still moving along a straight path, then the acceleration vector points exactly opposite the direction of the car's motion.

Figure 2-7 shows an acceleration scenario in which a car follows a curved, level road at a constant forward speed of 25 m/s. At A, we see three specific points, labeled X, Y, and Z. At B, we see the corresponding acceleration vectors x, y, and z. Because the forward speed never varies, the car accelerates only where the road bends. At point Y, the road is locally straight, so no acceleration occurs. That means **y** = **0**, where the bold numeral 0 denotes the *zero vector* (zero magnitude, no direction). In a vector graph, we illustrate the vector **0** by drawing a dot at the coordinate origin.

How to Determine Acceleration

Physicists express acceleration magnitude in units of distance per unit time *squared*, such as meters per second squared (m/s^2), also called meters per second per second. Imagine that you're in the driver's seat of a high-performance car that can go from zero to sixty miles per hour (0.00–60.0 mi/h) in 4 seconds flat (4.00 s). A speed of 60.0 mi/h equals about 26.8 m/s. Suppose that the car accelerates at a constant rate from the moment you first hit the gas pedal until you've attained a speed of 26.8 m/s on a level, straight racing track. You can calculate the acceleration magnitude a as follows:

$$a = (26.8 \text{ m/s}) / (4.00 \text{ s})$$
$$= 6.70 \text{ m/s}^2$$

We should expect the instantaneous acceleration magnitude to vary, at least a little bit, in a real-life test of a car's responsiveness. But the average acceleration magnitude will nevertheless work out as 6.70 m/s^2—a speed increase of 6.70 m/s with each passing second, assuming the vehicle's speed goes from 0.00 to 60.0 mi/h in 4.00 s.

We can measure the magnitude of an acceleration vector in terms of the force exerted on a known mass. This force might manifest itself as distortion in an elastic object such as a spring. We can adapt the "force meter" (Fig. 2-4) to obtain an "acceleration meter," technically known as an *accelerometer*, which will tell us the acceleration magnitude. We place an object having known mass in the device, and then calibrate the scale in a controlled lab environment. If we want the accelerometer to yield accurate readings, the acceleration vector's direction must correspond with the spring axis, and the acceleration vector must point outward from the fixed anchor toward the mass. This arrangement will produce a force on the mass that goes directly against the spring, as shown in Fig. 2-8.

Theoretically, you could employ a common weight scale to indirectly measure acceleration magnitude. When you stand on the scale, you compress a spring, or else balance a set of masses on a lever. The scale indicates the

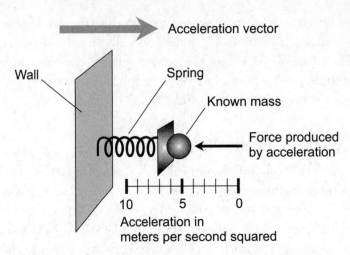

FIGURE 2-8 • An accelerometer measures acceleration magnitude, but it does not tell us the vector's direction. To get an accurate reading, we must orient the device properly.

downward force that the mass of your body exerts as a result of gravitational acceleration. The earth's gravitation acting on a mass produces the same effect as upward acceleration of 9.8 m/s². Force, mass, and acceleration interact, as we shall soon see.

Imagine that an object starts from a dead stop and accelerates at an average magnitude of a_{avg} in a straight line for a period of time t. Suppose that after this length of time, the magnitude of the displacement from the starting point equals q. Then the following formula applies:

$$q = a_{avg} \, t^2/2$$

In the foregoing example, suppose that the acceleration maintains a constant magnitude a. At time t, let's call the instantaneous speed v_{inst}. We can relate the instantaneous speed to the acceleration vector magnitude as follows:

$$v_{inst} = at$$

PROBLEM 2-9

Suppose that two objects, whose motions we express graphically as plots A and B in Fig. 2-9, both accelerate along straight-line paths. What's the instantaneous acceleration vector magnitude a_{inst} at $t = 4$ s for object A?

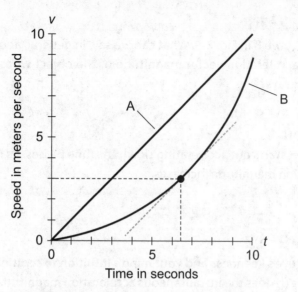

FIGURE 2-9 · Illustration for Problems 2-9 through 2-13.

SOLUTION

The acceleration depicted by plot A remains constant, because the speed increases at a constant rate with time. (That's why the plot shows up as a straight line.) The number of meters per second squared doesn't change throughout the time span shown. In 10 s, object A accelerates from 0 to 10 m/s; that's a rate of speed increase of 1 m/s^2. Therefore, the acceleration magnitude a_{inst} at time $t = 4$ s is equal to 1 m/s^2.

PROBLEM 2-10

What's the average acceleration magnitude a_{avg} of object A as shown in Fig. 2-9, during the time span from $t = 2$ s to $t = 8$ s?

SOLUTION

Because the graph is a straight line, the acceleration magnitude never varies. We've already calculated it as 1 m/s^2. Therefore, $a_{avg} = 1$ m/s^2 between the time points $t = 2$ s and $t = 8$ s, and in fact between any two time points within the domain of the graph.

PROBLEM 2-11

Examine curve B in Fig. 2-9. What can we say, in general, about the instantaneous acceleration vector magnitude of the object whose motion this curve portrays?

SOLUTION

The object starts out accelerating slowly. As time passes, its instantaneous acceleration magnitude increases.

PROBLEM 2-12

Use your eye's keenness and your mind's intuition to scrutinize Fig. 2-9. At what time (t) does the instantaneous acceleration magnitude a_{inst} of object B equal 1 m/s²?

SOLUTION

Take a ruler and find a straight line tangent to curve B whose slope equals the slope of line A. Then locate the point on curve B where the line just touches that curve. Finally, draw a line straight down, parallel to the speed (v) axis, until that line intersects the time (t) axis. Read the value directly from the t axis. It appears close to $t = 6.3$ s.

PROBLEM 2-13

Once again, use visual approximation on Fig. 2-9. Consider the object whose motion curve B defines. At the point in time t where the instantaneous acceleration magnitude a_{inst} equals 1 m/s, what's the instantaneous speed, v_{inst}, of object B?

SOLUTION

Locate the same point that you found in Problem 2-12, corresponding to the tangent point of curve B and the line parallel to A. Draw a horizontal line to the left until it intersects the speed (v) axis. Read the value off the v axis. In this example, it appears close to $v_{inst} = 3.0$ m/s.

Vectors in 3D

In rectangular 3D coordinates, we can define vectors as the sum of three components along mutually perpendicular x-, y-, and z-axes. The x-axis might run east and west, the y-axis north and south, and the z-axis up and down. Using velocity as an example, a moving object travels at a certain velocity parallel to each of these axes at every instant in time. When we add up the three *component velocity vectors*, we get an expression of the complete velocity vector in 3D.

We can also define 3D velocity in terms of one magnitude component and two direction angles. We denote the object's instantaneous speed as the length of the vector's line segment. Navigators and astronomers denote 3D vector direction in terms of points in the sky. On the earth's surface, we can define such points as *celestial latitude* and *celestial longitude*, or as *right ascension* and *declination*. Theoretical scientists and mathematicians sometimes use an alternative 3D coordinate scheme called *spherical coordinates* to determine vector magnitude and direction in space.

A third scheme exists for defining vectors in 3D space: *cylindrical coordinates*. In this system, we take the direction and the radial distance from the reference point (the *origin*) in a horizontal plane, and then we add a vertical elevation component, where up is positive and down is negative. In the case of a displacement vector, we might say, "Start here, then travel 10 km toward the northwest, and then go straight up 20 km."

Still Struggling

If the notion of 3D vectors baffles and frustrates you, don't feel bad about it. You have plenty of company. This stuff can get difficult! When I attended college, people warned me about how difficult "vector analysis" would be. I got lucky; I had a good professor, so I didn't have much trouble. If you need a refresher on vector topics, you should refer to a good pre-calculus textbook. I recommend *Pre-Calculus Know-It-All* (McGraw-Hill, 2010). When I wrote that book, I made a special effort to go into detail about vectors and 3D coordinates.

Newton's Laws of Motion

Three laws, credited to Isaac Newton, apply to the motions of objects in classical physics. These laws ignore the relativistic effects that become significant when speeds approach the speed of light, or when extreme gravitational fields exist. In his time, Newton did not know anything about relativity.

Newton's First Law

We can state Newton's first law in two parts. Both aspects of this law involve the tendency of an object to "keep doing the same thing" in the absence of interference from an unknown external source. The law can be summarized as follows:

- An object at rest will remain at rest unless an outside force acts on it.
- An object moving at a specific velocity will keep moving at that velocity unless an outside force acts on it.

Newton's Second Law

If a force of magnitude F (in newtons) acts on an object of mass m (in kilograms), then we can calculate the acceleration vector magnitude a (in meters per second squared) using the formula

$$a = F/m$$

The more "widely advertised" version of this formula is

$$F = ma$$

When we define force and acceleration as vector quantities, we can denote the complete description of the situation with the formula

$$\mathbf{F} = m\mathbf{a}$$

Newton's Third Law

Whenever we have an action, we also observe an equal and opposite reaction. In other words, if an object A exerts a force vector $\mathbf{F}$ on another object B, then object B exerts a force vector $-\mathbf{F}$ (the negative of vector $\mathbf{F}$) on object A. The negative of a vector $\mathbf{F}$ has the same magnitude as $\mathbf{F}$, but points in precisely the opposite direction.

PROBLEM 2-14

Imagine that a spacecraft of mass $m = 10,500$ (1.0500×10^4) kg, traveling through interplanetary space, is acted upon by a force vector $\mathbf{F} = 100,000$ (1.0000×10^5) N in the direction of Polaris, the North Star. Determine the magnitude and direction of the acceleration vector.

✔ SOLUTION

You can use the first formula stated under the header "Newton's second law." Plugging in the numbers for force magnitude F and mass m yields the acceleration magnitude a, as follows:

$$a = F/m$$
$$= 1.0000 \times 10^5 / 1.0500 \times 10^4$$
$$= 9.5238 \text{ m/s}^2$$

The direction of the acceleration vector **a** corresponds to the direction of the force vector **F** in this case; that is, toward the North Star. As an interesting aside, you might notice that this acceleration magnitude closely matches the gravitational acceleration at the earth's surface, 9.8 m/s^2. A person traveling in this spaceship would feel right at home in terms of body weight.

PROBLEM 2-15

According to Newton's first law, shouldn't the moon fly off in a straight line into interstellar space? Why does it orbit the earth instead?

✔ SOLUTION

The moon constantly experiences the effect of an acceleration vector that "tries" to pull it down to the earth. The moon's inertia "tries" to make it fly away in a straight line, but the inward force caused by the earth's gravitation keeps that from happening. A constant force vector points from the moon toward the earth, corresponding to the moon's instantaneous acceleration vector.

Still Struggling

In the scenario of Problem and Solution 2-15, don't get the moon's *instantaneous speed* mixed up with its *instantaneous velocity vector*. The moon's instantaneous speed remains nearly constant as time passes. (If the moon's orbit were a perfect circle, the moon's instantaneous speed would remain absolutely constant; but the slight elongation in the moon's orbit causes its speed to vary slightly.) The moon's instantaneous velocity vector changes (constant speed but rotating direction) because of the gravitational force between the moon and the earth.

QUIZ

Refer to the text in this chapter if necessary. A good score is eight correct. Answers are at the back of the book.

1. Which of the following parameters manifests itself as a vector quantity?
   A. Temperature
   B. Visible-light intensity
   C. Velocity
   D. Mass

2. Suppose that we're coasting along in a nonaccelerating ship somewhere in deep space, so that we experience constant weightlessness in the cabin. We look out one of the portholes and see an object of mass 2.0 kg traveling in a straight line at a constant speed of 1.0 m/s relative to our ship. From this observation, we know that the external force vector acting on that mass has a magnitude of
   A. 2.0 N.
   B. 0.50 N.
   C. 0.25 N.
   D. zero.

3. What would happen if the earth suddenly stopped moving in its orbit around the sun, but nothing else changed?
   A. Nobody knows.
   B. The earth would fly out of solar orbit.
   C. The earth would fall into the sun.
   D. The earth would remain fixed at the spot.

4. Imagine that you bring a small boat into port with the intention of docking. As you approach the dock, you shut off the motor, expecting the boat to come to a stop. Instead, the boat coasts along the water surface in a straight line and crashes into the dock. You've learned a practical lesson concerning
   A. Newton's first law.
   B. Newton's second law.
   C. Newton's third law.
   D. Newton's fourth law.

5. If we want to talk meaningfully about the velocity vector for an object at a certain moment in time, we must specify the object's instantaneous
   A. mass and speed.
   B. mass and displacement.
   C. speed and direction.
   D. mass and direction.

6. In 3D space, we can define the magnitude of a vector in terms of

    A. right ascension and declination.
    B. celestial latitude and longitude.
    C. the location of the coordinate origin.
    D. the length of the line segment representing it.

7. A force of 1 gram-meter per second squared represents 1/1000 of a

    A. watt.
    B. joule.
    C. newton.
    D. henry.

8. Imagine that we stand on top of a high cliff on a planet where the gravitational acceleration is 5.67 m/s². We drop a brick of mass 3.00 kg from the top of that cliff. After 2.00 s have passed, assuming the brick has sufficient density to overcome atmospheric resistance, its downward instantaneous speed will be

    A. 64.3 m/s.
    B. 34.0 m/s.
    C. 22.7 m/s.
    D. 11.3 m/s.

9. Suppose that we drop a second brick from the same cliff as the one described in Question 8, but this brick has a mass of 6.00 kg. After 2.00 s have passed, its instantaneous speed, compared with the speed of the first brick, will be

    A. the same.
    B. half as great.
    C. twice as great.
    D. four times as great.

10. While driving a truck along a straight, level highway on a windless night, you see an animal crossing the road. You press on the brake pedal but keep going straight, causing the truck's velocity vector to

    A. exactly reverse direction and decrease in magnitude.
    B. keep pointing in the same direction and decrease in magnitude.
    C. point straight down toward the pavement.
    D. vanish (become the zero vector).

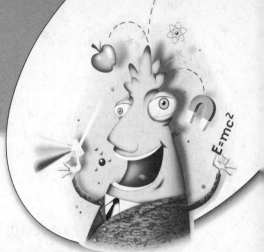

chapter 3

Momentum, Work, Energy, and Power

Classical mechanics describes the behavior of objects in motion. Any moving mass has *momentum* and *energy*. Whenever two objects collide, their momentum and energy values shift and change.

CHAPTER OBJECTIVES

In this chapter, you will

- Compare momentum with impulse.
- Observe conservation of momentum.
- Analyze collisions between objects.
- Compare work with energy.
- Calculate potential and kinetic energy.
- Distinguish between energy and power.

Momentum

When we multiply an object's mass by its velocity, we obtain a quantity known as *momentum*. As we've learned, the kilogram (kg) constitutes the base SI unit of mass, and the meter per second (m/s) constitutes the standard unit of speed. We can therefore express momentum magnitude in *kilogram-meters per second* (kg · m/s).

Momentum as a Vector

Suppose that an object having mass m (in kilograms) moves at speed v (in meters per second). We can calculate the magnitude p of the momentum vector using the formula

$$p = mv$$

If we want to fully describe the momentum of a moving object, then we must express it as a vector quantity. We must consider the velocity of the mass in terms of its speed and its direction. (A 2-kg brick flying at 5 m/s through your east window represents a different vector quantity than a 2-kg brick flying at 5 m/s through your north window.) If we let **v** represent the velocity vector and we let **p** represent the momentum vector, then

$$\mathbf{p} = m\mathbf{v}$$

Impulse

The momentum of a moving object can vary in any of three different ways:

- A change in the mass of the object.
- A change in the speed of the object.
- A change in the direction of the object's motion.

Let's consider the second and third of these possibilities together, so we have a change in the velocity.

Imagine a mass, such as a spaceship, coasting along a straight-line path in interstellar space. Consider a point of view, or *reference frame*, such that we can express the ship's velocity as a nonzero vector pointing in a certain direction. We might apply a force **F** to this vessel by firing one of its rocket engines. Imagine that our ship has several engines, one intended for driving the vessel forward

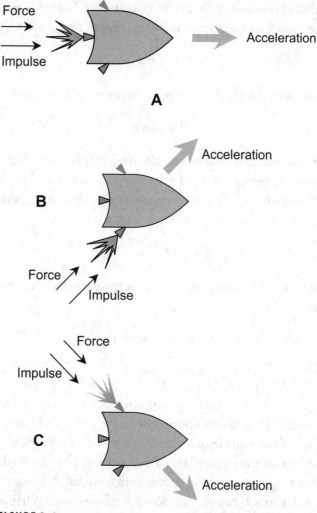

FIGURE 3-1 • Three ways that an impulse can accelerate a spaceship: forward (A), to the left (B), and to the right (C). We look "down" on the vessel in all three cases.

at increased speed, and others capable of changing the direction. Suppose that we fire one engine for t seconds, causing it to impose a force vector of $\mathbf{F}$ newtons (as shown by any of the examples in Fig. 3-1). When we multiply the force vector $\mathbf{F}$ by the scalar time t, we get the *impulse* vector, symbolized $\mathbf{I}$ and expressed in kilogram-meters per second (kg · m/s) as follows:

$$\mathbf{I} = \mathbf{F}t$$

Impulse always produces a change in velocity. In Chapter 2, we learned that mass m, force F, and acceleration a relate according to the formula

$$F = ma$$

When we substitute ma for F in the preceding formula for impulse, we get

$$I = (ma)t$$

Acceleration constitutes a change in velocity per unit time. Suppose that our spaceship travels at velocity v_1 before we fire the rocket, and the vessel travels at a different velocity v_2 after we fire the rocket. Assuming that the rocket engine produces a constant force while it burns, we have

$$a = (v_2 - v_1)/t$$

We can substitute the quantity $(v_2 - v_1)/t$ for the vector a in the previous equation to obtain

$$I = m[(v_2 - v_1)/t] \, t$$

$$= mv_2 - mv_1$$

This new equation tells us that the impulse vector equals the change in the momentum vector. (If you don't remember how vectors add and subtract, you might want to refresh your knowledge of basic vector arithmetic.)

We've derived an important law of Newtonian physics. Reduced to base units in SI, we can express impulse vector magnitude in kilogram-meters per second (kg · m/s), exactly as we can do with momentum. When we impart an impulse to an object, then that object's momentum vector p changes by an equivalent amount. The effect shows up as a change in the speed, the direction, or both.

PROBLEM 3-1

Suppose that an object of mass 2.00 kg moves at a constant speed of 50.0 m/s in a northerly direction. An impulse, acting in a southerly direction, slows this mass down to 25.0 m/s, but the mass still moves in a northerly direction. What's the impulse vector responsible for this change in momentum?

✔ SOLUTION

To calculate the original momentum, $\mathbf{p}_1$, we multiply the mass times the initial velocity to get

$$\mathbf{p}_1 = 2.00 \text{ kg} \times 50.0 \text{ m/s}$$
$$= 100 \text{ kg} \cdot \text{m/s}$$
$$\text{going north}$$

We calculate the final momentum, $\mathbf{p}_2$, using the figure for the final velocity, obtaining

$$\mathbf{p}_2 = 2.00 \text{ kg} \times 25.0 \text{ m/s}$$
$$= 50.0 \text{ kg} \cdot \text{m/s}$$
$$\text{going north}$$

The change in momentum equals $\mathbf{p}_2 - \mathbf{p}_1$, which we determine as follows:

$$\mathbf{p}_2 - \mathbf{p}_1 = 50.0 \text{ kg} \cdot \text{m/s} - 100 \text{ kg} \cdot \text{m/s}$$
$$= -50.0 \text{ kg} \cdot \text{m/s}$$
$$\text{going north}$$

which we can also express as a vector of magnitude 50.0 kg · m/s in a southerly direction. Because we describe impulse and momentum in the same units, we can say that the change in the object's momentum results from an impulse vector of 50.0 kg · m/s, acting in a southerly direction.

Still Struggling

If the result of Problem 3-1 baffles you, think back to your vector arithmetic courses. A vector having magnitude $-x$ in a certain direction has the same meaning, and the same effect, as a vector with magnitude x in the opposite direction. In physics, problems sometimes work out to yield vectors with negative magnitude. When that happens, you can reverse the vector's direction and then take the *absolute value* of the magnitude to get a positive value.

Collisions

When two moving objects hit each other because their paths cross at exactly the right (or wrong) instant in time, we say that a *collision* has taken place. The effects of a collision depend on whether the objects "stick together" or "bounce apart" after they hit each other.

Conservation of Momentum

Here's a principle that you should commit to your memory. Physicists call this rule the *law of conservation of momentum*:

- The total momentum contained in two objects does not change when the objects collide, unless we introduce some new mass or force into the scene.

The characteristics of the collision don't matter, as long as we have an *ideal system* in which no friction or other real-world imperfection occurs. The foregoing law applies not only to systems of two objects, but to systems having any number of objects. But it holds true only in a *closed system*—a system in which the total mass remains constant, and no forces are introduced from the outside.

A Note about Units

From now on in this course, if you don't see a specific unit given for a particular quantity, you should assume that we're working with standard SI units. Therefore, in the following examples, we'll discuss masses in kilograms, velocity vector magnitudes in meters per second, and momentum vector magnitudes in kilogram-meters per second.

Get into the habit of assuming that SI units are always implied, whether the units end up being important in the discussion, or not. Of course, if an author specifies some unit other than the standard SI unit, you should use it! But always make your calculations with units in mind. They must always agree throughout a calculation, or you'll risk getting an inaccurate or absurd result.

"Sticky" Objects

Figure 3-2A shows two objects having masses m_1 and m_2 moving at speeds v_1 and v_2, respectively. The velocity vectors $\mathbf{v}_1$ and $\mathbf{v}_2$ point in the directions indicated

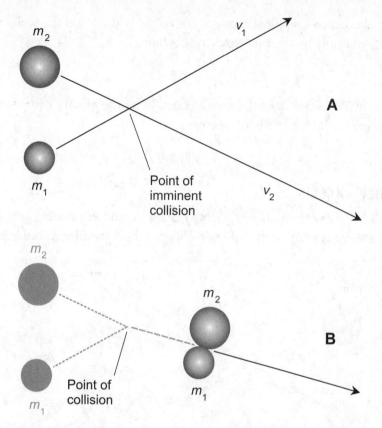

FIGURE 3-2 · At A, two "sticky" objects having different velocities approach each other on a collision course. At B, the collision has occurred.

by the arrows. The two objects travel on a collision course. The object with mass m_1 has momentum

$$\mathbf{p}_1 = m_1 \mathbf{v}_1$$

and the object with mass m_2 has momentum

$$\mathbf{p}_2 = m_2 \mathbf{v}_2$$

Figure 3-2B shows the objects after they have run into each other and stuck together. The composite object travels at a new velocity $\mathbf{v}$, different from either of the initial velocities. The new momentum vector (let's call it $\mathbf{p}$) equals the sum of the original momentum vectors. Mathematically, we have

$$\mathbf{p} = m_1 \mathbf{v}_1 + m_2 \mathbf{v}_2$$

We can also express the final momentum vector, **p**, if we add the scalar masses and then multiply by the final velocity, obtaining

$$\mathbf{p} = (m_1 + m_2)\,\mathbf{v}$$

If we divide each side of the above equation by the quantity $(m_1 + m_2)$, we can determine the final velocity vector

$$\mathbf{v} = \mathbf{p}\,/\,(m_1 + m_2)$$

"Bouncy" Objects

Figure 3-3A shows two objects having masses m_1 and m_2 moving at speeds v_1 and v_2, respectively. As before, the two objects follow a collision course, and the

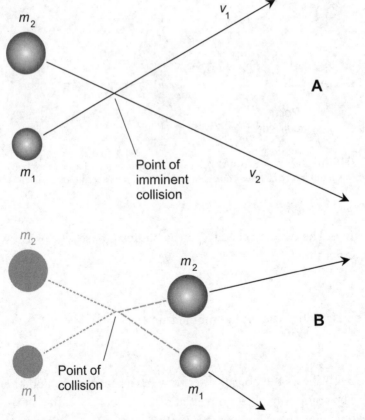

FIGURE 3-3 · At A, two "bouncy" objects having different velocities approach each other on a collision course. At B, the collision has occurred.

velocity vectors $\mathbf{v}_1$ and $\mathbf{v}_2$ point in the directions shown by the arrows. Also as before, the momentum of the object with mass m_1 is

$$\mathbf{p}_1 = m_1\mathbf{v}_1$$

and the momentum of the object with mass m_2 is

$$\mathbf{p}_2 = m_2\mathbf{v}_2$$

Prior to the collision, our situation here looks the same as the one portrayed by Fig. 3-2, but appearances deceive! This time, our objects consist of different stuff. When they hit each other, they'll bounce apart instead of sticking together.

In the scenario of Fig. 3-3B, the objects have just glanced off each other. Their individual masses have not changed, but their velocities have. Therefore, their individual momentum vectors have changed. According to the law of conservation of momentum, the total momentum of the system has not changed because of the collision. Suppose that the new velocity of m_1 is $\mathbf{v}_{1a}$, and the new velocity of m_2 is $\mathbf{v}_{2a}$. The new momentums of the objects are therefore

$$\mathbf{p}_{1a} = m_1\mathbf{v}_{1a}$$

and

$$\mathbf{p}_{2a} = m_2\mathbf{v}_{2a}$$

According to the law of conservation of momentum, we have

$$\mathbf{p}_1 + \mathbf{p}_2 = \mathbf{p}_1 + \mathbf{p}_{2a}$$

so it follows that

$$m_1\mathbf{v}_1 + m_2\mathbf{v}_2 = m_1\mathbf{v}_1 + m_2\mathbf{v}_{2a}$$

The examples shown in Figs. 3-2 and 3-3 represent idealized situations. In the physical world as we experience it, complications would arise. We've ignored such factors here for the sake of demonstrating the basic principles.

Still Struggling

Do you wonder whether or not the collisions shown in these drawings would impart *spin* to the composite mass (in Fig. 3-2) or to either or both masses (in Fig. 3-3)? In the real world, that would usually happen, and it would make our calculations vastly more complicated. In these idealized examples, we assume that no spin results from the collisions.

PROBLEM 3-2

Imagine that you have two electric toy trains set up on a long, straight track running east and west. Train A has mass 1.60 kg and travels east at 0.250 m/s. Train B has mass 2.50 kg and travels west at 0.500 m/s. The trains have "sticky pads" on the fronts of their engines so that if they crash, they won't bounce apart. You set the trains up with the intent of creating a head-on wreck! Suppose that the wheel bearings have no internal friction, and also that the wheels roll along the track with no friction. Further suppose that, at the instant the trains hit each other, you shut off the power to the engines. How fast, and in what direction, will the composite train move after the crash, assuming that neither train derails?

SOLUTION

First, calculate the momentum of each train. Call the masses of the trains m_a and m_b, respectively. Assign the direction east as positive, and west as negative. (You can do this because the two directions lie exactly opposite each other along a straight line.) Represent the velocity vector of train A as v_a. Represent the velocity vector of train B as v_b. For train A, you have

$$m_a = 1.60 \text{ kg}$$

and

$$v_a = +0.250 \text{ m/s}$$
$$\text{going east}$$

For train B, you have

$$m_b = 2.50 \text{ kg}$$

and

$$v_b = -0.500 \text{ m/s}$$
also going east

The trains' momentums, respectively, are therefore

$$p_a = m_a v_a$$
$$= (1.60 \text{ kg})(+0.250 \text{ m/s})$$
$$= +0.400 \text{ kg} \cdot \text{m/s}$$
going east

and

$$p_b = m_b v_b$$
$$= (2.50 \text{ kg})(-0.500 \text{ m/s})$$
$$= -1.25 \text{ kg} \cdot \text{m/s}$$
also going east

The sum total of their momentum vectors is

$$p = p_a + p_b$$
$$= +0.400 \text{ kg} \cdot \text{m/s} + (-1.25 \text{ kg} \cdot \text{m/s})$$
$$= -0.850 \text{ kg} \cdot \text{m/s}$$
going east

The mass m of the composite equals the sum of the masses of trains A and B, which remain the same throughout this experiment:

$$m = m_a + m_b$$
$$= 1.60 \text{ kg} + 2.50 \text{ kg}$$
$$= 4.10 \text{ kg}$$

Denote the final velocity as v. You know that momentum is conserved in this collision, as in all ideal collisions. Therefore, the final velocity equals the final momentum p divided by the final mass m, as follows:

$$v = p/m$$
$$= (-0.850 \text{ kg} \cdot \text{m/s}) / (4.10 \text{ kg})$$
$$\approx -0.207 \text{ m/s}$$
going east

That's the same as 0.207 m/s going west.

Arithmetic with Units

We can add, subtract, multiply, and divide units, just as we can do with plain numbers. For example, 0.850 kg · m/s divided by 4.10 kg causes kilograms to cancel out in the quotient, yielding m/s. You can keep the units in your calculations as you go through all the arithmetic, thereby ensuring that the units in the final result will make sense. If you had come up with, say, kg · m (kilogram-meters) in Solution 3-2, you'd know that you made a mistake somewhere, because kilogram-meters do not express speed.

Arithmetic with Vectors

We can multiply and divide vector quantities by scalar quantities. When we do either of those things, we always get another vector. For example, in Solution 3-2 we divided momentum (a vector) by mass (a scalar). However, we can't do addition or subtraction directly between quantities when one of them is a vector and the other is scalar. We can't multiply vectors by each other unless we know whether we should take the *dot product* (which yields a scalar) or the *cross product* (which yields a vector). If you don't remember how to find dot products and cross products, you might want to refresh your knowledge of pre-calculus.

 PROBLEM 3-3 _____

Imagine two electric toy trains set up as described in Problem 3-2. Suppose that train A has a mass 2.00 kg and travels east at 0.250 m/s, while train B has a mass of 1.00 kg and travels west at 0.500 m/s. How fast, and in what direction, will the composite train trundle along the tracks after the crash?

 SOLUTION _____

Call the masses of the trains m_a and m_b, respectively. Assign the direction east as positive, and west as negative. Represent the velocity vectors as v_a and v_b. For train A, you have the known quantities

$$m_a = 2.00 \text{ kg}$$

and

$$v_a = +0.250 \text{ m/s}$$

For train B, you have

$$m_b = 1.00 \text{ kg}$$

and

$$\mathbf{v}_b = -0.500 \text{ m/s}$$

The momentums, respectively, are

$$\mathbf{p}_a = m_a \mathbf{v}_a$$
$$= (2.00 \text{ kg})(+0.250 \text{ m/s})$$
$$= +0.500 \text{ kg} \cdot \text{m/s}$$

going east

and

$$\mathbf{p}_b = m_b \mathbf{v}_b$$
$$= (1.00 \text{ kg})(-0.500 \text{ m/s})$$
$$= -0.500 \text{ kg} \cdot \text{m/s}$$

also going east

The sum total of their momentums is

$$\mathbf{p} = \mathbf{p}_a + \mathbf{p}_b$$
$$= +0.500 \text{ kg} \cdot \text{m/s} + (-0.500 \text{ kg} \cdot \text{m/s})$$
$$= 0 \text{ kg} \cdot \text{m/s}$$

going nowhere

The mass m of the composite train is

$$m = m_a + m_b$$
$$= 2.00 \text{ kg} + 1.00 \text{ kg}$$
$$= 3.00 \text{ kg}$$

The final velocity $\mathbf{v}$ is therefore

$$\mathbf{v} = \mathbf{p}/m$$
$$= (0 \text{ kg} \cdot \text{m/s})/(3.00 \text{ kg})$$
$$= 0 \text{ m/s}$$

going nowhere

After the crash, your composite train will rest in place on the tracks, not moving either way. You can describe its velocity as **0** (the zero vector).

Still Struggling

Does the foregoing result seem impossible? If momentum is conserved—neither created nor destroyed—how can it vanish as the result of any collision, or for that matter, any event whatsoever? That's an excellent question, and the answer takes a little hard thought. The total momentum of this system actually equals zero before the crash, as well as after it! Remember that momentum is a vector quantity. Look at the above equations for momentum again:

$$\mathbf{p}_a = m_a \mathbf{v}_a$$
$$= (2.00 \text{ kg})(+0.250 \text{ m/s})$$
$$= +0.500 \text{ kg} \cdot \text{m/s}$$

and

$$\mathbf{p}_b = m_b \mathbf{v}_b$$
$$= (1.00 \text{ kg})(-0.500 \text{ m/s})$$
$$= -0.500 \text{ kg} \cdot \text{m/s}$$

The momentum vectors of the two trains have equal magnitude but opposite direction. So their vector sum equals **0** (the zero vector) before they collide, as well as afterward.

Work

In physics, *work* refers to a specific force applied over a specific distance. We do work when we lift an object of nonzero mass (which we might call a "weight" or "mass") directly against the force of gravity. We can calculate the work w done by a force of magnitude F over a displacement of magnitude q using the formula

$$w = Fq$$

The standard unit of work is the *newton-meter* (N · m), equivalent to a *kilogram-meter squared per second squared* (kg · m²/s²).

Work as a Dot Product of Vectors

If we want to express the above-described relation completely, we must use vectors to represent the force and displacement. How can we multiply two vectors? Fortunately, in this case, we don't have a complicated scenario. When we perform work on a mass, the force and displacement vectors usually point in the same direction. As things work out, the dot product, also known as the *scalar product*, provides our answer. We can therefore write the foregoing formula in more "powerful" form as

$$w = \mathbf{F} \bullet \mathbf{q}$$

where $\mathbf{F}$ is the force vector represented as newtons in a certain direction, and $\mathbf{q}$ is the displacement vector represented as meters in a certain direction. We use a heavy dot (•) to represent the dot product of two vectors. The use of the heavy dot distinguishes the dot product from the ordinary scalar product of two variables, units, or numbers (as we find, for example, in the unit expression kg · m²/s²), where we use a small elevated dot roughly the same size as a period or decimal point.

As long as the force and displacement vectors point in the same direction, we can simply multiply the vector magnitudes to calculate the amount of work done. Never forget the fact that work constitutes a scalar quantity, not a vector quantity.

Lifting an Object

Imagine a 1.0-kg object lifted upward against the earth's gravity with a rope-and-pulley system. Suppose the pulley has no friction, and the rope doesn't stretch. You stand on the floor, holding the rope, and pull downward. You exert a certain force over a certain distance. The force and displacement vectors through which your hands move point in the same direction. You can wag your arms back and forth while you pull, but in practice this won't make any difference in the amount of work required to lift the object a certain distance, so let's keep things simple and suppose that you pull in a straight line.

The force of your pulling downward translates to an equal force vector $\mathbf{F}$ directly upward on the mass, as shown in Fig. 3-4. The mass moves upward

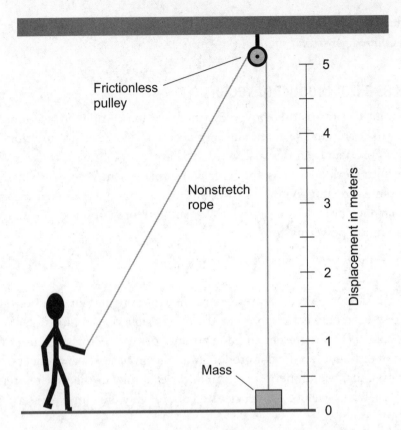

FIGURE 3-4 • We perform work when we apply a force over a specific distance. In this case, we apply the force upward against the earth's gravitation.

exactly as far as you pull the rope, that is, by displacement **q**. As you pull on the rope, you exert a force that precisely counteracts the force of gravitation on the mass. The force of gravity, $\mathbf{F}_g$, on the object equals the product of the object's mass m and the acceleration vector $\mathbf{a}_g$ of gravity. The value of $\mathbf{a}_g$ is approximately 9.8 m/s² acting downward. Therefore, as you lift the object, you impose upon the mass a force of

$$\mathbf{F} = m\mathbf{a}_g$$
$$= (9.8 \text{ m/s}^2)(1.0 \text{ kg})$$
$$= 9.8 \text{ kg} \cdot \text{m/s}^2$$
$$= 9.8 \text{ N upward}$$

PROBLEM 3-4

Consider the example described above and illustrated by Fig. 3-4. Imagine that you lift the object 1.5 m. How much work have you done?

SOLUTION

To move the object, the upward force **F** equals the product of the mass, $m = 1.0$ kg, and the acceleration of gravity, $a_g = 9.8$ m/s^2, as follows:

$$\mathbf{F} = m\mathbf{a}_g$$

$$= (1.0 \text{ kg})(9.8 \text{ m/s}^2)$$

$$= 9.8 \text{ kg} \cdot \text{m/s}^2$$

$$= 9.8 \text{ N upward}$$

You apply this force over a distance **q** = 1.5 m going straight up. Therefore, the work w equals the dot product **F** • **q**. Because **F** and **q** point in the same direction, you can simply multiply their magnitudes to get

$$w = Fq$$

$$= (9.8 \text{ kg} \cdot \text{m/s}^2)(1.5 \text{ m})$$

$$= 14.7 \text{ kg} \cdot \text{m}^2/\text{s}^2$$

You should round this off to 15 kg · m^2/s^2, because your input data only goes to two significant figures.

Still Struggling

Does this unit of work—the *kilogram-meter squared per second squared*—seem arcane and awkward? Well, it is! You might rather think of it as a *newton-meter*. However, physicists have another name for this unit, the *joule*, which we can abbreviate by writing J. You'll encounter the joule often if you do much study in any science or engineering field. When you lift the object in the example of Problem 3-4, you perform approximately 15 J of work.

Energy

Energy can exist in diverse forms. From time to time, we hear news about an "energy crisis." When people talk about that sort of thing, they usually refer to acute or anticipated shortages of the energy available from *fossil fuels* such as coal, oil, and (to a lesser extent) methane, also called "natural gas." Imagine that you find a lump of coal. Where's its energy? It looks like an inert rock. But if you light it on fire (don't try that at home!), you'll discover its energy content in a hurry. Scientists express and measure energy in joules, just as they do with work. In fact, one of the most well-known definitions of energy tells us that it's *the capacity to do work.*

Potential Energy in Mechanical Form

Look again at the situation shown by Fig. 3-4. When you raise an object of mass *m* upward through a displacement **q**, you apply an upward force **F** to it. Now imagine that you suddenly release the rope. The mass falls to the floor. Suppose that *m* = 5 kg, and that the mass is a hard, solid brick. If you raise the brick a couple of millimeters and then let it drop, it will strike the floor without much fanfare. If you raise the brick 2 m and then let it go, it will likely damage the floor when it hits. If you raise the brick 100 m and then let it go, you'll have big trouble when it reaches the floor! The dropping and landing of a heavy object can perform a useful task if controlled in the right way, such as pounding a stake into the ground. Such action can also cause considerable havoc, especially if you're the sort of person who has bad luck dropping massive bricks from great heights.

When you lift an object, you give it the ability to do work; you provide it with a certain amount of *potential energy*. In a mechanical sense, potential energy represents the same thing as work. If you impose a force vector of magnitude *F* upon an object against the earth's gravitation, and if by doing so you lift that object through a displacement vector of magnitude *q*, then you can calculate the potential energy, E_p, using the formula

$$E_p = Fq$$

Potential Energy in Other Forms

The foregoing formula provides us with a simplistic, mechanical view of potential energy. As we've discussed, potential energy can exist in a lump of coal (even if we don't lift it). When we burn the coal, we liberate a certain number of joules, mainly as thermal energy. Potential energy also exists in electrochemical cells or batteries, such as a lantern battery. When we connect a light bulb across an electrochemical battery, we liberate a certain number of joules, mainly in the form of visible light. Potential energy exists in wood, gasoline, oil, methane, propane, rocket fuel, and many other substances that seem "inert" when we merely look at them. With the help of the appropriate systems, we can turn such potential energy into heat, light, electricity, mechanical work, and other forms for our convenience.

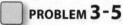

PROBLEM 3-5

Refer again to Fig. 3-4. If the object has a mass of 5.004 kg and you lift it 3.000 m, how much potential energy will it attain? Take the magnitude of earth's gravitational acceleration as $a_g = 9.8067$ m/s². You can neglect vectors here, because everything takes place along a single straight line.

SOLUTION

First, determine the force required to lift a 5.004-kg object in the earth's gravitational field, as follows:

$$F = ma_g$$
$$= (5.004 \text{ kg})(9.8067 \text{ m/s}^2)$$
$$= 49.0727268 \text{ N}$$

The potential energy is

$$E_p = Fq$$
$$= (49.0727268 \text{ N})(3.000 \text{ m})$$
$$= 147.2181804 \text{ J}$$

The least accurate input data has four significant figures, so you can conclude that the mass gains potential energy of $E_p = 147.2$ J.

Kinetic Energy

Suppose that you lift the object in the scenario of Fig. 3-4 by a certain distance, imparting to it a potential energy E_p. If you let go of the rope so that the object falls back to the floor, the object might do damage, either to the floor or to itself, when it hits. The object will move, and in fact will accelerate, from the moment you drop it until the moment it hits the floor. All of the potential energy that you imparted to the object by lifting it will instantly change to other forms: vibration, sound, heat, and possibly the outward motion of solid material fragments!

Now think of the situation an infinitesimal moment—an instant—before the object strikes the floor. At this point in time, the *kinetic energy*, E_k, in the mass will exactly equal the potential energy that you gave it when you lifted it. Mathematically, we have

$$E_k = Fq$$

$$= ma_g q$$

$$= 9.8 \, mq$$

where F is the force applied, q is the distance over which you raised the object (and thus the distance it falls), m is the mass of the object, and a_g is the acceleration of gravity. Here, we consider a_g as 9.8 m/s², accurate to only two significant figures.

We can express E_k for a moving object having a mass m in another way, as follows:

$$E_k = mv^2/2$$

where v represents the speed of the object at the instant before impact. We could use the formulas for displacement, speed, and acceleration from the previous chapter to calculate the instantaneous speed of the object when it hits the floor, but we don't have to do that. We already have a formula for kinetic energy in the example of Fig. 3-4. The mass-velocity formula applies to any moving object, even if we don't perform any work on it.

TIP   *Let's take careful note of the fact that we use the lowercase italic m for mass, and the lowercase nonitalic m for meter(s). Don't confuse these two! Mass expresses material quantity. The meter constitutes a physical unit. We symbolize them both in almost the same way, but in action they could hardly differ more. Other "dichotomies" of this sort occasionally arise in physics, engineering, and mathematics—so watch out for them!*

PROBLEM 3-6

Refer again to Fig. 3-4. Suppose the object has mass $m = 1.0$ kg and you raise it by a distance of 4.0 m. Determine its kinetic energy just before it strikes the floor, according to the force/displacement method. Use 9.8 m/s² as the value of a_g, the acceleration of gravity.

SOLUTION

You can take advantage of the formula $F = ma_g q$. When you set $m = 1.0$ kg, $a_g = 9.8$ m/s², and $q = 4.0$ m, you'll obtain

$$E_k = 1.0 \text{ kg} \times 9.8 \text{ m/s}^2 \times 4.0 \text{ m}$$

$$= 39.2 \text{ kg} \cdot \text{m}^2/\text{s}^2$$

$$= 39.2 \text{ J}$$

You know each input value to an accuracy of two significant figures, so you should round the answer off to $E_k = 39$ J.

PROBLEM 3-7

In the example of Problem 3-6, determine the kinetic energy of the object the instant before it strikes the floor, by using the mass/speed method. Demonstrate that it yields the same final answer, in the same units, as the method used in Problem 3-6.

SOLUTION

You need to use a little trick here: Calculate the object's speed at the instant before it strikes the floor. The process goes along straightforwardly enough, although you'll find it rather tedious. Are you ready?

First, let's figure out how long it takes for the object to fall. Recall from the previous chapter that

$$q = a_{avg} t^2/2$$

where q represents the displacement, a_{avg} represents the average acceleration, and t represents the elapsed time. In this situation, we can replace a_{avg} with a_g, because the earth's gravitational acceleration never changes;

it always stays exactly at its average value. We can manipulate the above formula to solve for time, as follows:

$$t = (2q/a_g)^{1/2}$$
$$= [(2 \times 4.0 \text{ m})/(9.8 \text{ m/s}^2)]^{1/2}$$
$$= (8.0 \text{ m} \times 0.102 \text{ s}^2/\text{m})^{1/2}$$

"Wait!" you say. "What have we done with a_g? Where does this s^2/m unit come from?" We multiply by the reciprocal of a_g, which constitutes the same action as dividing by a_g. When we take the reciprocal of a quantity expressed in terms of a unit, we must also take the reciprocal of the unit. That's where the s^2/m "unit" comes from. Continuing, we have

$$t = (8.0 \text{ m} \times 0.102 \text{ s}^2/\text{m})^{1/2}$$
$$= (0.816 \text{ m} \cdot \text{s}^2/\text{m})^{1/2}$$
$$= (0.816 \text{ s}^2)^{1/2}$$
$$= 0.9033 \text{ s}$$

Meters cancel out in the above process. Units, just like numbers and variables, cancel out to *unity* (that is, the exact value 1) when we divide them by themselves. Let's not round 0.9033 s off just yet. We still have some more calculations to do.

Recall from the previous chapter the formula for the relationship among instantaneous velocity v_{inst}, acceleration a, and time t for an object that accelerates at a constant rate:

$$v_{inst} = at$$

Here, we can replace a with a_g as before to get

$$v_{inst} = a_g t$$

We know both a_g and t already, so we can calculate

$$v_{inst} = (9.8 \text{ m/s}^2)(0.9033 \text{ s})$$
$$= 8.85234 \text{ m/s}$$

Don't worry about the fact that we keep getting more and more superfluous digits in our numbers. We'll round our answer off when we've finished all the arithmetic.

We now have only one more calculation to make. We must use the formula for E_k in terms of v_{inst} and m. We know that $v_{inst} = 8.85234$ m/s and $m = 1.0$ kg, so we can proceed as follows:

$$E_k = mv_{inst}^2/2$$
$$= (1.0 \text{ kg})(8.85234 \text{ m/s})^2/2$$
$$= (78.3639 \text{ kg} \cdot \text{m}^2/\text{s}^2)/2$$
$$= 39.18195 \text{ kg} \cdot \text{m}^2/\text{s}^2$$

The unit in this result, kg · m²/s², represents an alternative name for the joule (J). Because we can claim an accuracy of only two significant figures, we must round our answer off to 39 J. That's same final result, expressed in the same units, as you got when you used the force/displacement method to solve Problem 3-6.

Still Struggling

You might reasonably wonder why we would ever want to work out a problem with a scheme as convoluted Solution 3-7. Of course, given a choice, we would use the method of Solution 3-6 to determine kinetic energy in a scenario such as that illustrated by Fig. 3-4. We dragged ourselves through Problem 3-7 as an exercise—merely to verify the fact that either method will work.

Power

In the context of physics, *power* expresses the rate at which a system expends energy or converts it to another form. Mechanically, power quantifies the rate at which work is done. The standard unit of power is the *joule per second* (J/s), more commonly known as the *watt* (W). When we talk about power, we usually work with kinetic energy. Sometimes, power expresses the rate at which a system stores up or accumulates potential energy.

Mechanical Power

In the foregoing examples illustrated by Fig. 3-4, the object acquires potential energy when we lift it. All of that potential energy "morphs" into kinetic energy

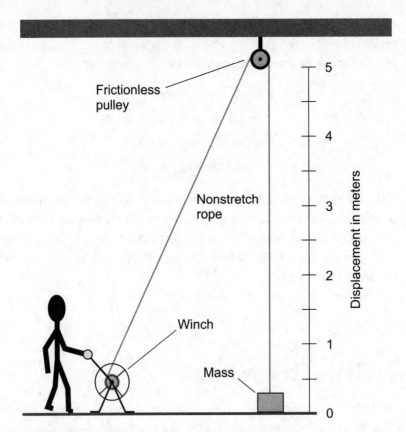

FIGURE 3-5 · We can use a winch and pulley to lift a heavy object, but we must expend power to do it.

if we let the object fall. The final burst of sound, shock waves, and perhaps outflying shrapnel manifests the last of the kinetic energy that we originally imparted to the object by lifting it up. "So," you wonder, "Where does power fit into this scenario?"

Imagine that, instead of the free end of a rope in your hand, you have a winch that you can turn to raise the object, as shown in Fig. 3-5. The object starts out sitting on the floor, and you crank the winch (or use a motor to crank it) to raise the object. You might actually need the winch, possibly in conjunction with a pulley system, if you want to lift a particularly massive object.

Let's Do It!

You'll need to expend energy to lift the object off the floor, no matter what sort of apparatus you use. Suppose that you crank the winch, imparting potential energy to the object. If you employ a complicated system of pulleys, you might not have to bear down on the winch with much force even if the object has considerable mass; but such a system will increase the number of times you'll have to turn the crank to move the object upward over a given distance.

You can express the rate at which you expend energy cranking the winch in watts, thereby describing the power that you impose on the system from moment to moment in time. The faster you crank the winch, for a given object mass, the more power you'll expend at any particular instant. The more massive the object, for a given winch-cranking speed, the more power you'll have expend at any given moment. However, the power you expend at any point in time does *not* depend on how high you ultimately raise the mass above the floor. In theory, you could expend a little bit of power for a long time and lift the mass by 15 m, or 150 m, or 1.5 km. At the other extreme, you could expend a lot of power for a short time and lift the mass by only 2.3 m, or 2.3 cm, or 2.3 mm.

Now imagine that you have an ideal system: The winch and pulley have no friction, and the rope doesn't stretch at all. Suppose that you crank the winch at a constant rotational rate, so you generate constant power. The power that you expend in terms of "strain and sweat," multiplied by the time that you spend applying that power, precisely equals the amount of potential energy that you impart to the object as time passes. (This simplistic relation holds only if the power remains constant, however.) If you let P represent the power in watts, and if you let t represent the time in seconds during which you apply the constant power P, then you can calculate the potential energy E_p that you impart to the object, in joules, using the formula

$$E_p = Pt$$

You can rearrange this formula to get

$$P = E_p/t$$

You know the potential energy equals the mass m times the acceleration of gravity a_g times the displacement q. So you can directly calculate the power with the formula

$$P = ma_g q/t$$

PROBLEM 3-8

Imagine that you want to lift a mass of 208 kg over a distance of 2.503 m in a time of 7.165 s, expending power at a constant rate. How much power will you expend? Take the acceleration of gravity as 9.8067 m/s².

SOLUTION

You can use the formula given above, letting a_g = 9.8067. When you do the arithmetic, you obtain

$$P = 9.8067 \ mq/t$$
$$= 9.8067 \times 208 \times 2.503 / 7.165$$
$$= 713 \ W$$

Still Struggling

Did you notice that in solving Problem 3-8, we didn't carry the units through the entire expression, multiplying and dividing them out along with the numbers? We don't have to do that once we know that a formula works, and as long as we keep our units consistent with each other in the context of the formula. In this case, we employ SI base units (meter, kilogram, second) exclusively, so we can have confidence that everything will come out okay in the end.

Electrical Power

You might wish to spare yourself the menial labor of cranking a winch to lift heavy objects over and over, merely to perform experiments that demonstrate the nature of power. Mechanical power tends to defy direct measurement anyhow, although we can calculate it theoretically as we've done in Problem 3-8.

Imagine that you connect an electric motor to the winch, as shown in Fig. 3-6. If you connect an *electrical wattmeter* between the power source and the motor, you can measure the mechanical power indirectly. For this method to provide a true indication of the power, however, you'll need an ideal system, where the motor functions at 100-percent efficiency, the rope does not stretch, and the pulley system has no friction. Of course a real-world pulley always exhibits

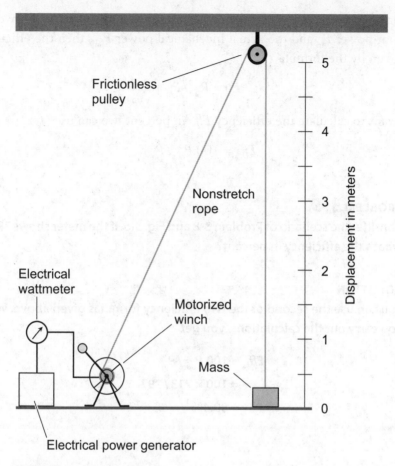

FIGURE 3-6 • We can indirectly measure electrical power by using a motor to drive a winch, lifting a heavy object.

friction; a real-world rope inevitably stretches a little, and a real-world electric motor never converts every watt of electrical input power into mechanical power. Therefore, if you were to actually perform the experiment portrayed in Fig. 3-6, you'd observe a wattmeter reading slightly greater than the reading you'd get if you calculated the mechanical power directly as described in Solution 3-8.

System Efficiency

Suppose we connect the apparatus of Fig. 3-6 and do the experiment described in Problem 3-8. The wattmeter might show something like 797 W. In that case, we can calculate the efficiency of the whole system by dividing the actual mechanical power (713 W) by the measured input power (797 W). If we call the input power P_{in} and the actual mechanical power P_{out}, then the efficiency *Eff* is given by the formula

$$Eff = P_{out} / P_{in}$$

If we want to calculate the efficiency $Eff_\%$ in percent, we can use

$$Eff_\% = 100 \, P_{out} / P_{in}$$

PROBLEM 3-9

Consider the scenario of Problem 3-8 and Fig. 3-6. If the meter shows 797 W, what's the efficiency in percent?

SOLUTION

You can use the second of the two efficiency formulas given above. When you carry out the calculations, you get

$$Eff_\% = 100 \, P_{out} / P_{in}$$
$$= 100 \times 713 / 797$$
$$= 89.4\%$$

QUIZ

Refer to the text in this chapter if necessary. A good score is eight correct. Answers are at the back of the book.

1. We can represent impulse in terms of
   A. a constant force applied for a certain length of time in a specific direction.
   B. a constant speed for a certain length of time in a specific direction.
   C. a constant acceleration for a certain length of time in a specific direction.
   D. a constant displacement over a certain span of time in a specific direction.

2. Suppose that you lift an object of mass 5.00 kg straight upward over a distance of 4.00 m on a planet where the gravitational acceleration is 2.00 m/s². How much mechanical energy must you expend to do this?
   A. 2.50 J
   B. 10.0 J
   C. 40.0 J
   D. 160 J

3. Imagine that you fire a rocket engine to push a spacecraft through interstellar space, so that the ship accelerates at a constant rate. When you multiply the vessel's mass by the length of time for which the rocket engine pushes it, you get
   A. momentum.
   B. velocity.
   C. impulse.
   D. None of the above

4. Envision a minor motor-vehicle accident. Car X backs out of a parking space at 1.000 m/s toward the east. Car Y, whose driver searches for a place to park, travels north at 1.000 m/s. Neither driver sees the other car, and the cars collide. Suppose that each car (including its driver) has a mass of 1000 kg. The total system momentum vector before the collision is approximately
   A. 1000 kg · m/s toward the northeast.
   B. 1414 kg · m/s toward the northeast.
   C. 2000 kg · m/s toward the northeast.
   D. zero, because the vehicles haven't hit each other yet!

5. Consider a hockey player who skates across the ice at a constant speed of 5.000 m/s in a straight line toward the goal. He has a total body mass (including uniform, skates, and hockey stick) of 80.00 kg. How much kinetic energy does he have?
   A. 400.0 J
   B. 1000 J
   C. 1.600×10^4 J
   D. 3.200×10^5 J

6. Imagine that a 4000-kg motorboat sits still on a frictionless lake on a windless day. The captain starts the motor and runs it steadily for 5.000 s in a forward direction. Then she shuts the motor down, and the boat coasts forward at a speed of 5.000 m/s. How much forward impulse did the motor impose on the boat?

   A. 2.000×10^4 kg · m/s
   B. 1.000×10^5 kg · m/s
   C. 160 kg · m/s
   D. We need more information to answer this.

7. We can define power as

   A. the amount of energy that we expend, dissipate, consume, or supply, multiplied by the total elapsed time.
   B. the rate, at any given moment in time, at which we expend, dissipate, consume, or supply energy.
   C. the average amount of energy that we we expend, dissipate, consume, or supply over a defined period of time.
   D. the sum of the potential energy and the kinetic energy that we expend, dissipate, consume, or supply over a defined period of time.

8. Suppose that two objects, both moving at constant velocity, collide in a closed system, sticking to each other after the impact. How does the total system momentum after the collision compare with the total system momentum before the collision?

   A. It depends on the relative masses of the objects.
   B. It depends on the relative velocities of the objects.
   C. It stays the same.
   D. It increases.

9. When we make a calculation involving quantities expressed in incompatible units (such as kilograms, feet, and minutes),

   A. we can directly multiply or directly divide the units by each other.
   B. we can directly multiply the units by each other, but we can't directly divide them.
   C. we can directly divide the units by each other, but we can't directly multiply them.
   D. we can neither directly multiply nor directly divide the units by each other.

10. We can theoretically employ units of gram-centimeters per second to quantify

   A. velocity.
   B. acceleration.
   C. kinetic energy.
   D. momentum.

chapter **4**

Particles of Matter

Centuries ago, a few *alchemists* (the forerunners of today's chemists and physicists) hypothesized that all matter must consist of tiny, invisible particles, which they called *atoms*. The *atomic theory of matter* allowed scientists to explain why some substances have greater density than others, some materials retain their shape while others flow freely, and we can see some substances but not others.

CHAPTER OBJECTIVES

In this chapter, you will

- Discover the nature of matter.
- Explore the interior of an atom.
- Observe subatomic particles in action.
- See how matter can become energy.
- Learn the truth about antimatter.
- Split and combine atoms.

Early Theories

Every atom, no matter what other properties it has, comprises *subatomic particles* of incredible density and minuscule volume. We perceive matter as "continuous" because the atoms, and the subatomic particles within them, are too small to see, even with ordinary microscopes. The particles move so fast that we could never resolve them individually even if we could see them.

More "Space" than "Stuff"

The distances between the particles in an atom dwarf the diameters of the particles themselves. Have you heard scientists say that matter contains mostly empty space? In a literal sense, that's true. If we could shrink ourselves down to the subatomic scale and slow time in proportion, a piece of metal might look like a huge swarm of gnats. We would gape at the scene around us, and find it difficult to believe that full-size people really see and feel that "stuff" as a solid object.

Before microscopes existed, scientists deduced the particle nature of matter by observing the behavior of things like water, rocks, and metals. These substances present themselves in myriad different forms. But any given material, such as copper, water, or glass, appears the same at room temperature and normal atmospheric pressure, no matter where we find it. Even without doing any complicated experiments, early physicists figured out that substances could exhibit these consistent behaviors only if they possessed diverse particle arrangements.

The Elements

Until about the year 1900, plenty of respectable people refused to believe the atomic theory of matter. Today, almost everyone accepts that notion; rational people dismiss those who don't as cranks. The atomic theory explains the behavior of matter better than any other model.

Eventually, scientists identified 92 unique substances that occur in nature, and called them the *chemical elements*. Later, scientists discovered that they could synthesize additional elements. Using machines known as *particle accelerators*, sometimes called "atom smashers," nuclear physicists have fabricated human-made elements that can't exist in nature—at least not under conditions resembling anything we would imagine as normal.

Uniqueness

When you have samples of two different elements, their atomic structures differ. The slightest change in the internal structure of an atom can make a tremendous difference in the behavior of the corresponding substance on a large scale.

You can live by breathing pure oxygen gas, but you can't survive on an atmosphere of pure nitrogen gas. Oxygen will cause iron to corrode, but nitrogen will not. Wood will burn furiously in an atmosphere of pure oxygen, but will not even ignite in an atmosphere of pure nitrogen. Nevertheless, both of these two gases look, smell, and feel the same at normal temperature and pressure. Both gases are invisible, colorless, and odorless.

Still Struggling

You might wonder how two substances, such as oxygen and nitrogen, can look, feel, and smell identical, yet exhibit such different effects. Oxygen and nitrogen influence their surroundings differently because their atoms don't have quite the same internal structure. An oxygen atom has a few more particles than a nitrogen atom.

The Nucleus

An element maintains its identity on the basis of its *atomic nucleus*, a cluster of particles at the "center" of every atom. An atomic nucleus can contain one or more subatomic particles known as the *proton* and the *neutron*.

The Proton

All protons carry a small *positive electric charge*. Any individual proton carries the same *charge quantity* as every other individual proton. Every proton at rest has the same mass as every other proton at rest. Most scientists accept the proposition that all protons are identical, at least in our part of the universe, although they, like all other particles, gain mass if accelerated to extreme speeds. (You'll learn about that effect toward the conclusion of this course.)

While we can't see a single proton, and we can't detect its mass under ordinary conditions, a high-speed barrage of these particles can have dramatic effects.

If you could scoop up a teaspoon full of protons the way you scoop up a teaspoon full of sugar, the resulting sample would weigh tons. If you could hold a "stone" made of solid protons in your hand and then drop it, that "stone" would fall to the earth's center as easily as a lead shot falls to the bottom of a lake.

The Neutron

A neutron has a mass slightly greater than that of a proton. Neutrons have no electrical charge. An individual neutron has roughly the same density as an individual proton. But, while protons can exist intact for a long time all by themselves in free space, neutrons cannot. A single, isolated neutron has a so-called *mean life* (a fancy term for average life) of only about 15 minutes.

If you gathered up a batch of 1,000,000 neutrons and let them float around in space separately from each other, you'd find only about 500,000 neutrons after a quarter of an hour. After half an hour, you'd have approximately 250,000 neutrons. After three-quarters of an hour, about 125,000 neutrons would remain. After an hour, you'd have roughly 62,500 neutrons, and so on.

The Two Most Abundant Elements

The simplest chemical element, hydrogen, has a nucleus made up of a lone proton. Hydrogen constitutes the most abundant element in the universe. Sometimes a nucleus of hydrogen contains a neutron or two along with the proton, but we don't encounter such a thing very often. These mutant forms of hydrogen do, nevertheless, play significant roles in atomic physics.

Helium exists as the second most abundant element in the universe. In most cases, a helium atom has a nucleus with two protons and two neutrons. In the cores of all active stars including our own sun, hydrogen "morphs" into helium, liberating energy. This process, called *atomic fusion* or *nuclear fusion*, also makes possible the terrific power of a *hydrogen bomb*. Some scientists believe that we might eventually manage to harness nuclear fusion to generate useful energy for humankind.

Atomic Number

According to modern atomic theory, if we take any two individual protons and isolate them, we'll find them identical in every way. The same holds true for neutrons. The number of protons in an element's nucleus, the *atomic number*, gives that element its identity.

We've already learned about the elements whose nuclei contain one or two protons. The element with three protons is *lithium*, a light metal. The element with four protons is *beryllium*, also a metal. Carbon has six protons in its nucleus, nitrogen has seven, and oxygen has eight. In general, as the number of protons in an element's nucleus increases, the number of neutrons also increases. Also in general, elements with high atomic numbers, such as lead, have greater density than elements with low atomic numbers, such as carbon. Perhaps you've compared a lead shot with a lump of coal of similar size, and noticed this difference.

Changing an Atom's Identity

If you could add two protons to the nucleus of every atom in a sample of carbon, you'd end up with an equal number of oxygen atoms. That theoretical principle seems simple enough, but such processes have proven difficult to carry out in practice, even in well-equipped modern laboratories.

In ancient times, alchemists tried to make elements change. The most well-known of their pursuits was the quest to turn lead (atomic number 82) into gold (atomic number 79). As far as anyone knows, nobody has ever done it.

In the 1940s, when physicists and engineers built and tested the first atomic bombs, humans finally succeeded in "morphing" one chemical element into another. The end product, of course, bore no resemblance to the "golden ideal" for which the alchemists strove.

Isotopes

Within the atoms of any element, the number of neutrons can vary. But regardless of how many neutrons the nucleus contains, the element keeps its identity based on the atomic number. Differing numbers of neutrons produce various *isotopes* of the element.

Each element has one particular isotope that occurs most often in nature, but every element possesses more than one possible isotope. Changing the number of neutrons in an element's nucleus results in a difference in the mass, and also a difference in the density, of the element. For example, scientists sometimes call hydrogen containing a neutron or two in the nucleus, along with the proton, *heavy hydrogen*. The naturally occurring isotope of uranium has three more neutrons in its nucleus than the isotope used to make atomic weapons.

Adding or taking away neutrons from an atom's nucleus constitutes a somewhat less difficult feat than adding or taking away protons, but all such tasks involve high-energy physics technology. We can't simply take a balloon filled with nitrogen and expect to turn the gas into pure oxygen by blowing neutrons and protons in!

Atomic Mass

The *atomic mass*, sometimes called the *atomic weight*, of an element, approximately equals the sum of the number of protons and the number of neutrons in the nucleus. Scientists formally express this quantity in *atomic mass units* (AMU), where one AMU represents precisely 1/12 of the mass of the nucleus of the carbon isotope having six neutrons. That's the most common isotope of carbon, symbolized C-12, C12, or ^{12}C (yes, the superscript 12 appears *before* the C). Any individual proton or neutron has a mass of approximately 1/12 AMU, but an isolated neutron has a little more mass than an isolated proton.

We can uniquely identify any element by its atomic number, but the atomic mass of an element depends on the isotope. A well-known isotope of carbon, C-14, exists in trace amounts in virtually all carbon-containing substances. This fact has proven useful to geologists and archeologists. Samples of C-14 emit high-energy radiation, while samples of C-12 do not. The *radioactivity* of C-14 diminishes with time according to a well-known, predictable mathematical function, allowing geologists to determine when carbon-containing compounds formed, and thereby to determine the ages of the artifacts they find in their "dig sites."

In nuclear reactions capable of producing energy, such as the reactions that take place inside stars, atomic bombs, and nuclear power plants, a certain amount of mass is always given up—and converted into energy—in the transactions between the atoms. Even a tiny mass can liberate an enormous burst of energy. In his theory of relativity, Albert Einstein formalized this relation with the now-famous equation

$$E = mc^2$$

where E represents the energy produced (in joules), m represents the total mass (in kilograms) lost during the reaction, and c represents the speed of light in a vacuum (in meters per second). The value of c^2 boggles the imagination, amounting to approximately 90 quadrillion meters squared per second squared, which we can write in scientific notation as 9×10^{16} m²/s². The magnitude of c^2 explains why vast amounts of energy can come out of an atomic reaction involving samples having modest mass.

As of this writing, you can find information concerning all the known elements, including atomic number, atomic mass, and various other characteristics, on the Internet at www.webelements.com.

You can also enter "table of the elements" into your favorite search engine using the phrase mode.

Still Struggling

If you're astute, you might ask, "If neutrons decay so fast, how can matter exist in stable form?" As things work out in nature, neutrons die quickly when they're "all alone," but they last almost forever in an atom's nucleus when they're "with friends." If that weren't the case, matter as we know it couldn't exist. Neutrons can also survive for a long time when a huge number of them accumulate and fall in upon one another as a result of mutual gravitation. When a large star uses up all of its fuel and then explodes, the remaining matter can collapse into a "clump" of neutrons called a *neutron star*.

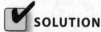

PROBLEM 4-1

Suppose that we could split the nucleus of an oxygen-16 (O-16) atom, which has eight protons and eight neutrons, exactly in two without liberating or absorbing any energy. What element would result? How many atoms of the element would we have?

SOLUTION

This reaction would produce two atoms of beryllium-8 (Be-8), each with four protons and four neutrons. The most common isotope of beryllium (Be-9), however, has five neutrons in its nucleus.

Outside the Nucleus

Surrounding the nucleus of an atom, we find *electrons* having electric charge opposite the protons' charge. Physicists arbitrarily call the electron charge negative, and the proton charge positive.

The Electron

An individual electron has the same charge quantity as an individual proton, but with opposite polarity. An electron has far less mass than a proton. You would need to gather up about 2000 electrons to get an object having the same mass as a single proton.

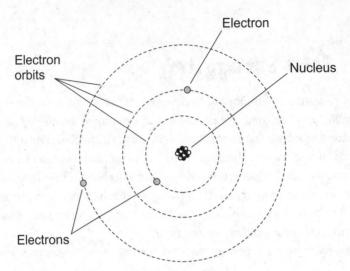

FIGURE 4-1 · An early model of the atom, developed around the year 1900.

One of the earliest theories concerning the structure of the atom pictured the electrons embedded in the nucleus, like raisins in a cake. Later, scientists envisioned the electrons "in orbit" around the nucleus. According to this model, every atom resembled a miniature "solar system" with the nucleus playing the role of the "sun" and the electrons playing the roles of the "planets," as shown in Fig. 4-1.

In today's model of the atom, the electrons surround the nucleus, but they follow "orbits" so complex that we can't pinpoint the location of any individual electron at any given instant of time. However, we can say that, at any particular time point, an electron will just as likely exist inside a certain sphere as outside. We call a sphere of this sort, centered at the nucleus, an *electron shell* or simply a *shell*. The greater a particular shell's radius, the more energy an electron in that shell has. All atoms have multiple, concentric shells. Figure 4-2 illustrates (in greatly simplified form) what happens when an electron gains exactly the right amount of energy to jump from one shell to a larger shell representing more energy.

In some materials, electrons can roam easily among the atoms. We call such substances *electrical conductors*. In other materials, the electrons don't "want" to move among atoms. We call these substances *electrical insulators*. In any case, we'll find it far easier to move electrons than to move protons or neutrons. The phenomenon that we call *electricity* almost always results, in some way, from the motion of electrons.

Generally, the number of electrons in an atom equals the number of protons. The negative charges therefore exactly cancel out the positive charges, giving

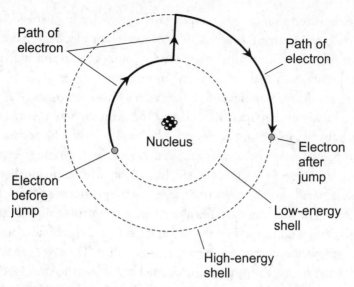

FIGURE 4-2 · Electrons "orbit" the nucleus of an atom at defined levels. Within an atom, an electron can jump or fall from one level to another.

us an *electrically neutral* atom. Under some conditions, an atom can have an excess or shortage of electrons. High levels of radiant energy, extreme heat, or the presence of an *electric field* (to be discussed later) can "knock" electrons loose from atoms, upsetting the charge balance.

Ions

If an atom has more or fewer electrons than protons, then the atom carries an electrical charge. A shortage of electrons results in positive charge; an excess of electrons produces a negative charge. An element's identity remains the same, no matter how great the excess or shortage of electrons. In the extreme case, an atom can lose all of its electrons, leaving only the nucleus, but even then, the element retains its identity.

When an atom has an electric charge, we call it an *ion*. If the atom lacks electrons so that it carries a positive charge, we can call it a *cation* (pronounced "cat-eye-on" or "catty-on"). If the atom has extra electrons so that it carries a negative charge, we can call it an *anion* (pronounced "an-eye-on" or "annie-on"). When a substance contains many ions, we say that the material is *ionized*.

We can cause ionization to take place when we heat a material sample to high temperatures, or when we place a sample between two regions of space having opposite electric charge. Ionization can also occur in a substance as a result of exposure to *ultraviolet*, *x rays*, *gamma rays*, or high-speed subatomic particles such as neutrons, protons, helium nuclei, or electrons. So-called *ionizing radiation*, more often called *radioactivity*, ionizes the atoms in living tissue, and can cause illness, death, and genetic mutations.

Lightning results from ionization in the earth's lower atmosphere. A vast electric charge builds up between a cloud and the ground, between two different clouds, or between two different regions of a single cloud. The accumulation of charge produces a powerful *electric field* that exerts forces on the electrons in the intervening air. These forces pull the electrons away from individual atoms. Ionized atoms generally conduct electric currents with greater ease than electrically neutral atoms. The ionization occurs along a jagged, narrow *channel* through the air. When a certain number of atoms become ionized, the channel conducts well enough to allow the charge discrepancy to equalize. Then we see a lightning *stroke*. After the stroke, the atomic nuclei quickly recover the stray electrons, and the air becomes electrically neutral for awhile. In a thundershower, however, the same sort of charge buildup will likely recur before long.

The atmosphere of our planet grows less dense with increasing altitude. Because of this, the ultraviolet and x-ray energy received from the sun gets more intense as we go higher. At certain altitudes, the gases in the atmosphere become ionized by solar radiation. These regions form the *ionosphere*. The ionosphere has many interesting properties. In particular, it can dramatically influence the propagation of some radio waves originating from transmitting stations on the earth's surface. The ionized layers refract the waves, bending them back toward the surface instead of allowing them to escape into outer space. This effect makes long-distance communication possible at the so-called shortwave frequencies. If you've ever heard a broadcast from a distant country on a "shortwave" radio receiver, you've directly witnessed the effects of *ionospheric radio-wave propagation*.

Still Struggling

Can an element exist as an ion and also as an isotope different from the usual isotope? The answer is yes. For example, an atom of carbon might have eight neutrons rather than the usual six, thus constituting the isotope C-14, and it might have one electron missing, giving it a positive unit electric charge and making it a cation.

PROBLEM 4-2

Imagine that we split the nucleus of an oxygen atom exactly in two, just as we did in Problem 4-1. Suppose that the original oxygen atom carries no electric charge, and no electrons are gained or lost during the reaction. Can both of the resulting atoms end up electrically neutral?

SOLUTION

Yes. The original oxygen atom must have eight electrons in order to exist in an electrically neutral state. If these eight electrons divide equally between the two beryllium atoms, each of which has four protons in its nucleus, then both beryllium atoms will end up with four electrons, rendering them electrically neutral.

PROBLEM 4-3

Consider the above scenario, except that this time, suppose that the oxygen atom has two electrons missing, making it a cation. Can the resulting two beryllium atoms both be electrically neutral, assuming no electrons are gained or lost during the reaction?

SOLUTION

No. We must have eight electrons, in total, for both beryllium atoms to end up neutral. It's possible for one beryllium atom to lack an electric charge, but in that case, the other beryllium atom will exist as a cation.

Energy from Matter

If we can manage to split an atomic nucleus into two smaller nuclei, we carry out a process called *nuclear fission*. In a sense, fission constitutes the opposite of nuclear fusion, which occurs inside the sun and other stars. The very first atomic bombs, developed in the 1940s, made use of fission reactions to produce energy. More powerful weapons, created in the 1950s, used atomic fission bombs to produce the high temperatures necessary to produce hydrogen fusion reactions.

Human-Caused and Natural Fission

A physicist can't snap an atomic nucleus apart as if it were a toy. Nuclear reactions can occur only under special conditions. Things don't take place in as straightforward a manner as the foregoing problems might suggest. To split atomic nuclei in the laboratory, physicists employ a *particle accelerator*. This machine takes advantage of electrical charges, magnetic fields, and other effects to hurl subatomic particles, at extreme speeds, at the nuclei of atoms to cause nuclear fission, often attended by the liberation of energy.

Some fission reactions occur spontaneously. Such a reaction can occur atom-by-atom over a long period of time, as with the decay of radioactive minerals in the environment such as C-14. A fission reaction may occur rapidly but under controlled conditions, as in a nuclear power plant. In the extreme, nuclear fission takes place almost instantaneously and out of control, as in an atomic bomb when we press two sufficiently massive samples of certain radioactive materials together.

Matter and Antimatter

The proton, the neutron, and the electron each have their own "nemesis" particle that occurs in the form of *antimatter*. It should not surprise you to learn that we call such particles *antiparticles*, naming the "big three" as follows:

- The proton has its counterpart in the *antiproton*.
- The neutron has its counterpart in the *antineutron*.
- The electron has its counterpart in the *positron*.

An antiproton has the same mass as a proton, but in a negative sort of way, and it carries a negative electric charge equal but opposite to the positive charge of the proton. The antineutron has the same mass as the neutron, but again, in a negative sense. Neither the neutron nor the antineutron have any electrical charge. The positron has same mass as the electron, but in a negative sense, and it holds a positive electric charge equal in quantity to the negative charge on an electron.

Have you read or heard that when a particle of matter collides with its nemesis, they annihilate each other? It's true—almost. The particles don't cease to exist altogether; their "essences" remain as they change from matter into energy. When a particle combines with its antiparticle, we can calculate the released energy according to the modified Einstein formula

$$E = (m_+ + m_-) \, c^2$$

where E represents the liberated energy (in joules), m_+ represents the mass of the matter particle (in kilograms), m_- represents the mass of the antimatter particle (in kilograms or, if you prefer, in "antikilograms"), and c^2 represents the speed of light squared, that gigantic constant equal to approximately 9×10^{16} m²/s².

Unimaginable Power

In theory, if we bring equal masses of matter and antimatter together, all the mass "morphs" into energy. If we have a little more matter than antimatter, then we'll have some matter leftover after the encounter. Conversely, if we start with a little more antimatter than matter, some antimatter will remain after the reaction.

In a nuclear reaction such as fusion or fission involving matter, only a tiny fraction of the total mass converts directly to energy. Plenty of matter always remains, although it can (and often does) end up in modified form. You might push together two chunks of U 235, the isotope of uranium having atomic mass 235 AMU; if they have enough combined mass, an atomic explosion will take place. But after the reaction, most of the matter will remain. We can therefore say that a fission reaction has low "matter-to-energy conversion efficiency."

We can get a higher "matter-to-energy conversion efficiency" with a hydrogen bomb, in which nuclear fusion produces the explosive force. The same holds true for controlled nuclear fusion reactions, such as those sought after by scientists and engineers for the purpose of obtaining energy for future generations. But even a fusion reaction has low "matter-to-energy conversion efficiency" compared with a matter-antimatter reaction! Even when hydrogen atoms fuse to produce helium atoms and energy, the overall mass does not decrease very much.

If we can ever produce a sizable matter-antimatter reaction between equal masses of particles and antiparticles, we will achieve a "matter-to-energy conversion efficiency" of practically 100 percent. As you can imagine, a destructive weapon of this sort would make a hydrogen bomb seem like a little firecracker by comparison. A single matter-antimatter weapon of modest size could wipe out all life on earth. A large one could vaporize the whole planet.

Still Struggling

Why don't we see antimatter floating around in the universe? Why are the moon and all the extraterrestrial planets, as well as all the known asteroids and comets in the solar system, made of matter rather than antimatter? (If any celestial object were made of antimatter, then the instant that a matter object landed on it, that object would vanish with a flash that we could see from trillions of kilometers away.) This is an interesting question. Who really knows whether or not all of the stars and galaxies that we can see actually consist of matter?

Evidently, if any antimatter existed in the neighborhood of our sun when it was born, that antimatter combined with nearby matter and disappeared long ago. If both matter and antimatter existed in the primordial solar system, the mass of the matter must have exceeded the mass of the antimatter, because matter prevailed after the contest. Most astronomers doubt that our Milky Way galaxy contains much antimatter. If it did, we would observe periodic explosions of unimaginable brilliance, or else a continuous flow of energy that we couldn't attribute to anything other than ongoing matter-antimatter annihilation.

PROBLEM 4-4

Suppose that we bring a 1.00-kg block of matter and a 1.00-kg block of antimatter into direct contact. How much energy will result from the annihilation? Will any matter or antimatter remain afterwards?

SOLUTION

Let's answer the second question first. No matter or antimatter will remain, because the two blocks have equal (and, in a sense, opposite) mass. As for the first question, the total mass involved in this encounter equals 2.00 kg, so we can use the Einstein formula to calculate the liberated energy E as follows:

$$E = mc^2$$
$$= 2.00 \times 9.00 \times 10^{16}$$
$$= 1.80 \times 10^{17} \text{ J}$$

PROBLEM 4-5

We know that 1 W = 1 J/s. How long would the energy produced in the above-described matter/antimatter reaction, assuming that we could control and harness it, illuminate a "never-burn-out" 50-W light bulb?

SOLUTION

We divide the amount of energy, in joules, by the wattage of the bulb, in joules per second. We know that this approach will work because, in terms of units, we have

$$J/W = J / (J/s) = J \cdot (s/J) = s$$

The joules cancel out. When we divide joules by watts, we get seconds. We use the small elevated dot (·) to represent multiplication when dealing with units, as opposed to the slanted cross (×) that we use with numerals. Getting down to the actual numbers, let P represent the power consumed by the bulb (50 W), let t_s represent the number of seconds that the bulb will burn, and let E represent the total energy produced by the matter-antimatter reaction, 1.80×10^{17} J. Then

$$t_s = E/P$$
$$= 1.80 \times 10^{17}/50$$
$$= 3.60 \times 10^{15} \text{ s}$$

A minute contains 60.0 seconds, an hour contains 60.0 minutes, a mean solar day consists of 24.0 hours, and, on the average, a year lasts for 365 mean solar days (rounded off to the nearest day). That gives us approximately 3.15×10^7 seconds in a year. Let t_{yr} be the time, in years, that the light bulb burns. Then we can calculate

$$t_{yr} = t_s/(3.15 \times 10^7)$$
$$= (3.60 \times 10^{15})/(3.15 \times 10^7)$$
$$= 1.14 \times 10^8 \text{ yr}$$

That's 114 million years.

PROBLEM 4-6

Suppose that we double the amount of matter in the previous two problems to 2.00 kg, but we leave the amount of antimatter at 1.00 kg. How much energy will result from the encounter in this case? Will any matter or antimatter remain afterward?

SOLUTION

We will get the same amount of energy as we did in Solutions 4-4 and 4-5, that is, 1.80×10^{17} J. We will find 1.00 kg of matter leftover (the difference between the masses) after the reaction. However, assuming the encounter produces an explosion, the matter won't remain in the form of a solid block. It will scatter, particle by particle, into space.

Compounds

Atoms of two or more elements can join together, sharing electrons. When this happens, we get a molecule of a chemical *compound*. On our planet, a common compound forms when two hydrogen atoms join with a single atom of oxygen. That process yields H_2O, usually called water. Thousands of different chemical compounds exist in nature; humans can synthesize many more.

More than a Mixture

A compound differs from a plain mixture of elements. We can put hydrogen gas into a balloon along with oxygen gas, but that doesn't automatically give us water. We must supply a "triggering event" to get the atoms of the two gases to form molecules of a compound. Under the right conditions, an explosion occurs as part of this *chemical reaction*, because these two particular elements combine with gusto when sufficiently provoked! Figure 4-3 illustrates a water molecule.

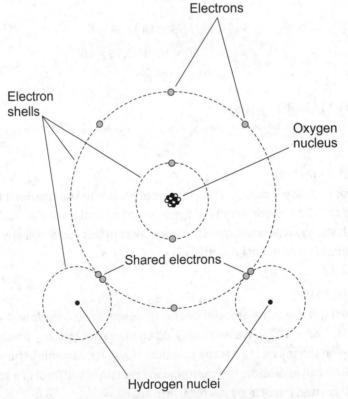

FIGURE 4-3 · Simplified diagram of a water molecule.

The oxygen and hydrogen atoms share electrons in a peculiar arrangement known as a *covalent bond*. In this way, the atoms "link up," making the water molecule fundamentally different from a mere threesome consisting of two hydrogen atoms and one oxygen atom in close proximity.

Compounds often, but not always, appear different from any of the elements that make them up. At room temperature and pressure, both hydrogen and oxygen occur in the *gaseous phase*. Under the same conditions, water appears mainly in the *liquid phase*. The heat of the reaction between hydrogen and oxygen yields water vapor initially; that's a colorless, odorless gas. Some of this vapor will condense into liquid if the temperature gets low enough. A further decrease in temperature will cause the liquid to freeze into the *solid phase*.

WARNING   *Do not try to make water by igniting hydrogen gas in combination with air or oxygen gas. You could get severely burned. If you inhale hydrogen gas along with air and the mixture ignites, your lungs could sustain fatal injury.*

Common rust provides another example of a chemical compound. It forms when iron joins with oxygen. Under ordinary temperature and pressure conditions, pure elemental iron appears as a dull gray solid, while pure oxygen manifests as a clear, odorless gas. Iron rust, however, looks dramatically different from either of its constituents, appearing as a reddish-brown, flaky solid or powder. The reaction between iron and oxygen takes place slowly, unlike the rapid combination of hydrogen and oxygen when ignited. We can speed up the rate of the iron–oxygen reaction by exposing the mixture to water vapor, as anyone who lives in a humid climate knows.

Compounds Torn Asunder

The opposite of the element-combination process can occur with many compounds. Water, again, provides us with a convenient example. We can separate liquid water into hydrogen and oxygen atoms by means of a process called *electrolysis*.

You can *electrolyze* water in your kitchen. Obtain two large metal nails and some *bell wire* (ordinary insulated copper wire, available at nearly all hardware stores). Cut off two 3-ft (approximately 1-m) lengths of the wire, and strip approximately 2 in (5 cm) of the insulation off of each end of both pieces of the wire. Wrap one stripped end of wire around each nail near the nail head in a tight helix. Add a measuring cupful (approximately 0.25 L) of table salt to a bucket full of water, and dissolve the salt thoroughly to get a salt-water solution. Using the free ends of the wires, connect the two electrodes to opposite poles of a 12-volt (12-V) battery made from two 6-V "lantern batteries" or eight ordinary dry cells connected in series. (Do not use an automotive battery for this experiment.) Insert the electrodes into the water a few centimeters apart, as shown in Fig. 4-4. You'll see bubbles rising up from both electrodes. The bubbles coming from the negative electrode contain pure hydrogen gas, while the bubbles coming from the positive electrode contain pure oxygen gas. You should see more hydrogen bubbles than oxygen bubbles.

Be careful when doing this experiment. Don't reach into the bucket and grab the electrodes. In fact, you shouldn't grab the electrodes or the battery terminals at all. A 12-V battery can give you a nasty (and potentially dangerous) shock when you have wet hands.

If you leave the apparatus shown in Fig. 4-4 running for awhile, you'll notice corrosion forming on the exposed wire and the electrodes, and especially on the positive electrode where oxygen bubbles come off. Remember that you've added table salt (sodium chloride) to the water! Therefore, the positive electrode will attract chlorine ions as well as oxygen ions. Both oxygen and chlorine combine readily with the copper in the wire and the iron in the nail. The resulting compounds manifest as solids that eventually coat the wire and the nail. If you continue the electrolysis process for a long enough time, this coating will form an electrical insulator and hinder the flow of current through the salt-water solution.

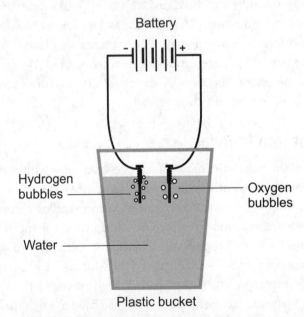

FIGURE 4-4 · Electrolysis of water, in which molecules of the compound separate into individual hydrogen and oxygen atoms.

The Restless Molecule

Looking back at Fig. 4-3, you see a simplified "schematic diagram" of a water molecule. It contains three atoms of two different elements, linked together by electron-sharing. Sometimes, molecules also form as the result of two or more atoms of a single element joining together.

Oxygen atoms tend to "pair up" in the earth's atmosphere. For this reason, you'll often see an oxygen molecule denoted as O_2. The O represents oxygen, and the subscript 2 indicates two atoms per molecule. You symbolize water as H_2O because each molecule has two atoms of hydrogen and one atom of oxygen. (You don't have to write a subscript numeral 1 for the oxygen. The absence of a subscript translates to 1 by default.)

Sometimes oxygen atoms occur singly. Then you denote it by writing O without any subscript. Sometimes, three atoms of oxygen group together to form a molecule of *ozone*, symbolized as O_3. You've doubtless heard and read about this gas! Its presence in the upper atmosphere protects earthly life from excessive solar radiation, but in the lower atmosphere, ozone acts as an irritant or pollutant.

Molecules constantly move; their speed depends on the temperature. As the temperature increases, the molecules in any particular substance move with increasing speed. In a solid, the molecules interlock in a rigid pattern, although they vibrate continuously (Fig. 4-5A). In a liquid, the molecules slither and slide past each other (Fig. 4-5B). In a gas, the molecules fly all around (Fig. 4-5C), bouncing off each other and crashing into internal or surrounding barriers. On a large scale, we witness the force of these collisions as *gas pressure*.

Still Struggling

Does a single atom constitute a molecule, or not? Several readers brought this question to my attention after the publication of the first edition of this book. A few sources will tell you that a single atom constitutes a "monatomic molecule." However, most sources, and most of the texts that I've seen, define a molecule as two or more atoms joined together by chemical bonds. Some chemical elements occur naturally as groups of single atoms; examples include the so-called *noble gases* helium, neon, argon, krypton, xenon, and radon. Some scientists call them *monatomic gases*.

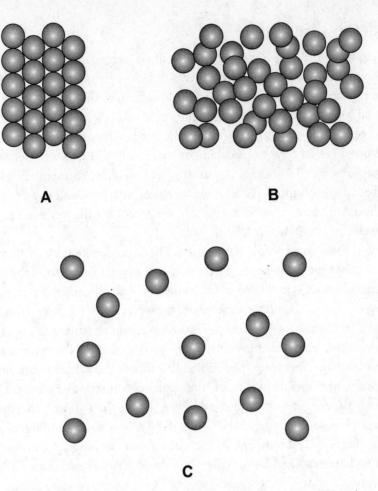

FIGURE 4-5 · Simplified rendition of molecules in a solid (A), a liquid (B), and a gas (C).

QUIZ

Refer to the text in this chapter if necessary. A good score is eight correct. Answers are at the back of the book.

1. Suppose that an isotope of sodium contains 11 protons and 12 neutrons in the nucleus. What, approximately, is the atomic mass?
    A. 11 AMU
    B. 12 AMU
    C. 23 AMU
    D. We need more information to answer this question.

2. When 3 kg of matter collide with 3 kg of antimatter, we get
    A. a burst of energy with nothing left over.
    B. a burst of energy with some matter left over.
    C. a burst of energy with some antimatter left over.
    D. a simple mixture unless a spark or flame causes a chemical reaction.

3. If we could isolate a neutron from an atomic nucleus and let it "sit" all by itself in free space, what would happen to it?
    A. Nothing
    B. It would probably lose its identity after a short while.
    C. It would turn into an electron.
    D. It would turn into antimatter.

4. When two or more atoms join to form a molecule,
    A. the atomic nuclei merge.
    B. the atoms share some electrons.
    C. the atoms release energy according to the Einstein formula.
    D. the electrons in the whole molecule all orbit around the same nucleus.

5. Examine Fig. 4-3. How many electrons exist in the inner shell of the oxygen atom of this molecule?
    A. Two
    B. Six
    C. Eight
    D. We can't tell from this illustration.

6. Two individual atoms representing different isotopes of a given element can never have the same number of
    A. neutrons.
    B. protons.
    C. electrons.
    D. nuclei.

7. When scientists first began to refine the atomic theory of matter, they found dozens of different elements in nature. Later, scientists created elements that never occur in nature with the help of

    A. an electrolysis machine.
    B. a particle accelerator.
    C. a fusion reactor.
    D. a fission reactor.

8. The number of protons in an atomic nucleus defines the associated element's

    A. isotope.
    B. ion charge.
    C. atomic number.
    D. atomic mass.

9. An electron has a mass

    A. slightly greater than the mass of a proton.
    B. much greater than the mass of a proton.
    C. slightly less than the mass of a proton.
    D. much less than the mass of a proton.

10. Suppose that an atom of helium, whose atomic number is 2, has one electron "orbiting" its nucleus. This atom constitutes

    A. a positive isotope.
    B. a negative isotope.
    C. an anion.
    D. a cation.

chapter **5**

Basic States of Matter

In the ancient Greek and Roman civilizations, scientists believed that all things comprised combinations of four "elements": *earth*, *water*, *air*, and *fire*. According to this theory, different proportions of these "elements" gave materials their unique properties. After the Roman civilization declined, the Western world came under a collective trance in which superstition prevailed for centuries. When scientific reasoning regained respect, physicists defined three basic *states* or *phases* of matter: *solid* (the analog of earth), *liquid* (the analog of water), and *gaseous* (the analog of air). The ancient notion of fire evolved into the phenomenon we now call *energy*.

CHAPTER OBJECTIVES

In this chapter, you will

- Compare solids, liquids, and gases.
- Learn how diverse substances behave.
- Define the hardness of a solid sample.
- Discover Hooke's law for solids.
- Observe Pascal's law for incompressible liquids.
- Calculate pressure and density.

The Solid Phase

A sample of matter in the solid phase will retain its shape unless subjected to violent impact, placed under stress, or heated enough to affect the way its molecules behave. Examples of solids at room temperature and ordinary pressure include rock, steel, salt, wood, and glass.

The Electric Force

When you place a concrete block on a concrete floor, why doesn't the block sink into or meld with the floor? Why, if you strike a brick wall with your fist, are you likely to hurt yourself, rather than having your fist pass painlessly into the bricks? Internally, atoms contain mostly space—even in dense solids. Why can't solid objects pass through one another the way galaxies can in outer space, or the way dust clouds do in the atmosphere? They're mostly space as well!

The answers to these questions lie in the nature of the electrical forces within and around atoms. Shells of negatively charged electrons surround most atomic nuclei. Every electron carries a negative charge. Objects with electrical charges of the same polarity (negative-negative or positive-positive) always repel. As two objects with like charge approach each other, they repel with increasing force. Therefore, even when an atom has an equal number of electrons and protons, the charges exist in different distributions: The positive charge concentrates in the nucleus, and the negative charge surrounds the nucleus in one or more concentric spheres.

Imagine that you could shrink down to submicroscopic size and stand on the surface of an aluminum platter. Beneath your feet, the surface would look something like a huge field full of soccer balls, as shown in Fig. 5-1. You would find it

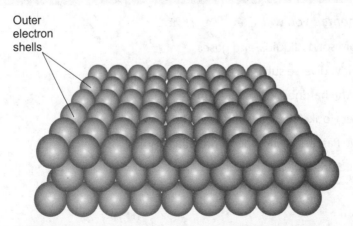

Outer electron shells

FIGURE 5-1 · In a solid, the outer electron shells of the atoms pack tightly together.

difficult to walk on this surface because of its irregularity. Nevertheless, every ball would resist penetration by other balls. All of the balls would carry negative electric charge, so they would all repel each other. They could not pass through each other, unless you could apply tremendous force to them. The mutual repulsion would maintain the aluminum surface in a stable, fixed state. The balls, even though nearly hollow inside, would exist in a tightly packed formation.

Of course, the foregoing description represents an oversimplification, but it should give you an idea of why solids don't normally pass through each other, and in fact why many solids even resist penetration by liquids or gases.

Brittleness, Malleability, and Ductility

The atoms of elemental solids can "stack up" in various ways. This fact makes itself evident in the shapes of the *crystals* we observe in some solid substances. Salt, for example, has a characteristic cubical crystalline shape. The same is true of sugar. Water ice crystals can appear in a fantastic variety of shapes, but they always have six sides, axes, or facets. Some substances, such as iron, don't form visibly apparent crystals under normal circumstances. Some materials, such as glass, fracture along smooth, but curved, boundaries. We can grind up some solid materials into fine powder, while others defy all attempts to pulverize them.

Crystalline solids generally exhibit a property called *brittleness*. If you use a hammer to strike a sample of crystalline material with enough force, the sample will crack or shatter. You cannot stretch, compress, or bend such a sample very much without breaking it. Glass nevertheless has a little bit of "give"; it's not *totally* rigid. You can see the flexibility of glass if you watch the reflections from large window panes during a full gale. But no one would call glass "flexible" in the everyday sense. You can't bend a straight glass rod into a donut shape, or a flat window pane into a cylinder, at the temperatures and pressures normally prevailing on the surface of the earth.

Copper utility wire, in contrast to glass, exhibits *malleability*. That means we can mold it or pound it flat. Copper wire also has a property called *ductility*, meaning that we can stretch it and bend it without difficulty, at least to some extent. Many other metals behave this way as well. Gold represents one of the most malleable known metals; crafts people can pound it into sheets thin enough to coat the cupolas of cathedrals without breaking the budget. Aluminum has greater malleability and ductility than glass, but not to the extent of soft copper or gold. Wood exhibits some flexibility, but practically no malleability or ductility. We can bend a strip of wood to a variable extent (depending on its water content and the type of tree from which it came), but we can't pound it into a thin sheet or stretch it to make a wire.

For some solids, the brittleness, ductility, and malleability vary with the temperature. We can make samples of glass, copper, and gold more malleable and ductile by heating them. The professional glass blower takes advantage of this phenomenon, as does the coin minter and the wire manufacturer. A person who works with wood has no such good fortune. If you heat wood, it gets drier and less flexible. Ultimately, if you heat glass, copper, or gold enough, it will turn into a liquid. As you subject wood to increasing temperature, it will remain solid up to a certain point, and then it will undergo *combustion*, a rapid form of *oxidation*. In other words, it will catch on fire.

Hardness

Some solids present themselves to us as "more solid" than others. A quantitative means of expressing hardness, known as the *Mohs scale*, classifies solids from 1 to 10. The lower numbers represent softer solids, and the higher numbers represent harder ones. Table 5-1 lists the standard substances used in the Mohs scale, along with their hardness numbers. When you want to test the hardness of a substance, you must take note of two facts:

- A sample *always* scratches something softer than itself.
- A sample *never* scratches anything harder than itself.

Talc, which you can crumble in your hand, represents a soft solid. Old-fashioned blackboard chalk provides another example. Wood exhibits hardness

TABLE 5-1 The Mohs scale of hardness. As the number increases, the hardness also increases.

Hardness Number	Standard Substance
1	Talc
2	Gypsum
3	Calcite
4	Fluorite
5	Apatite
6	Orthoclase
7	Quartz
8	Topaz
9	Corundum
10	Diamond

somewhat greater than talc or chalk, but limestone has greater hardness than wood. Then, in increasing order of hardness, we encounter glass, quartz, and diamond. We can always determine the approximate hardness of a particular solid sample by seeing how easily it scratches one of the standard Mohs solids, and by seeing how easily one of the standard Mohs solids scratches it.

Many substances have hardness numbers that change with temperature. In general, cold temperatures harden these materials. Ice constitutes a common example. On a skating rink where the temperature hovers near the freezing point of water, water ice appears rather soft. On the surface of *Charon* (the bitterly cold companion of *Pluto*, the outermost large object in our solar system) where temperatures plummet into the hundreds of degrees below zero, water ice exhibits hardness comparable to granite.

Chemists and earth scientists maintain lab samples of each of the 10 substances noted in Table 5-1. A "scratch" must show up as a permanent mark or disfigurement, not merely a few particles transferred from one substance to the other. Many materials have hardness values that fall between two whole numbers on the scale. Therefore, the Mohs hardness scale doesn't provide a very precise method for measuring hardness.

Density

We can measure the *density* of a solid in terms of the mass (in kilograms) that a cubic meter of it has. Density varies in direct proportion to mass divided by volume. Scientists express density in units of *kilograms per meter cubed* (kg/m^3 or $kg \cdot m^{-3}$) in SI. That constitutes an awkward unit in most real-life situations. Imagine trying to determine the density of sandstone by cutting out a cubical block measuring 1 m on an edge, and placing it on a laboratory scale! You'd need a construction crane to lift the block, and when you finally got it onto the scale, it would crush the instrument beyond recognition.

Because of the impracticality of measuring density in kilograms per meter cubed, scientists sometimes use the centimeter-gram-second (cgs) system of units instead. We can figure out the density of a sample by taking a cubic-centimeter sample of the material in question, and then finding its mass in grams. This gives us a unit called the *gram per centimeter cubed* (g/cm^3 or $g \cdot cm^{-3}$). We can convert this unit to SI form or vice versa as follows:

- To convert from g/cm^3 to kg/m^3, multiply by 1000.
- To convert from kg/m^3 to g/cm^3, multiply by 0.001.

You can doubtless think of extremely dense solids. How about lead? Gold and iron also have relatively high density. Aluminum exhibits far lower density than lead, gold, or iron. Most rocks have lower density than most metals. Glass has roughly the same density as silicate rock (from which glass derives). Wood, and most plastics, have low density values.

PROBLEM 5-1

A sample of solid matter has a volume of 20.7 cm³ and a mass of 0.211 kg. What is the density in grams per centimeter cubed?

SOLUTION

This problem presents a trick, because it involves two different systems of units: cgs for the volume and SI for the mass. To get a meaningful answer, we must remain consistent with our units. The problem requires that we express the answer in the cgs system, so let's begin by converting kilograms to grams. To do that, we multiply the mass figure by 1000, which tells us that the sample masses 211 g. If we let d represent the density, m represent the mass, and v represent volume, then

$$d = m/v$$

In this case, we have

$$d = 211/20.7$$
$$= 10.2 \text{ g/cm}^3$$

rounded to three significant figures.

PROBLEM 5-2

Calculate the density of the sample from Problem 5-1 in kilograms per meter cubed. Do not use the conversion factor or the result of Problem 5-1. Work out the whole problem from scratch.

SOLUTION

Let's start by converting the volume to meters cubed. A cubic meter contains 1,000,000, or 10^6, cubic centimeters. Therefore, to convert the cgs volume to SI volume, we must divide by 10^6 or multiply by 10^{-6}. When we do that, we get 20.7×10^{-6} m³ (which we should express as 2.07×10^{-5} m³ in standard scientific notation). Now we can divide the mass in kilograms by the volume in cubic meters to obtain the density as follows:

$$d = m/v$$
$$= 0.211 / (2.07 \times 10^{-5})$$
$$= 0.102 \times 10^5$$
$$= 1.02 \times 10^4 \, \text{kg/m}^3$$

We've obtained an answer 1000 times the size of the value we got when we calculated the density in grams per centimeter cubed, as the conversion scheme would have told us if we'd taken the "easy way out."

Measuring Solid Volume

Now imagine that we have an irregular object instead of a perfect cube. How can we figure out its volume? We could easily calculate it if the object were a perfect sphere, a rectangular prism, or a pyramid of known dimensions. But what if it's a gnarled and knobby nugget?

Scientists have come up with a clever way of measuring the volumes of irregular solids: immerse them in liquid. First, we measure the amount of liquid in a container, as shown in Fig. 5-2A. Then we measure the amount of liquid

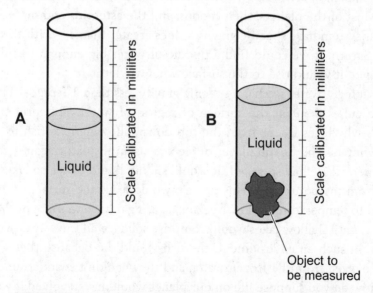

FIGURE 5-2 · We can determine the volume of a solid by measuring the amount of liquid that it displaces. At A, a container with liquid but without a sample; at B, a container with a completely submerged sample.

that the object displaces when we completely submerge it. This will show up as an increase in the apparent amount of liquid in the container (Fig. 5-2B). One *milliliter* (1 ml) of pure liquid water has a volume of precisely 1 cm³. In any decent chemistry lab, we'll find at least one container marked off in milliliters. As long as none of the solid dissolves in the liquid, and as long as the solid doesn't absorb any of the liquid, we can use this technique to measure the volume of a reasonable-sized sample.

Specific Gravity

The density of a solid relative to the density of pure liquid water at 4°C (about 39°F) constitutes another important characteristic. Under normal atmospheric pressure conditions, water exhibits its greatest density at this temperature, and we assign it a relative density value of exactly 1. Substances with relative density greater than 1 will sink in pure water at 4°C, and substances with relative density less than 1 will float in pure water at 4°C. We call the relative density of a solid, defined in this way, the *specific gravity*. You'll often see it abbreviated sp gr. Some scientists call it the *relative density*.

You should have no trouble thinking of substances whose specific gravity numbers far exceed 1. Examples include most rocks and virtually all metals. However, *pumice*, a volcanic rock typically filled with air pockets, floats on water. Most of the planets, their moons, and the asteroids and meteorites in our solar system have specific gravity values greater than 1—with the exception of Saturn. If we could find a puddle of water big enough to submerge that planet, it would rise to the surface when we let it go.

Interestingly, water ice has a specific gravity less than 1 (sp gr < 1). That's why ice cubes float on the surface of a glass of water. This property also explains why lakes freeze from the top down, allowing lake fish to survive severe winters. The floating layer of ice acts as an insulator against the cold atmosphere. If ice had sp gr > 1, it would sink in liquid water, so lakes would freeze from the bottom up. This process would leave the surfaces constantly exposed to temperatures below freezing, causing more and more of the water to freeze, until shallow lakes would comprise solid ice all the way down to the bottom. In such an environment, the fish would die because they couldn't extract oxygen from the frozen water, and they couldn't swim around to get food. How do you suppose life on our planet would have evolved if water ice had sp gr > 1?

Elasticity

We can stretch some solid samples easily, while other samples resist stretching and might break instead. We can stretch a long span of soft-drawn copper wire; a similar length of rubber would stretch a lot more. However, the "stretchiness" of these two substances differs qualitatively as well as quantitatively. If you let go of a rubber cable after stretching it, it will spring back to its original length immediately. If you let go of a stretched-out length of soft-drawn copper wire, it will "stay stretched."

We can define the *elasticity* of a substance as the extent to which a sample of it returns to its original dimensions after stretching or compression. According to this definition, rubber has high elasticity, and copper has low elasticity. This definition is qualitative (it says something about how a substance behaves) but not quantitative (we don't assign specific numbers to it).

Still Struggling

Scientists sometimes talk or write about *ideal substances* (meaning theoretically perfect samples of matter) as if to imply that they can actually exist. Don't get theoretical ideals and physical reality mixed up! In the real world, we'll never encounter a *perfectly elastic* substance, that is, one that returns precisely to its original size and shape after stretching or compression. Neither will we ever find a sample of *perfectly inelastic* material, that is, something that totally defies any stretching or compression. Both of these extremes constitute theoretical ideals, which we might informally define as elasticity factors of 1 (perfectly elastic) and 0 (no elasticity whatsoever).

Hooke's Law

For a moment, imagine that we manage to find a perfectly elastic sample of material. It will obey a strict principle called *Hooke's law* concerning how much we can stretch or compress it:

- The extent to which we can deform a perfectly elastic solid sample, either by stretching or compression, varies in direct proportion to the external force that we apply.

Let's state this principle mathematically. If we let F represent the magnitude of the applied force (in newtons) and we let s represent the amount of stretching or compression expressed as a displacement (in meters), then

$$s = kF$$

where k represents a physical constant that depends on the substance. We can write this equation in vector form as follows:

$$s = k\mathbf{F}$$

The stretching or compression displacement vector always points in the same direction as the applied force vector.

Even though perfectly elastic solids don't exist in our physical world, plenty of materials come close enough to that ideal so that we can use Hooke's law to describe their behavior when we apply external force, as long as we don't exert so much force that we break, crush, or permanently deform the sample under test.

PROBLEM 5-3

Suppose an elastic cord has theoretically perfect elasticity, provided that we don't apply more than 5.00 N of force. When we apply no force at all, the cord measures 1.00 m long. When we apply 5.00 N of stretching force, the length increases to 2.00 m. What length should we expect the cord to attain if we apply only 2.00 N of stretching force?

✔ SOLUTION

Applying 5.00 N of force causes the cord to become 1.00 m longer than its natural length (no applied force). We know that the cord behaves as a perfectly elastic material as long as the force doesn't exceed 5.00 N. We can calculate the value of the constant k, called the *spring constant*, in meters per newton (m/N) directly from the formula

$$s = kF$$

If we apply at least a minimal force (to obtain a nonzero value for F so we don't end up dividing by 0), then we can rearrange the above equation to get

$$k = s/F$$

Now we can input the numbers that we know for this particular cord, getting

$$k = (1.00 \text{ m}) / (5.00 \text{ N})$$
$$= 0.200 \text{ m/N}$$

This value of k will work as long as $F \leq 5.00$ N. (It might work for greater forces as well, but we'd have to conduct experiments to verify that hypothesis.) Therefore, the formula for displacement as a function of force becomes

$$s = 0.200F$$

If $F = 2.00$ N, then we have

$$s = 0.200 \text{ m/N} \times 2.00 \text{ N}$$
$$= 0.400 \text{ m}$$

That's the *additional length* by which the cord will expand when we apply 2.00 N of external stretching force. With no applied force, the cord measures 1.00 m long, so when we exert 2.00 N of stretching force, we get a total cord length of 1.00 m + 0.400 m, or 1.400 m, which we can round off to 1.40 m.

We can graphically illustrate the behavior of this cord, for stretching forces between zero and 5.00 N, as shown in Fig. 5-3. This plot portrays a *linear function*, because it appears as a straight line in rectangular coordinates. If the force exceeds 5.00 N, according to the specifications for our cord, we have no assurance that the displacement-versus-force function will remain linear.

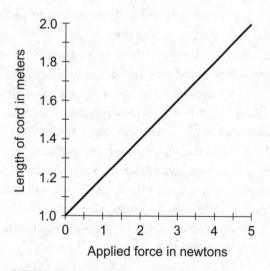

FIGURE 5-3 • Illustration for Problem 5-3. The function is linear (its graph appears as a straight line) within the range of forces shown here.

The Liquid Phase

In the liquid state or phase, a substance has two properties that distinguish it from the solid phase. First, a liquid can "shape-shift" so that it conforms to the inside boundaries of a container. Second, when we place a liquid in an open container (such as a bucket) in an environment with gravitation, the liquid flows to the bottom of the container and develops a defined, flat surface on top.

Diffusion

Imagine that we're riding on a vessel coasting through interplanetary space, so that no acceleration force exists within. Suppose that we fill a jar halfway with a certain liquid, and then we introduce another liquid that doesn't react chemically with the first sample. Gradually, the two liquids will blend together so that we have a *homogeneous* (uniform, evenly distributed) mixture. We call this blending process *diffusion*.

In a liquid, diffusion takes place rather slowly unless we "shake things up." However, some combinations of liquids diffuse faster than others. Alcohol diffuses into water at room temperature more quickly than heavy motor oil diffuses into light motor oil under the same conditions. But eventually, whenever we mix any two liquids that don't chemically react with each other, the mixture will become homogeneous throughout any container of finite size. We don't have to disturb the container for this process to happen, because the molecules of a liquid constantly move. They push and jostle each other until they attain uniform distribution throughout.

Now suppose that we conduct the same experiment in a bucket on the earth's surface where gravitational acceleration exists. In this case we'll still observe diffusion, but "heavier" liquids will sink toward the bottom of the bucket while "lighter" liquids will rise toward the surface. Alcohol, for example, will "float" to some extent on water. (We won't find a sharply defined "surface" between the alcohol and the water, as we see between the liquid and the air, but the force imposed by gravitation will keep the mixture from becoming perfectly homogeneous throughout the bucket.) We will observe the same "partial mixing" effect with any two nonreacting liquids *unless* they have equal density. We'll discuss the meaning of the term *density* for liquids shortly.

Viscosity

Some liquids flow more easily than others. Some seem more "slimy" than others. You can easily observe the difference in behavior at room temperature between water and molasses, for example. If you fill a glass with water and another glass with an equal amount of molasses and then pour the contents of both glasses into your kitchen sink, the glass containing the water will empty in a hurry, while the molasses will take awhile to get out of its glass. To describe this difference in "runniness," we say that the molasses has higher *viscosity* than the water at room temperature. In a hot kitchen, you'll observe less difference in viscosity between molasses and water than you'll see a cold kitchen, but you'll never get your kitchen so torrid that the molasses flows as fast as the water does.

Some liquids exhibit far more viscosity than molasses. Consider hot tar as construction workers pour it down to make the surface of a new highway, or the oil in the engine of your truck on a bitter-cold winter morning! These substances theoretically qualify as liquids, but as the temperature falls they behave increasingly less like liquids and more like solids.

We cannot "draw an exact line" between the liquid and the solid phases for tar or oil. They don't freeze and change state from liquid to solid in an obvious way, as water does. As hot tar cools, do we ever reach a point in time where we can say "This stuff is liquid," and then one second later say, "Now, this stuff is solid"? Can you say that on a morning when the temperature is −15°C, the motor oil in your truck constitutes a solid, but if you wait a few minutes until the temperature rises to −14°C, the oil transitions into a liquid?

Still Struggling

We can't always give a clear answer to the question, "Does this substance constitute a solid or a liquid?" It depends on the observer's point of reference, and to some extent on judgment as well.

We can consider some substances as solid in a short-term time sense, but liquid in a long-term sense. For example, think about the mantle of the earth—the layer of rock between the crust and the core. In a long-term time sense, pieces of the crust, known as *tectonic plates*, float around on top of the mantle like scum on the surface of a liquid. Over eons of time, this motion shows up as so-called continental drift. But from one moment (as we perceive it) to the next, and even from hour to hour or from day to day, the crust seems rigidly fixed on the mantle. The earth's mantle behaves like a solid in the short-term sense, but like a liquid in the long-term sense.

Imagine that we could turn ourselves into creatures whose life spans extended over trillions (units of 10^{12}) of years, so that 1,000,000 years seemed to pass like a moment. Then from our point of view, the earth's mantle would behave like a liquid with low viscosity, just as water presents to us in our actual state of time-awareness. If we could "morph" into beings whose lives lasted for only a tiny fraction of a second, then water would seem to take forever to get out of a glass tipped on its side; we would call it a solid substance, or maybe a liquid with extremely high viscosity.

Density

We can define the density of a liquid in three different ways: *mass density*, *weight density*, and *particle density*. The distinction between these quantities might seem theoretically subtle, but in practical situations, a significant difference exists:

- We express *liquid mass density* in terms of the number of kilograms per meter cubed (kg/m^3) in a sample of the liquid.
- We define *liquid weight density* in newtons per meter cubed (N/m^3), or the mass density times the acceleration in meters per second squared (m/s^2) to which we subject a sample of the liquid.
- We define *liquid particle density* as the number of moles of atoms or molecules per meter cubed (mol/m^3) in a sample of the liquid, where 1 mol represents approximately 6.02×10^{23} atoms or molecules.

Let d_m represent the mass density of a liquid sample (in kilograms per meter cubed), let d_w represent the weight density (in newtons per meter cubed), and let d_p represent the particle density (in moles per meter cubed). Let m represent the mass of the sample (in kilograms), let V represent the volume of the sample (in meters cubed), and let N represent the number of moles of atoms in the sample. Let a represent the acceleration (in meters per second squared) to which we subject the sample. Once we've defined the quantities in the preceding way, we find that the following three relations hold among them:

$$d_m = m/V$$
$$d_w = ma/V$$
$$d_p = N/V$$

TIP   *Alternative definitions for mass density, weight density, and particle density make use of the* liter, *which equals a thousand centimeters cubed (1000 cm³) or one-thousandth of a meter cubed (0.001 m³), as the standard unit of volume. Once in awhile you'll see the centimeter cubed (cm³), also known as the* milliliter *because it equals 0.001 liter, used as the standard unit of volume. Scientists base these definitions on the assumption that the liquid has uniform density throughout the sample.*

PROBLEM 5-4

A sample of liquid measures 0.275 m³ in volume, and masses 300 kg. What's its mass density in kilograms per meter cubed?

SOLUTION

We have the input values in SI, making our task easy. We can simply divide the mass by the volume, obtaining

$$d_m = m/V$$
$$= 300 \text{ kg} / 0.275 \text{ m}^3$$
$$= 1091 \text{ kg/m}^3$$

We're entitled to only three significant figures because that's as far as our input values go. Therefore, we can say that $d_m = 1.09 \times 10^3$ kg/m³.

PROBLEM 5-5

If we approximate the acceleration of gravity at the earth's surface as 9.81 m/s², what's the weight density of the sample of liquid described in Problem 5-4?

SOLUTION

We multiply our mass density answer by 9.81 m/s² to obtain

$$d_w = 1.09 \times 10^3 \text{ kg/m}^3 \times 9.81 \text{ m/s}^2$$
$$= 10{,}693 \text{ N/m}^3$$
$$= 1.07 \times 10^4 \text{ N/m}^3$$

Measuring Liquid Volume

A laboratory scientist can measure the volume of a liquid sample using a test tube or flask marked off in milliliters or liters. But there's another way to measure the volume of a liquid sample, provided we know its chemical composition and the weight density of the substance in question. We can weigh the sample of liquid, and then divide the weight by the weight density. We must, of course, pay careful attention to the units when we do this. In particular, we must express the weight in newtons, which equals the mass in kilograms times the acceleration of the earth's gravity (approximately 9.81 m/s²).

Let's do a mathematical exercise to show why we can measure volume in this way. Let d_w represent the known weight density of a huge sample of liquid, too large for us to measure its volume using a flask or test tube. Suppose that this substance has a weight of w (in newtons). If V represents the volume in meters cubed, we know from the above formula that

$$d_w = w/V$$

because $w = ma$, where a represents the acceleration of gravity. If we divide both sides of this equation by w, we get

$$d_w/w = 1/V$$

Now we can invert both sides of this equation, and exchange the left-hand and the right-hand sides, to obtain

$$V = w/d_w$$

If we want all of this algebra to work, we must assume that none of the quantities V, w, and d_w can ever attain a value of 0. In the real world, this assumption presents no trouble for us; all materials occupy at least some volume, have at least some weight because of gravitation, and have finite, nonzero density because we'll always find at least a little of it in any discernible space.

Pressure

Have you ever heard, or read in a text, that you can't compress liquid water? In a simplistic sense, that's true, but liquid water can nevertheless exert pressure, as anyone who has lived through a flash flood or tsunami will verify. You can experience "water pressure" by diving down several feet in a swimming pool and noting the sensation that the water produces as it presses against your eardrums.

In a fluid, the pressure, as force per unit area, varies in direct proportion to the distance we go below the surface. At any fixed depth, the pressure varies in direct proportion to the weight density of the liquid. If d_w represents the weight density of a liquid (in newtons per meter cubed) and s represents the depth below the surface (in meters), then we can calculate the pressure P (in newtons per meter squared) using the formula

$$P = d_w s$$

If we know the mass density d_m (in kilograms per meter cubed) rather than the weight density, the formula becomes

$$P = a_g d_m s$$

where a_g represents the acceleration of gravity (in meters per second squared).

PROBLEM 5-6

Pure water at room temperature has a mass density of 1000 kg/m³. Suppose that we submerge a tiny cube measuring 1.000 mm along each edge in a body of water, so that the cube's center lies 1.000 m below the surface. How much force does the cube's surface experience as a result of water pressure?

SOLUTION

First, we must figure out the total surface area of the cube. It measures 1.000 mm, or 1.000×10^{-3} m, along each edge. Therefore, each of its faces has a surface area of $(1.000 \times 10^{-3})^2$ m² $= 1.000 \times 10^{-6}$ m². A cube has six faces, so the total surface area of the cube equals 1.000×10^{-6} m² $\times 6.000 = 6.000 \times 10^{-6}$ m². (Don't let the "extra" zeroes in these expressions irritate you. They indicate that we know the length of the cube's edges to an accuracy of four significant digits.)

Next, we calculate the weight density of water in newtons per meter cubed. On the earth's surface, it's 9.81 times the mass density of water, which is 1000 kg/m³. When we multiply out, we obtain 9.81×10^3 N/m³. We've submerged the center of our cube to a depth of 1.00 m. The water pressure at that depth is 9.81×10^3 N/m³ $\times 1.00$ m, or 9.81×10^3 N/m². The force F (in newtons) on the cube equals this number times the surface area of the cube, as follows:

$$F = 9.81 \times 10^3 \text{ N/m}^2 \times 6.00 \times 10^{-6} \text{ m}^2$$
$$= 58.9 \times 10^{-3} \text{ N}$$
$$= 5.89 \times 10^{-2} \text{ N}$$
$$= 0.0589 \text{ N}$$

Pascal's Law for Incompressible Liquids

Imagine a watertight, rigid container. Suppose that two pipes of unequal diameters run straight upward out of this container. Imagine that you fill the container with an *incompressible liquid* such as water, overflowing to the extent that the water rises partway up into the pipes. Now suppose that you place pistons inside the pipes so that they form perfect water seals. You let both pistons rest on the water surface, as shown in Fig. 5-4.

Because the pipes have unequal diameters, the pistons' faces have different surface areas. One of the pistons' faces has area A_1 (in meters squared), and the other has area A_2. Suppose that you push downward on piston number 1 (the one whose face has area A_1) with force F_1 (in newtons). You wonder: How much upward force, F_2, will this action produce at piston number 2 (the one whose face has area A_2)? *Pascal's law* provides the answer. The forces vary in direct proportion to the areas of the piston faces in contact with the liquid.

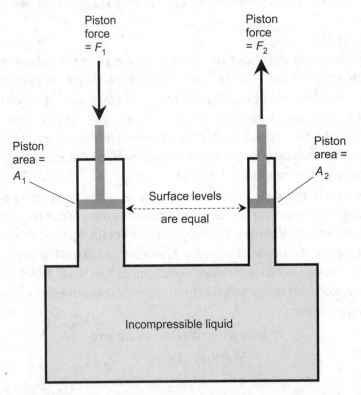

FIGURE 5-4 • Pascal's law for confined, incompressible liquids. The forces vary in direct proportion to the surface areas of the pistons' faces.

In the example shown by Fig. 5-4, piston number 2 has a smaller diameter than piston number 1, so the force F_2 is proportionately less than the force F_1. Mathematically, the following equations both hold:

$$F_1/F_2 = A_1/A_2$$

and

$$A_1 F_2 = A_2 F_1$$

Still Struggling

If we use either of the above equations, we must remain consistent with our units throughout the calculations. In addition, we must realize that the top equation has meaning only as long as piston number 2 has a nonzero surface area!

PROBLEM 5-7

Suppose that the faces of the pistons shown in Fig. 5-4 have surface areas of

$$A_1 = 12.0 \text{ cm}^2$$

and

$$A_2 = 8.00 \text{ cm}^2$$

in contact with the liquid. If you press down on piston number 1 with a force of 10.0 N, how much upward force will result at piston number 2?

SOLUTION

At first, you might suspect that we must convert the areas of the pistons' faces to meters squared to solve this problem. But actually, we need only determine the *ratio* of the areas, because we know them both in the same units. We calculate this ratio as

$$A_1/A_2 = 12.0 \text{ cm}^2/8.00 \text{ cm}^2$$
$$= 1.50$$

From Pascal's law, we know that $F_1/F_2 = 1.50$. We know that $F_1 = 10.0$ N, so we can solve for F_2 by setting up the ratio

$$10.0/F_2 = 1.50$$

When we multiply this equation through by F_2, we get

$$10.0 = 1.50 \, F_2$$

Dividing through by 1.50 yields the answer

$$F_2 = 10.0 / 1.50$$
$$= 6.67 \, \text{N}$$

The Gaseous Phase

The gaseous phase of matter resembles the liquid phase, insofar as a gas conforms to the boundaries of a container or enclosure. But gravitation and acceleration affect gases much less than they affect liquids. If you fill up a bottle with a gas, you won't observe any discernible "gas surface" as a result of the earth's gravitation. Gases differ from liquids in another important way as well. While we can rarely compress a liquid such as water, we can almost always compress a gas such as air.

Density

Three distinct definitions exist for gas density, following the fashion of the definitions for liquid density, as follows:

- We express *gas mass density* in terms of the number of kilograms per meter cubed (kg/m^3) in a sample of the gas.
- We define *gas weight density* in newtons per meter cubed (N/m^3), or the mass density times the acceleration in meters per second squared (m/s^2) to which we subject a sample of the gas.
- We define *gas particle density* as the number of moles of atoms or molecules per meter cubed (mol/m^3) in a sample of the gas, where 1 mol represents approximately 6.02×10^{23} atoms or molecules.

As with liquid density, we base these definitions on the assumption that the gas maintains uniform density throughout the sample.

Diffusion in Small Containers

Imagine a rigid enclosure, such as a glass jar, from which we've removed all the air to get a perfect vacuum inside. Suppose that we place this jar somewhere out in space, far away from the gravitational effects of stars and planets, and where space itself constitutes a nearly perfect vacuum. Suppose that the temperature way out there happens to match the temperature in a typical household, say 20°C. We have brought with us a tank of compressed elemental gas such as nitrogen or oxygen. We pump some of that gas into the jar, keeping the whole assembly sealed so that the gas can't escape. As we would expect, the gas diffuses quickly and evenly throughout the interior of the jar.

Now suppose that we introduce a certain amount of a second gas—something that won't react chemically with the first gas—into the jar. The second gas mixes with the first. The diffusion process occurs rapidly, so the mixture attains uniform density throughout the enclosure after a short while (Fig. 5-5A). What do you think would happen if we did the same experiment

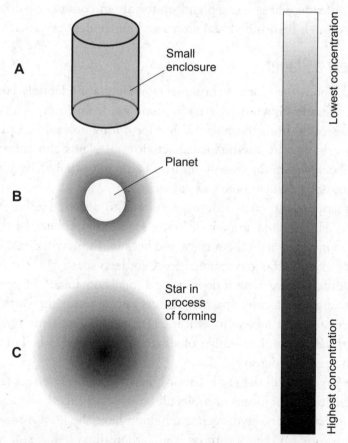

FIGURE 5-5 · At A, distribution of gas inside a container. At B, distribution of gas around a planet with an atmosphere. At C, distribution of gas in space as a star forms. Darkest shading indicates highest concentration.

in the presence of a gravitational field? As you can probably guess, the gases would still mix uniformly inside the jar. The force of gravity wouldn't prevent the gases from mixing. This phenomenon happens with all gases in containers of reasonable size. Gravitation doesn't keep gases from mixing up—at least, not at temperatures we normally encounter, and not as long as the gases don't differ in density by an extreme factor.

Planetary atmospheres, such as that of our own earth, comprise mixtures of various gases. On our planet, nitrogen makes up approximately 78 percent of the gas in the atmosphere at the surface. Oxygen accounts for about 21 percent. The remaining 1 percent consists of many other gases including argon, carbon dioxide, hydrogen, helium, ozone (oxygen molecules with three atoms rather than the usual two), and tiny quantities of some gases that would kill humans in high concentrations, such as chlorine, methane, and carbon monoxide. These gases blend uniformly in containers of reasonable size, even though their individual atoms and molecules vary greatly in mass.

Gases Near a Planet

Let's think about the gaseous shroud surrounding a moderately large planet such as the earth. Gravitation attracts some gas from interplanetary space. Other gases come from the planet's interior during volcanic eruptions. Still other gases result from the biological activities of plants and animals (if the planet harbors life). In the case of our own earth, industrial activity and fossil-fuel combustion produce some gases as well.

All the gases in the earth's atmosphere tend to diffuse. However, because there exists an unlimited amount of "outer space" around our planet and only a finite amount of gas near the surface, and because the gravitational force near the surface exceeds the gravitational force in deep space, the diffusion takes place in a different way than it does inside a small container. The greatest concentration of gas molecules (particle density) occurs near the earth's surface, and it decreases with increasing altitude (Fig. 5-5B). The same phenomenon occurs with respect to the number of kilograms per meter cubed of the atmosphere (the gas mass density).

On the large scale of the earth's atmosphere, yet another effect takes place. For a given number of atoms or molecules per meter cubed, some gases have more mass than others. Hydrogen and helium have the least mass density at a given pressure. Oxygen has more, and carbon dioxide has still more. The most mass-dense gases tend to sink toward the surface, while the least mass-dense gases rise up high, and some of their atoms escape into outer space.

Still Struggling

Why do we observe no distinct boundaries, or layers, from one type of gas to another in the earth's atmosphere? The answer lies in the fact that in a mixture of gases at a reasonable temperature, the molecules and atoms diffuse; they move around with relative freedom. Therefore, instead of finding gases stacked one on top of the other like a layer cake (with the heaviest gases on the bottom, of course), we see gradual and vague transitions. That's fortunate, because if the gases of the atmosphere were rigorously stratified, we would likely find insufficient oxygen to sustain most forms of life at sea level. We'd suffocate in a deadly layer of some noxious gas such as carbon monoxide, chlorine, or methane. We might do okay if we climbed a mountain to exactly the optimum altitude, so we got to a layer where oxygen prevailed—but what a chore!

Gases in Outer Space

Scientists once thought that interstellar space contained no matter whatsoever. But today we know that plenty of atoms and molecules exist out there, largely hydrogen and helium. Trace amounts of heavier gases also exist. All the atoms in outer space interact gravitationally with all the others. Even a single atom exerts some gravitation on every other atom, no matter how great the distance between them.

The motion of atoms in outer space occurs almost at random, but not quite. The slightest perturbation in the randomness of the motion gives gravitation a chance to help the gas congeal or "clump up" in defined locations. Once this process begins, it can continue until a "fuzzy globe" of gas forms with significant particle density at the center, as shown in Fig. 5-5C. As gravitation continues to pull the atoms in toward the center, the mutual attraction among the atoms increases. If the gas cloud has some spin, it flattens into an oblate spherical shape and eventually into a disk with a bulge at the center. A "vicious circle" ensues; the density in the central region skyrockets. The gas pressure in the center rises, heating the gas there. Ultimately the central region gets so hot that *nuclear fusion* begins, and we have a new star! Similar events among the atoms of the gas on a smaller scale can result in the formation of asteroids, planets, and planetary moons.

Pressure

We can compress most gases to an almost unlimited extent. That's why we can fill up hundreds of party balloons from a single, small tank of helium gas. That's why a SCUBA diver can breathe for a long time from a single small tank of air.

Imagine a container whose volume (in meters cubed) equals V. Suppose that N moles of atoms of a particular gas exist inside this container, which is surrounded by a perfect vacuum. We can say certain things about the pressure P, in newtons per meter squared, that the gas exerts outward on the walls of the container. First, P varies in direct proportion to N, provided that we hold V constant. Second, if V increases while N remains constant, P will decrease.

Another important factor—the absolute temperature—comes into play with gases under pressure as they expand and contract. We express absolute temperature in kelvins, representing degrees Celsius above absolute zero (the absence of all heat). When we compress a parcel of gas comprising a fixed number of atoms or molecules, the parcel heats up; when we decompress the same parcel, it cools off. If we heat up a parcel of gas and keep the volume and the number of particles constant, the pressure rises; if we cool the same parcel down, the pressure drops.

In the next chapter, we'll study the behavior of matter, especially liquids and gases, under conditions of varying temperature and pressure.

QUIZ

Refer to the text in this chapter if necessary. A good score is eight correct. Answers are at the back of the book.

1. If we can easily stretch a sample of solid material into a long, thin wire, we might say that the substance exhibits exceptional
   A. hardness.
   B. ductility.
   C. brittleness.
   D. viscosity.

2. Suppose that a sample of substance has a mass density of 4.0×10^3 kg/m³ at the earth's surface. If we take the sample to the surface of Mars, where the acceleration of gravity is 37 percent as strong as on earth, we will find that
   A. the mass density has remained the same at 4.0×10^3 kg/m³.
   B. the mass density has decreased to 1.5×10^3 kg/m³.
   C. the mass density has increased to 1.1×10^4 kg/m³.
   D. we need more information to determine the mass density.

3. Suppose that a sample of substance has a weight density of 1.5×10^4 N/m³ at the earth's surface. If we take the sample to the surface of Mars, where the acceleration of gravity is 37 percent as strong as on earth, we will find that
   A. the weight density has remained the same at 1.5×10^4 N/m³.
   B. the weight density has decreased to 5.6×10^3 N/m³.
   C. the weight density has increased to 4.1×10^4 N/m³.
   D. we need more information to determine the weight density.

4. Refer to Fig. 5-6. Suppose the pistons have areas of 0.040 m² and 0.010 m², as shown. With how much force must you press down on the left-hand piston to cause the right-hand piston to exert an upward force of 12 N?
   A. 48 N
   B. 24 N
   C. 6.0 N
   D. 3.0 N

5. Suppose that a rubber cord has a spring constant of 0.20 m/N for stretching forces ranging from 0 to 10 N. If the cord measures 1.0 m long when we apply no stretching force, how long will it get when we apply 3.0 N of stretching force?
   A. 3.2 m
   B. 2.8 m
   C. 1.6 m
   D. 1.2 m

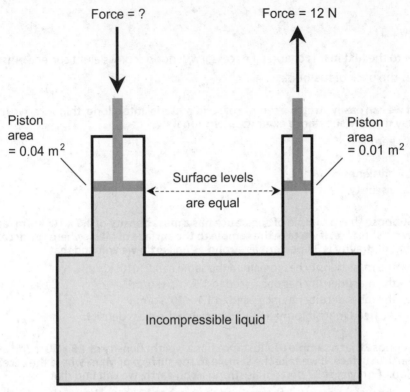

Force = ?   Force = 12 N

Piston
area
= 0.04 m²

Piston
area
= 0.01 m²

Surface levels

are equal

Incompressible liquid

FIGURE 5-6 · Illustration for Quiz Question 4.

6. Suppose that we pour 80.00 ml of liquid into a flask, and we know that this liquid
    has a mass density of 1500 kg/m³. How many grams of liquid have we poured
    into the flask?

    A. 18.75 g
    B. 53.33 g
    C. 120.0 g
    D. 1200 g

7. Suppose that we confine a sample of a gas whose atoms occur singly (rather
    than grouping together into molecules having two or more atoms) to a cubical
    container measuring 10.0 cm along each edge. Somehow, we count the atoms
    in the container and find 3.01×10^{22} of them, uniformly distributed. What's the
    particle density of the gas?

    A. 0.500 mol/m³
    B. 20.0 mol/m³
    C. 50.0 mol/m³
    D. 200 mol/m³

8. **Why can't gases having nearly identical mass densities, such as oxygen and nitrogen, diffuse at room temperature (say, 20°C)?**
   A. The premise in this question is wrong! They can and do diffuse.
   B. The heavier gas sinks, while the lighter gas rises.
   C. Room temperature is too warm for diffusion to occur.
   D. Room temperature is too cold for diffusion to occur.

9. **What will happen to a solid object having a specific gravity of 1.2 if we place it in a bucket of pure liquid water? Assume that the object doesn't dissolve in the water.**
   A. It will float on the surface.
   B. It will hover at a fixed depth.
   C. It will sink to the bottom.
   D. We need more information to answer this question.

10. **We might use the Mohs scale to quantify the**
    A. ductility of a metal.
    B. particle density of a gas.
    C. mass density of a liquid.
    D. hardness of a solid.

chapter **6**

Temperature, Pressure, and Changes of State

When we heat a confined sample of gas, the pressure increases. When we increase the pressure of a confined parcel of gas, the temperature rises. What do we mean, exactly, when we talk about *heat* and *temperature*? In this chapter, we'll define those parameters, and learn how they affect matter.

CHAPTER OBJECTIVES

In this chapter, you will

- Discover how energy can "travel."
- See how scientists measure temperature.
- Learn how temperature, volume, and pressure interact.
- Watch matter expand and contract as the temperature rises and falls.
- Analyze solidification, liquefication, condensation, and vaporization.

What Is Heat?

The term *heat* refers to the transfer of energy from one material object, place, or region to another. Three distinct heat-transfer processes (or *modes*) commonly occur: *conduction, radiation,* and *convection.* Examples of each mode follow:

- When you place a kettle of water on a hot stove, energy transfers by conduction from the burner to the water as shown in Fig. 6-1A.

- When an infrared (IR) lamp, sometimes called a "heat lamp," shines on your skin, energy transfers by radiation from the lamp to your body as shown in Fig. 6-1B.

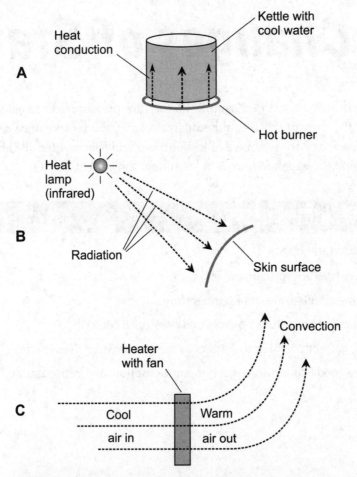

FIGURE 6-1 • Examples of energy transfer as heat by conduction (A), radiation (B), and convection (C).

- When an electric heater-fan warms up a room, the heated air from the fan rises and mixes by convection with rest of the air in the room as shown in Fig. 6-1C.

Heat and energy differ in a subjective way, even though we define both phenomena in the same dimensions. Heat does not comprise energy itself, but the *transfer of energy* that occurs by conduction, radiation, and/or convection.

The Calorie

Physicists sometimes use a unit called the *calorie* to quantify heat. You've heard and read this word many times (maybe too many!). When a physicist talks about calories, she refers to a different unit and a different process than does the dietician when he uses the same word. A "physics calorie" represents only 1/1000 (0.001 or 10^{-3}) of a "diet calorie." The scientific term usually refers to matter only, while the nutritional term refers to biological processes.

The calorie (cal) that interests us, as physicists, quantifies the energy transfer that raises or lowers the temperature of 1 gram (1 g) of pure liquid water by 1 degree Celsius (1°C). The *kilocalorie* (kcal), equivalent to the "diet calorie," equals the amount of energy transfer that alters the temperature of 1 kg, or 1000 g, of pure liquid water by 1°C. These definitions hold valid *if and only if* the water remains liquid during the entire process. If any of the water freezes, thaws, boils, or condenses, the foregoing definitions won't work. At standard atmospheric pressure at the earth's surface, in general, these definitions apply at temperatures between approximately 0°C (the freezing point of water) and 100°C (the boiling point of water).

Specific Heat

Pure liquid water requires 1 calorie per gram (1 cal/g) to warm or cool by 1°C (as long as it's not exactly at the melting/freezing temperature or the vaporization/ condensation temperature, as we shall shortly see). What about other liquids such as oil, alcohol, or salt brine? What about solids such as steel or wood? What about gases such as air? Things don't work out so simply for those substances. Some matter takes more than 1 cal/g to warm or cool by 1°C; some matter takes less. Pure liquid water takes precisely 1 cal/g to change temperature by precisely 1°C for a trivial reason: *scientific convention*. We base our definition of the calorie on the behavior of pure liquid water at everyday temperature and atmospheric pressure.

Imagine that we have found a sample of a mysterious liquid in a remote cave. We call this liquid "substance X." We measure out 1 gram (1.00 g), accurate to three significant figures, of this liquid by pouring some of it into a test tube placed on a laboratory balance. Then we transfer 1 calorie (1.00 cal) of energy to substance X, noting that it remains entirely liquid. Suppose that, as a result of this energy transfer, our sample of substance X increases in temperature by 1.20°C. Obviously, substance X isn't pure liquid water, because it behaves differently than water when it receives a transfer of energy. In order to raise the temperature of 1.00 g of this liquid by 1.00°C, it takes somewhat less than 1.00 cal of heat. To be exact, at least insofar as the rules of significant figures allow, it will take 1.00/1.20 = 0.833 cal to raise the temperature of this sample of substance X by 1.00°C.

Now consider a sample of another material from another remote cave, this time a solid. Let's call it "substance Y." We carve a chunk of it down until we have a piece that masses 1.0000 g, accurate to five significant figures. We transfer 1.0000 cal of energy to our sample; its temperature rises by 0.80000°C, and it remains entirely solid. This material accepts heat energy in a manner different from either liquid water or substance X. It takes a little more than 1.0000 cal of heat to increase the temperature of 1.0000 g of this material by 1.0000°C. Calculating to the allowed number of significant figures, we can determine that it takes 1.0000/0.80000 = 1.2500 cal to raise the temperature of this sample of substance Y by 1.0000°C.

In the performance of our tests on substances X and Y, we have witnessed a special property of matter called the *specific heat*, defined in units of *calories per gram per degree Celsius* (cal/g/°C). Suppose that it takes c calories of heat to raise the temperature of 1 g of a substance by 1°C. For water, we already know that $c = 1$ cal/g/°C, to however many significant figures we want. For substance X described above, $c = 0.833$ cal/g/°C (to three significant figures). For substance Y described above, $c = 1.2500$ cal/g/°C (to five significant figures).

Alternatively, we can express c in *kilocalories per kilogram per degree Celsius* (kcal/kg/°C), and the value for any given substance will turn out the same as it does when we use units of calories per gram per degree Celsius (cal/g/°C). For water, $c = 1$ kcal/kg/°C, to however many significant figures we want. For substance X above, $c = 0.833$ kcal/kg/°C (to three significant figures), and for substance Y above, $c = 1.2500$ kcal/kg/°C (to five significant figures).

The British Thermal Unit (Btu)

In some applications, people use a completely different unit of heat: the *British thermal unit* (Btu). Civil engineers and scientists define the Btu as the amount of heat that will raise of lower the temperature of exactly 1 pound (1 lb) of pure liquid water by 1 degree Fahrenheit (1°F).

Think about the foregoing definition for a little while. Does something seem flawed here? If you're uneasy about it, you have a good reason. What's a "pound"? It depends on your location. How much water weighs 1 lb? On the earth's surface, 1 lb of matter masses approximately 0.454 kg or 454 g. But on Mars, you would need about 1.23 kg of mass to weigh 1 lb. In a "weightless" environment, such as on board a space vessel orbiting the earth or coasting through deep space, the strict definition of Btu loses meaning altogether, because no amount of mass will give you an object that weighs 1 lb! To get rid of this inherent lack of clarity in the definition of Btu, we can define 1 lb as the equivalent of 0.454 kg and leave it at that, no matter where we go.

Despite its "character defects," you'll hear or read about the Btu quite often, especially if you live in the United States, so you should remain familiar with this unit. You might also read or hear specific heat quantified in terms of *Btu per pound per degree Fahrenheit* (Btu/lb/°F). In general this unit does not yield the same number, for any given substance, as the specific heat figure in calories per gram per degree Celsius (cal/g/°C) or kilocalories per kilogram per degree Celsius (kcal/kg/°C).

Still Struggling

Have you heard the Btu mentioned, or seen it quoted, in advertisements for furnaces and air conditioners? When people talk about "Btu" alone in regards to the heating or cooling capacity of a furnace or air conditioner, they technically misuse the term. They really mean to quote the *rate* of energy transfer in *Btu per hour*, not the total amount of energy transfer. British thermal units per hour express *thermal power*, not thermal energy!

PROBLEM 6-1

Suppose that we have 3.00 g of a certain substance. We transfer 5.0000 cal of energy to it, and the temperature goes up uniformly throughout the sample by 1.1234°C. The substance does not boil, condense, freeze, or thaw during this process. What is the specific heat of this material?

SOLUTION

Let's see how much energy 1.00 g of the matter accepts. We have 3.00 g of the material. The entire sample receives 5.0000 cal. Therefore, each gram receives 1/3 of the 5.0000 cal, or 1.6667 cal.

We know that the temperature rises uniformly throughout the sample. In other words, it gets hotter to exactly the same extent everywhere. Therefore, 1.00 g of this substance rises in temperature by 1.1234°C when we transfer 1.6667 cal to it. How much heat will raise the temperature by 1.0000°C? That's the number c we seek, the specific heat. To get c, we must divide 1.6667 cal/g by 1.1234°C. This gives us $c = 1.4836$ cal/g/°C. Because we know the mass of the sample to only three significant figures, we must round our final answer off to 1.48 cal/g/°C.

Temperature

In the scientific sense, the term *temperature* refers to the amount of kinetic energy in a sample of matter. In general, as the temperature of any given material sample rises, its atoms or molecules move faster. We can express temperature in another, less familiar way. To ascertain how hot distant stars, planets, and nebulae in outer space are, astronomers look at the way they emit *electromagnetic* (EM) energy in the form of radio waves, infrared (IR) rays, visible light rays, ultraviolet rays, x rays, and gamma rays. By examining the intensity of the EM radiation as a function of the wavelength, astronomers can determine a value for the *spectral temperature* of the distant celestial object.

Heat Entropy

When energy can move freely from one material object or medium into another as heat, the temperatures of the two media "try to equalize." Ultimately, if the energy transfer process continues for a long enough time, the temperatures of the two media approach the same value unless one of them departs the scene (for example, some steam boils away from a kettle of water).

On the largest imaginable scale, the kinetic energy of everything in the universe "tries to level off" to a state of equilibrium. The cosmos won't come near completing this "quest" in our lifetimes, or even during the lifetime of our planet. Nevertheless, the process will continue indefinitely, just as rivers flow to the seas, just as mountains erode down to the plains. Some cosmologists call this inexorable process *heat entropy*. A few people refer to the hypothetical end product of heat entropy as the *heat death of the universe*.

The Celsius (or Centigrade) Scale

Up until now, we've expressed temperature in terms of the Celsius or centi-grade scale (°C), which derives from the behavior of water at the surface of the earth, under normal atmospheric pressure, and at sea level.

If we have an extremely cold sample of ice and we begin to warm it up, it will start to melt as it accepts heat from the environment. We assign the ice, and the liquid water produced as it melts, a temperature value of 0°C by convention (Fig. 6-2A). As we continue to pump energy into the chunk of ice, more and more of it will melt, and its temperature will stay at 0°C. It won't get any hotter because it's not yet all liquid, and doesn't yet obey the rules for pure liquid water.

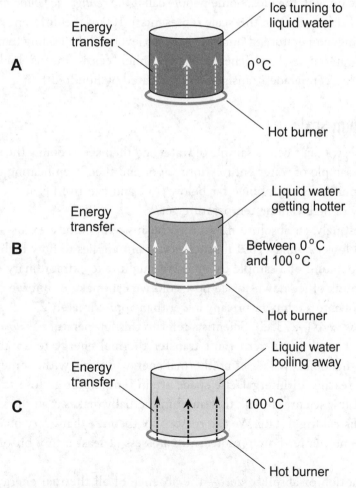

FIGURE 6-2 · Ice melting into liquid water (A), liquid water warming up without boiling (B), and liquid water starting to boil (C).

Once all the water has become liquid, and as we keep pumping energy into it, its temperature will start to increase (Fig. 6-2B). For awhile, the water will remain liquid and will get steadily warmer, obeying the 1 cal/g/°C rule. But eventually, we'll reach a point where the water starts to boil, and some of it changes to the gaseous state. We then assign to the liquid water, and the vapor that boils away from it, a temperature value of 100°C by convention (Fig. 6-2C).

Now we have two definitive points—the *freezing point* of water and the *boiling point*—at which there exist two specific numbers for temperature, 0°C and +100°C. We can define a scheme to express temperature based on these two points: the *Celsius temperature scale*, named after the scientist who supposedly first came up with the idea. Some people call it the *centigrade temperature scale*, because 1 degree (1°) on this scale represents 1/100 of the difference between the melting temperature of pure water at sea level and the boiling temperature of pure water at sea level. The prefix multiplier "centi-" means "hundredths," so the term "centigrade" translates to "graduated by hundredths."

The Kelvin Scale

Of course, we can freeze a sample of water and then keep cooling the ice down, or boil a sample of water entirely into vapor and then keep heating the vapor up. Temperatures can plunge far below 0°C, and can rise far above +100°C. How low or high can the temperature get?

Interestingly, an absolute minimum Celsius temperature exists—a sort of "thermal lower bound"—but no such constraint applies to how high the Celsius temperature of a sample can get. We might take extraordinary efforts to cool a chunk of ice down to see how cold we can make it, but we can never chill it down to a temperature any lower than approximately 273 degrees Celsius below zero (−273°C). Scientists refer to this temperature as *absolute zero*. An object at absolute zero can't transfer thermal energy to anything else, because that object possesses no thermal energy to begin with. Some atoms in the vast reaches of intergalactic space attain temperatures close to absolute zero, but few scientists think that anything actually makes it all the way down to that dismal frigid state. We can reasonably theorize that every object in the universe, no matter how remote, must possess at least a tiny bit of thermal energy.

The notion of absolute zero—the absence of all thermal energy—forms the basis for the *Kelvin temperature scale*, named after *Lord Kelvin*, a British physicist who lived from 1824 to 1907. A temperature of −273.15°C, or absolute zero, represents 0 K. The size of the kelvin "degree" (which scientists simply call the *kelvin* and symbolize by writing K without the little "degree" sign) exactly equals the size of the Celsius "degree," so 0°C = 273.15 K, and +100°C = 373.15 K. We don't have to bother with plus or minus signs when we express temperatures in kelvins, because the values never get negative.

On the high end, we can theoretically keep heating matter up forever. In the cores of stars, temperatures rise to millions of kelvins. Regardless of the actual temperature, the kelvin temperature always exceeds the Celsius temperature by 273.15 degrees. However, in extremely hot environments, we can often consider the kelvin and Celsius temperatures equivalent. When you hear someone say that a particular star's core has a temperature of 30,000,000 K, it means the same thing as +30,000,000°C for the purposes of most discussions, because ±273.15 constitutes a negligible difference value relative to 30,000,000.

The Rankine Scale

An interesting but antiquated temperature scale, called the *Rankine scale* (°R), assigns the value zero to the coldest possible temperature, just as the Kelvin scale does. The difference between the two scales involves the size of the "degree" increment. A temperature increase or decrease of 1 Rankine degree translates to an increase or decrease of 5/9 of a kelvin. Conversely, 1 kelvin increment measures 9/5 (or 1.8 times) the size of 1 Rankine increment. As with kelvins, we don't have to use plus or minus signs when we work in degrees Rankine, because the values never get negative.

If we have an object at a temperature of 50 K, then it's at 90°R. A temperature of 360°R equals 200 K. We can convert easily between kelvins and degrees Rankine using the following two rules:

- To convert degrees Rankine to kelvins, multiply by 5/9.
- To convert kelvins to degrees Rankine, multiply by 9/5 or 1.8.

Readings in the Kelvin and Rankine scales always differ significantly, even at extreme temperatures. If you hear someone say that a star's core has a temperature of 30,000,000°R, they're talking about the equivalent of approximately 16,700,000 K. However, you will not likely hear any serious physicist talk about Rankine temperatures these days!

The Fahrenheit Scale

In much of the English-speaking world, and especially in the United States, lay people use the *Fahrenheit temperature scale* (°F). A Fahrenheit "degree" is the same size as a Rankine "degree," but the scales are offset. The melting temperature of pure water ice at sea level is +32°F, and the boiling point of pure liquid water is +212°F. Absolute zero lies in "far negative territory" at approximately −459.67°F.

The most common temperature conversions that you'll likely perform involve changing a Fahrenheit reading to Celsius, or vice versa. Convenient formulas exist for this purpose. Let F represent the temperature in degrees Fahrenheit, and let C represent the temperature in degrees Celsius. If you want to convert from Celsius to Fahrenheit, use the formula

$$F = 1.8C + 32$$

If you want to convert from Fahrenheit to Celsius, use the formula

$$C = (5/9)(F - 32)$$

Although we express the constants in the above formulas to only one or two significant figures (1.8, 5/9, and 32), we can consider them mathematically exact for calculation purposes. Figure 6-3 shows a side-by-side scale comparison, called a *nomograph*, that you can use if you want to make quick, approximate temperature conversions for values in the range from −50°C to +150°C.

 Still Struggling

Will the universe actually undergo a "heat death" of the sort described earlier in this chapter? No one knows for certain. Some academics will tell you that good evidence exists. However, because of the depressing tenor of that scenario, some scientists have come up with creative theories in an effort to get away from it. One of the most fascinating of these hypotheses, the "oscillating universe theory," suggests that our entire cosmos alternately explodes and collapses in a never-ending series of "big bangs" and "big crunches." If that theory represents reality, then we'll never have to worry about an end to the action!

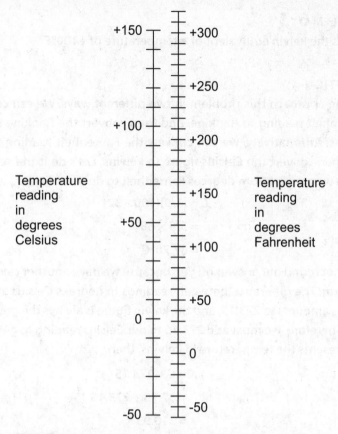

FIGURE 6-3 · You can use this nomograph to convert temperature readings between degrees Fahrenheit (°F) and degrees Celsius (°C).

PROBLEM 6-2

What's the Celsius equivalent of +50°F?

SOLUTION

We can use the above formula for converting Fahrenheit to Celsius temperatures:

$$C = (5/9)(F - 32)$$

In this case, we have

$$C = (5/9)(50 - 32)$$
$$= 5/9 \times 18$$
$$= +10°C$$

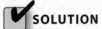

PROBLEM 6-3
_____

What's the kelvin equivalent of a temperature of +100°F?

SOLUTION
_____

We can approach this problem in two different ways. We can convert the Fahrenheit reading to Rankine, and then convert the Rankine reading to kelvins. Alternatively, we can convert the Fahrenheit reading to Celsius, and then convert the Celsius figure to kelvins. Let's do it the second way. When we convert from degrees Fahrenheit to degrees Celsius, we get

$$C = (5/9)(100 - 32)$$
$$= 5/9 \times 68$$
$$= 37.78°C$$

Let's not round our answer off yet, because we have another calculation to perform. The difference between readings in degrees Celsius and kelvins always amounts to 273.15, and the kelvin figure is always the greater of the two. Therefore, we must add 273.15 to our Celsius reading to get kelvins. If K represents the temperature in kelvins, then

$$K = C + 273.15$$
$$= 37.78 + 273.15$$
$$= 310.93 \text{ K}$$

Now we can round our answer off. Because we know our input data to three significant figures, we can say that in this situation, we have an absolute temperature of 311 K.

Some Effects of Temperature

Temperature can affect the volume of, or the pressure exerted by, a sample of matter. You know that most materials expand when heated, although some things expand more than others. Let's take a quantitative look at how temperature influences the behavior of matter.

Temperature, Volume, and Pressure

A sample of gas, confined to a rigid container, will exert increasing pressure on the walls of the container as the temperature rises. If we have a flexible container such as a balloon, the volume of the gas will increase as the temperature rises. Similarly, if we take a container with a certain amount of gas in it and suddenly make the container bigger without adding any more gas, the drop in pressure will cause the temperature to fall. If we have a rigid container with gas in it and then we let some of the gas escape, the drop in pressure will chill the container. This fact explains, for example, why a compressed-air canister gets cold when you use it to blow dust out of your computer keyboard.

Liquids behave a little more strangely. The volume of the liquid water in a kettle, and the pressure it exerts on the kettle walls, don't change when the temperature goes up and down, unless the water freezes or boils. But some liquids, unlike water, expand when they heat up. Mercury constitutes a good example. Thermometer manufacturers exploited this property of mercury for many years, until scientists discovered that mercury presents a biological toxicity danger and an environmental hazard.

Solids, in general, expand when the temperature rises, and contract when the temperature falls. In many cases you don't notice this expansion and contraction. Does your study desk look bigger at +30°C than it does at +20°C? Of course not. But, in fact, the size of your desk does vary with the temperature! The difference escapes your notice simply because it's too small for you to see. However, the bimetallic strip in the thermostat that controls your furnace or air conditioner bends considerably when one of its metals expands or contracts a little more than the other. If you hold such a strip near a hot flame, you can actually watch it curl up or straighten out.

Standard Temperature and Pressure

To set a reference for temperature and pressure for experimentation and measurement, scientists define *standard temperature and pressure* (STP). A condition of STP represents a more-or-less typical state of affairs at sea level on the earth's surface on a dry day.

We define standard temperature as 0°C (+32°F), which represents the freezing point or melting point of pure liquid water. We define standard pressure as the air pressure that will support a column of mercury 0.760 m (just a little less than 30 in) high. This pressure constitutes the well-known value of 14.7 pounds per inch squared (lb/in²), which translates to approximately 1.01×10^5 newtons per meter squared (N/m²).

Thermal Expansion and Contraction

Imagine a sample of solid material that expands when the temperature rises. Some solids expand to a greater extent per degree Celsius than others. We call the extent to which the height, width, or depth of a solid (its *linear dimension*) changes per degree Celsius the *thermal coefficient of linear expansion*.

For most materials, within a reasonable range of temperatures, the thermal coefficient of linear expansion remains constant. If the temperature changes by 2°C, for example, the linear dimension will change twice as much as it would do if the temperature variation were only 1°C. But this rule has limits. If you heat a metal to a high enough temperature, it will melt, burn, or vaporize. If you cool the mercury in an old-fashioned thermometer down enough (assuming you can actually find one), the liquid will freeze. In cases like these, we can't apply the simple length-versus-temperature rule.

In general, if *s* represents the difference in linear dimension (in meters) produced by a temperature change of *T* (in degrees Celsius) for an object whose original or "starting" linear dimension (in meters) equals *d*, then we can calculate the thermal coefficient of linear expansion, symbolized by the lowercase Greek letter alpha (α), using the equation

$$\alpha = s/(dT)$$

When the linear dimension increases, we consider *s* to have a positive value. When the linear dimension decreases, we consider *s* to have a negative value. For most materials, rising temperatures produce positive values of *T*, while falling temperatures produce negative values of *T*.

Scientists quantify the thermal coefficient of linear expansion in *meters per meter per degree Celsius*. Because this expression divides meters by themselves, the meters cancel out, leaving us with units called *per degree Celsius*, symbolized /°C.

 PROBLEM 6-4

Imagine a metal rod that measures 10.000 m long at a temperature of +20.00°C. Suppose that this rod expands to a length of 10.025 m at +25.00°C. What's the thermal coefficient of linear expansion for the material that composes this rod?

 SOLUTION

The rod increases in length by 0.025 m for a temperature increase of 5.00°C. Therefore, we know the following three quantities:

$$s = 0.025$$
$$d = 10.000$$
$$T = 5.00$$

Plugging these numbers into the formula above, we get

$$\alpha = 0.025 / (10.000 \times 5.00)$$
$$= 0.00050$$
$$= 5.0 \times 10^{-4}/°C$$

PROBLEM 6-5

Suppose that the thermal coefficient of linear expansion equals $2.50 \times 10^{-4}/°C$ for a certain substance. Imagine a perfect geometric cube of this substance with a volume V_1 of 8.000 m³ at a temperature of +30.0°C. What volume V_2 will cube the have if the temperature falls to +20.0°C?

✔ SOLUTION

In this context, we must ensure that we know what *linear dimension* means in relation to volume. The length of each edge of the cube will vary directly according to the thermal coefficient of linear expansion $\alpha = 2.50 \times 10^{-4}/°C$. The volume will vary according to the cube (third power) of the extent to which the length of each edge increases or decreases.

We can rearrange the earlier general formula for α so that it solves for the change in linear dimension, s, as follows:

$$s = \alpha dT$$

where T represents the temperature change (in degrees Celsius) and d represents the initial linear dimension (in meters). Because our object constitutes a perfect geometric cube, the initial length, d, of each edge equals 2.000 m (the cube root of 8.000, or $8.000^{1/3}$). Because the temperature falls, we can say that $T = -10.0$. Therefore

$$s = 2.50 \times 10^{-4} \times 2.000 \times (-10.0)$$
$$= -0.00500 \text{ m}$$

The length of each side of the cube at 20.0°C will measure

$$2.000 - 0.00500 = 1.995 \text{ m}$$

in length. The volume of the cube at +20.0°C will therefore equal

$$1.995^3 = 7.940149875 \text{ m}^3$$

Because we know our input data to only three significant figures, we must round our final answer off to 7.94 m³.

Still Struggling

How "heavy" is the air at the earth's surface? Think about it for a minute. Do you really have a clue? Most people don't think of air as "heavy" because we're all immersed in it. The same delusion can occur with water. When you dive a couple of meters down in a swimming pool, you don't feel a lot of pressure and the water doesn't feel massive, but if you calculate the huge amount of mass above you, it might scare you out of the water! The density of dry air at STP equals approximately 1.29 kg/m³. A parcel of air measuring 4.00 m high by 4.00 m deep by 4.00 m wide, the size of a "high-ceiling" bedroom, masses 82.6 kg. In the earth's gravitational field, that translates to 182 lb, the weight of a good-sized, full-grown man.

Temperature and States of Matter

When we heat or cool a sample of matter, it can do things besides simply expanding or contracting, or exerting increased or decreased pressure. It might undergo a *change of state*, such as when solid ice melts into liquid water, or when water boils into vapor.

Thawing and Freezing

Once again, let's think about ordinary water. Imagine a late winter day at a lakeside cabin in northern Wisconsin, where winter temperatures can fall far below freezing. Suppose that the temperature of the ice on the lake surface equals exactly 0°C. You can't safely walk or skate on the ice as you could do in the middle of the winter, because with the approach of spring, the ice has softened into a slushy mixture, part solid and part liquid. Nevertheless, the temperature of this soft ice equals 0°C.

The weather grows warmer hour by hour. As the temperature continues to rise, the slush gets softer. It becomes proportionately more liquid water and less solid ice. You might leave for school or work one morning and see the lake nearly "socked in" with slush, and return in the evening to find it almost entirely thawed. Now you can think about getting the boat out for some serious fishing! But you won't want to go swimming in that water. The liquid water will stay at 0°C until all of the ice has vanished. Only then will the water temperature begin to rise.

Consider now what happens in the late autumn. The weather, and the water, grows colder with each passing week. The water finally drops to 0°C. The lake surface begins to freeze. The temperature of the new ice equals 0°C. The weather grows colder until the whole lake surface has frozen solid. Then, as the temperature continues to fall day by day, the temperature of the ice drops below 0°C (although it remains at 0°C at the boundary just beneath the surface where solid ice meets liquid water). The layer of ice thickens. The ice near the surface can get much colder than 0°C. How much colder depends on various factors, such as the severity of the winter and the amount of snow that falls on top of the ice and insulates it against the bitter chill of the air.

The temperature of water does not follow exactly along with the air temperature when heating or cooling takes place in the vicinity of 0°C. Instead, the water temperature follows a curve something like that shown in Fig. 6-4. At A, the *air temperature* steadily rises; at B, the *air temperature* steadily falls. However, the *water temperature* "stalls" as it thaws or freezes. Most other substances exhibit this same property when they thaw or freeze, although the "stalling-out" temperature can vary considerably for different substances.

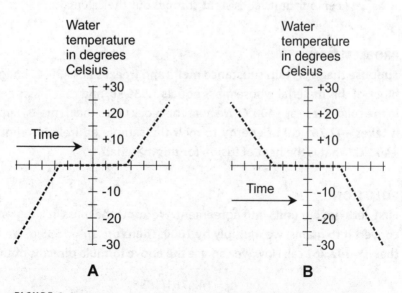

FIGURE 6-4 · Water as it thaws and freezes. At A, the temperature rises and the ice thaws. At B, the temperature falls and the liquid freezes.

Heat of Fusion

The temperature of a sample "levels off" for awhile as it thaws because it takes a certain amount of energy transfer to change a sample of solid matter to its liquid state, assuming the matter can exist in either of these two states. In the case of water, it takes 80 cal to convert 1 g of ice at 0°C to 1 g of pure liquid water at 0°C. We call it the *heat of fusion*. In the reverse scenario, if 1 g of pure liquid water at 0°C freezes completely solid and becomes ice at 0°C, it gives up 80 cal.

We can express heat of fusion in *calories per gram* (cal/g) or in *kilocalories per kilogram* (kcal/kg). For any particular substance, either unit will give us exactly the same number. Of course, if we're working with something other than water, then we must substitute the freezing/melting temperature of that substance for 0°C in the discussion. Once in awhile, you'll hear or read about heat of fusion in terms of *calories per mole* (cal/mol) rather than in calories per gram. But unless an author specifically expresses the units in calories per mole, you should assume that the discussion involves calories per gram (cal/g) or kilocalories per kilogram (kcal/kg).

If we symbolize the heat of fusion as h_f, the heat added or given up by a sample of matter as h, and the mass of the sample as m, then

$$h_f = h/m$$

as long as we keep our units consistent throughout the calculation.

PROBLEM 6-6

Suppose that a certain substance melts and freezes at +400°C. Imagine a block of this material whose mass equals 1.535 kg, and that exists entirely in the solid phase at +400°C. We heat the block up so that it melts. Suppose it takes 142,761 cal of energy to melt the sample entirely into liquid at +400°C. What is the heat of fusion for this material?

SOLUTION

First, let's get our units into agreement. We know the mass in kilograms; to convert it to grams, we multiply by 1000. Therefore, $m = 1535$ g. We know that $h = 142,761$ cal. Now we can use the above formula directly, obtaining

$$h_f = 142,761/1535$$

$$= 93.00 \text{ cal/g}$$

Boiling and Condensing

Let's return to the stove, where a kettle of water heats up. Suppose that we're at sea level and the temperature of the water equals exactly +100°C, but it has not yet begun to boil. As heat keeps going into the water, boiling begins. The water becomes proportionately more vapor and less liquid, so the mass of the liquid in the kettle gradually decreases. Nevertheless, the temperature of the liquid stubbornly remains at +100°C. Eventually, all of the liquid boils away, leaving only vapor (or "steam"). Imagine that we have managed to capture all of the vapor from the entire water sample in an enclosure, and in the process of the water's boiling away, we've driven all the air out of the enclosure so as to have pure water vapor. Suppose that we allow the stove burner to keep on heating the sample, even after all of it has boiled into vapor.

At the moment when the last of the liquid vanishes, the temperature of the vapor equals +100°C. Once all the liquid has boiled off, the vapor can grow hotter than +100°C. The ultimate extent to which we can heat up the vapor depends on the power of our stove burner, and on the quality of the chamber's thermal insulation.

Now let's think about what happens if we take the enclosure, along with the kettle, off of the stove and put it into a refrigerator. The environment, and the water vapor, begins to grow colder. The vapor temperature eventually drops to +100°C. The vapor begins to condense. The temperature of the "new" liquid water equals +100°C. Condensation takes place until all the vapor has condensed. (Hardly any of the vapor condenses back in the kettle. Instead, "dew" forms all around inside the chamber.) We allow a small amount of air into the chamber near the end of this experiment to maintain a reasonable pressure inside. The chamber keeps growing colder. Once all the vapor has condensed, the temperature of the liquid begins to fall below +100°C.

As is the case with melting and freezing, the temperature of water does not follow exactly along with the air temperature when heating or cooling takes place in the vicinity of +100°C. Instead, the water temperature follows a curve something like that shown in Fig. 6-5. At A, the air temperature rises; at B, the air temperature falls. The water temperature "stalls" as it boils or condenses. Most other substances exhibit this same property when they boil or condense, although the "stalling-out" temperature may vary, just as it does for heat of fusion.

Heat of Vaporization

It takes a certain amount of energy transfer to change a sample of liquid to its gaseous state, assuming the matter can exist in either of these two states. In the

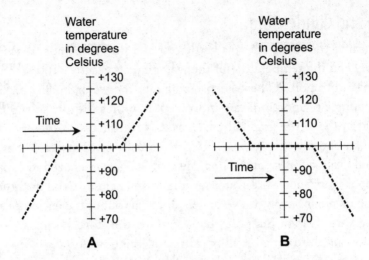

FIGURE 6-5 • Water as it boils and condenses. At A, the temperature rises and the liquid boils. At B, the temperature falls and the vapor condenses.

case of water, it takes 540 cal to convert 1 g of liquid at +100°C to 1 g of pure water vapor at +100°C. We call this quantity the *heat of vaporization*. In the reverse scenario, if 1 g of pure water vapor at +100°C condenses completely and becomes liquid water at +100°C, it loses 540 cal of heat.

We typically express heat of vaporization in the same units as heat of fusion: calories per gram (cal/g). We can also express heat of vaporization in kilocalories per kilogram (kcal/kg), and we'll get the same number as we get for the cal/g figure. When we work with a substance other than water, then we must, of course, substitute the boiling/condensation point of that substance for +100°C. Occasionally we might see the heat of vaporization for a certain substance quoted in calories per mole (cal/mol) rather than in cal/g.

If we symbolize the heat of vaporization (in calories per gram) as h_v, the heat added or given up by a sample of matter (in calories) as h, and the mass of the sample (in grams) as m, then

$$h_v = h/m$$

If you're astute, you'll notice that this formula matches the formula for heat of fusion, except that h_v appears in place of h_f.

PROBLEM 6-7

Imagine that a certain substance boils and condenses at +500°C. Imagine a beaker of this material whose mass is 67.5 g, and it exists entirely in the liquid phase at +500°C. It has a heat of vaporization specified as 845 cal/g. How much heat, in calories and in kilocalories, does it take to completely boil away this liquid?

SOLUTION

Our units already agree: We have grams for m and calories per gram for h_v. We can use simple algebra to manipulate the above formula to express the heat in terms of the other given quantities, obtaining

$$h = h_v m$$

Now we merely put in the numbers and do the arithmetic, getting

$$h = 845 \times 67.5$$
$$= 5.70 \times 10^4$$
$$= 57.0 \text{ kcal}$$

QUIZ

Refer to the text in this chapter if necessary. A good score is eight correct. Answers are at the back of the book.

1. If we transfer 850 kcal of heat energy to 3 kg of water vapor in a sealed chamber, the vapor will
   A. all condense into liquid.
   B. all freeze into ice.
   C. remain in the gaseous state.
   D. partially condense into liquid, and partially freeze into ice.

2. We can express thermal power in terms of
   A. British thermal units per hour (Btu/h).
   B. calories (cal).
   C. calories per gram (cal/g).
   D. kelvins per kilogram (K/kg).

3. Suppose that a rod made of some strange-looking material measures 2.0000 m long at a temperature of +30.0°C. We heat the rod up to +70.0°C, and then measure its length as 2.0800 m. What is the thermal coefficient of linear expansion for this substance?
   A. 2.00×10^{-3} /°C
   B. 1.00×10^{-3} /°C
   C. 5.00×10^{-4} /°C
   D. 2.00×10^{-5} /°C

4. Imagine that a certain liquid boils and condenses at +150.5°C. We have a flask containing 100.000 g of this substance. We measure its temperature as +150.5°C. Then, using a stove burner, we transfer heat energy to the liquid until it boils away. At the moment the last bit of liquid disappears into vapor at +150.5°C, we determine that we've transferred 300.00 kcal of heat to the substance. What is its heat of vaporization?
   A. 3.000 kcal/kg
   B. 3000 kcal/kg
   C. 0.3333 kcal/kg
   D. 333.3 kcal/kg

5. When we discuss the amount of thermal energy that it takes to convert 1 g of liquid to 1 g of the same substance in the form of a gas, we're talking about the
   A. thermal expansion coefficient.
   B. thermal transfer ratio.
   C. state conversion index.
   D. heat of vaporization.

6. Suppose that an astronomer says the core of a certain star maintains a constant temperature of $+2.00 \times 10^7$ degrees Celsius, accurate to three significant figures. That's essentially the equivalent of

   A. 3.60×10^7 degrees Rankine.

   B. $+3.60 \times 10^7$ degrees Fahrenheit.

   C. 2.00×10^7 kelvins.

   D. All of the above

7. Imagine a vessel containing 2.000 kg of liquid whose specific heat equals 1.50 kcal/kg/°C. Suppose that all of the liquid exists at its vaporization temperature of 450 K. We transfer 150 cal of heat energy to the liquid. What is the temperature of the liquid in the vessel after we've transferred this energy to it?

   A. 540 K

   B. 495 K

   C. 450 K

   D. We can't answer this question without more information.

8. After transferring the heat energy to the liquid as described in Question 7, we will find that the liquid in the vessel has a mass of

   A. zero.

   B. less than 2.000 kg.

   C. 2.000 kg.

   D. more than 2.000 kg.

9. The coldest possible temperature is

   A. 0°C.

   B. −459.67°C.

   C. −273.15°C.

   D. nonexistent, because there's no limit to how cold things can get.

10. The calorie per gram per degree Celsius (cal/g/°C) quantifies

   A. specific heat.

   B. heat of fusion.

   C. thermal coefficient of linear expansion.

   D. thermal power.

Test: Part I

Don't look back at any of the text while taking this test. The correct answer choices appear at the back of the book. Consider having a friend check your score the first time you take this test, without telling you which questions you got right and which ones you missed. That way, you won't subconsciously memorize the answers in case you want to take the test again later.

1. Suppose that a sample of a certain substance melts and freezes at −15°F. We carve out a solid 1.3-kg block of this material in a deep-freeze chamber at −45°F. Then we warm the chamber to 70°F and wait until the block melts. At the moment the last of the solid liquefies, we measure the temperature of the liquid. It's still −15°F. Suppose that the sample absorbed 6.5×10^5 cal as it melted. What's the heat of fusion for this substance, expressed in calories per gram?

   A. 0.0020 cal/g
   B. 2.0×10^{-6} cal/g
   C. 500 cal/g
   D. 5.0×10^5 cal/g
   E. Zero, because the temperature of the sample has remained constant.

2. What's the heat of fusion for the substance described in Question 1, expressed in kilocalories per kilogram?

   A. 2.0 kcal/kg
   B. 2000 kcal/kg
   C. 50×10^{-4} kcal/kg
   D. 500 kcal/kg
   E. Zero, because the temperature of the sample has remained constant.

3. We lift 3.50-kg brick straight up by 7.00 m on the surface of a planet where the gravitational acceleration is 5.00 m/s². This action causes the brick to gain potential energy. How much?

   A. 2.50 J
   B. 4.90 J
   C. 10.0 J
   D. 24.5 J
   E. 123 J

4. Imagine an airtight chamber whose size we can change at will. We evacuate the chamber and then introduce a certain amount of nitrogen gas into it. We allow the temperature of the gas to stabilize. Then we suddenly double the chamber's volume without introducing or removing any gas. What happens?

   A. The gas pressure and temperature both fall.
   B. The gas pressure falls but the temperature remains constant.
   C. The gas pressure falls but the temperature rises.
   D. The gas pressure remains constant but the temperature falls.
   E. The gas pressure remains constant but the temperature rises.

5. Imagine that you and I stand atop a cliff on Mars, where the gravitational acceleration amounts to approximately 3.5 m/s². You drop a rock of mass 150 g off the cliff. At the same instant, I drop a rock of mass 600 g. How will the instantaneous acceleration vectors of the rocks compare after 2 s of time, assuming that neither rock has hit the surface yet? (Let's neglect the effect of the small amount of friction caused by the thin atmosphere of Mars.)

   A. The two instantaneous acceleration vectors will coincide, in terms of both the magnitude and the direction.
   B. The more massive rock's instantaneous acceleration vector magnitude will be 16 times the less massive rock's instantaneous acceleration vector magnitude, but the vector directions will coincide.

C. The more massive rock's instantaneous acceleration vector magnitude will be four times the less massive rock's instantaneous acceleration vector magnitude, but the vector directions will coincide.

D. The more massive rock's instantaneous acceleration vector magnitude will be twice the less massive rock's instantaneous acceleration vector magnitude, but the vector directions will coincide.

E. The two acceleration vector magnitudes will coincide, but the directions will differ.

6. **In the scenario of Question 5, suppose that we wait until both rocks strike the surface. We measure their respective kinetic energy levels at the instant of impact. How will these levels compare?**

A. They will be identical.

B. The more massive rock will hit the surface with 16 times the kinetic energy of the less massive rock.

C. The more massive rock will hit the surface with four times the kinetic energy of the less massive rock.

D. The more massive rock will hit the surface with twice the kinetic energy of the less massive rock.

E. To answer this question, we must know how long it takes each rock to fall.

7. **When we talk about the specific heat of a certain substance, we refer to**

A. the absolute temperature of the sample in kelvins, multiplied by its mass in grams, divided by the energy in calories that it contains.

B. the amount of energy, in calories, required to raise the temperature of a 1-g sample of that substance by 1°C.

C. the amount of energy, in calories, given up by a 1-g sample of that substance if it cools down to absolute zero.

D. the amount of energy, in calories, required to raise or lower the temperature of a 1-g sample of that substance to 0°C.

E. The absolute temperature of the sample in kelvins, divided by its mass in grams, multiplied by the energy in calories that it contains.

8. **In theory, we can define potential energy in terms of**

A. foot-pounds per second.

B. gram-centimeters squared per second squared.

C. kilogram-meters per second.

D. inch-ounces squared per second.

E. gram- seconds squared per meter.

9. **Suppose that a woman masses 44.0 kg. She weighs herself on the surface of a planet that has no atmosphere (and therefore no air resistance) where, if she fell freely, she would accelerate downward at 11.0 m/s². How much does she weigh on that planet?**

A. We need more information to answer this question.

B. 146 N

C. 484 N

D. 5.32×10^3 N

E. 2.13×10^4 N

10. **Which of the following nonstandard "units" could we theoretically use to express the frequency of a sound wave?**

   A. The British thermal unit per minute
   B. The joule per minute
   C. The hertz per minute
   D. The kilogram-meter per minute
   E. The cycle per minute

11. **Imagine a motor vehicle (car) traveling at a constant speed of 20 m/s along a curving road as shown in Fig. Test I-1. Based on the appearance of this illustration, at which of the three points does the car's acceleration vector appear to have the greatest magnitude?**

   A. Point X
   B. Point Y
   C. Point Z
   D. We can't answer this question based on the appearance of the illustration alone. We need quantitative information about the contour of the road.
   E. The car's acceleration vector magnitude remains constant at all points because the car's speed never changes.

12. **Which, of any of the following statements accurately describes the motion of a satellite in a circular orbit around the earth?**

   A. The satellite's instantaneous speed remains constant.
   B. The satellite's instantaneous speed increases at a constant rate.
   C. The satellite's instantaneous speed increases at an ever-increasing rate.
   D. The satellite's instantaneous velocity vector remains constant.
   E. The satellite's instantaneous acceleration vector remains constant.

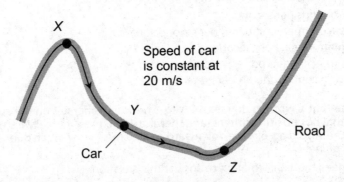

FIGURE TEST I-1 • Illustration for Part I Test Question 11.

13. Consider a solid, rigid rod that measures 20.0 m long at a temperature of 300 K. We cool the rod down to 200 K, and then find that it has shortened to a length of 19.9 m. What is the thermal coefficient of linear expansion for this substance?

    A. 2.00×10^{-4} /°C

    B. 1.00×10^{-4} /°C

    C. 5.00×10^{-5} /°C

    D. 2.00×10^{-5} /°C

    E. We must convert kelvins to degrees Celsius before we can calculate the answer to this question.

14. The term *heat of fusion* refers to

    A. the amount of heat necessary to convert a fixed mass of liquid entirely to the gaseous state without any increase in the sample's temperature.

    B. the amount of heat given up by a fixed mass of gas as it condenses entirely into the liquid state without any decrease in the sample's temperature.

    C. the energy produced when two atomic nuclei combine to form a new atomic nucleus.

    D. the extent to which the temperature in kelvins rises or falls when two atomic nuclei fuse to form a new atomic nucleus.

    E. None of the above

15. If we want to rigorously define one of the following parameters, we must describe it as a vector quantity. Which one?

    A. Speed

    B. Time

    C. Absolute temperature

    D. Acceleration

    E. Mass

16. Imagine that we ride in a van having a mass of 8000 kg (including the body masses of the passengers), traveling along a straight, level highway at a constant speed of 24.00 m/s. How much kinetic energy does the van have?

    A. 4.608×10^6 J

    B. 2.304×10^6 J

    C. 1.920×10^5 J

    D. 9.600×10^4 J

    E. We need more information to answer this question.

17. In the United States of America, people often talk about British thermal units per hour (Btu/h) to quantify

    A. energy.

    B. condensation rates.

    C. power.

    D. specific heat.

    E. heat of fusion.

18. One meter represents approximately 39.37 in. One foot represents exactly
    12 in. One statute mile represents exactly 5280 ft. Based on this information, we
    can calculate that a kilometer represents approximately

    A. 1.732 statute miles.
    B. 0.5773 statute miles.
    C. 1.609 statute miles.
    D. 0.6214 statute miles.
    E. None of the above

19. Suppose that the air in a room cools by 54°C. How much cooler has the air
    become in kelvins?

    A. 97 K
    B. 65 K
    C. 62 K
    D. 54 K
    E. 30 K

20. Which, if any, of the following statements A., B., C., or D. sometimes fails to hold
    true for an object moving along a straight-line path?

    A. The instantaneous acceleration vector magnitude equals the rate at which the
        speed changes at that instant.
    B. The instantaneous velocity vector always points in the same direction as the
        instantaneous displacement vector, or else in the exact opposite direction.
    C. The instantaneous velocity vector equals the rate at which the displacement
        vector changes at that instant.
    D. The instantaneous acceleration vector always points in the same direction as
        the instantaneous velocity vector, or else in the exact opposite direction.
    E. Every one of the above statements A., B., C., and D. always holds true for an
        object moving along a straight-line path.

21. A freight train travels a total distance of 180 km between 1:00 p.m. and 5:00 p.m.
    on a certain day. The train remains in the same time zone for the duration of the
    journey. Over that period of time, the train's speed averages

    A. 80.0 m/s.
    B. 125 m/s.
    C. 8.00 m/s.
    D. 12.5 m/s.
    E. None of the above

22. At a temperature of 4°C, we can define the specific gravity of a solid material
    sample as

    A. the ratio of its mass density relative to the mass density of pure liquid water.
    B. the pressure that it exerts relative to that of dry air at the earth's surface.
    C. its mass relative to an equal volume of hydrogen.
    D. its weight relative to an equal volume of carbon.
    E. the force that the earth's gravity exerts on a single atom of that substance.

23. We can define the atomic number of a chemical element as the

    A. sum of the number of protons and neutrons in the nucleus of one of its atoms.
    B. number of protons in the nucleus of one of its atoms.
    C. number of neutrons in the nucleus of one of its atoms.
    D. number of electrons "orbiting" one of its atoms.
    E. net electric charge on one of its atoms.

24. Imagine a material that industrial manufacturers can easily stretch and form into thin, flexible wire. A scientist would call this substance

    A. temperable.
    B. reactive.
    C. ductile.
    D. volatile.
    E. fissionable.

25. Suppose that an electric motor drives a rope-and-pulley system to lift heavy objects. The motor draws a constant 1.500 kW of power from the electric utility. The end of the rope, tied to a heavy block, imparts potential energy to the block at the rate of 1200 joules per second as the block rises into the air. What is the efficiency of this system?

    A. 80.00 percent
    B. 64.00 percent
    C. 125.0 percent
    D. 156.3 percent
    E. We must know the mass of the block to answer this question.

26. Suppose that we know the temperature of a certain sample in degrees Celsius. If we want to know the temperature in kelvins, we should

    A. multiply by 9/5.
    B. multiply by 5/9.
    C. multiply by 9/5 and then add 32.
    D. subtract 273.15.
    E. add 273.15.

27. Refer to Fig. Test I-2. Suppose that the pistons have areas of 0.500 m^2 and 0.100 m^2, as shown. We press down on the left-hand piston with a force of 125 N, causing the right-hand piston to move upward. How much force do we observe at the right-hand piston?

    A. 5.00 N
    B. 25.0 N
    C. 625 N
    D. 3.13×10^3 N
    E. We need more information to answer this question.

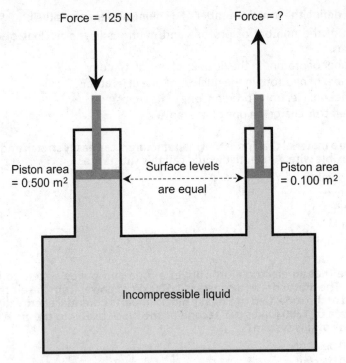

Force = 125 N   Force = ?

Piston area   Surface levels   Piston area
= 0.500 m²   are equal   = 0.100 m²

Incompressible liquid

FIGURE TEST I-2 · Illustration for Part I Test Question 27.

28. **We can use the newton to quantify an object's**
    A. acceleration.
    B. weight.
    C. gravitation.
    D. kinetic energy.
    E. potential energy.

29. **If we add an electron to an atom of carbon (atomic number 6) that has seven electrons to begin with, we get**
    A. a different isotope of carbon.
    B. a carbon anion.
    C. an oxygen atom (atomic number 8).
    D. a carbon cation.
    E. a negative nitrogen ion (atomic number 7).

30. **Suppose that a race car, initially at rest, accelerates along a straight, level road at 5.000 m/s². How far will the car travel in the first 10.00 s after it begins to move?**
    A. We need more information to answer this question.
    B. 2500 m
    C. 1250 m
    D. 500.0 m
    E. 250.0 m

31. Suppose that a cubical chunk of matter measuring 7.000 cm along each edge has a mass of 343.0 g. What is its density?

    A. 49,000 kg/m³
    B. 7000 kg/m³
    C. 1000 kg/m³
    D. 142.9 kg/m³
    E. 20.41 kg/m³

32. Suppose that we use a rope-and-pulley system to raise a 3.500-kg object by a distance of 7.000 m on the surface of a planet where the gravitational acceleration is 8.000 m/s². How much work does this task require?

    A. 1568 N · m
    B. 196.0 N · m
    C. 28.00 N · m
    D. 16.00 N · m
    E. 3.500 N · m

33. As long as we don't accelerate with respect to a source of visible light, the speed of the light coming to us from that source through a vacuum depends on

    A. the direction from which it arrives.
    B. its wavelength.
    C. its distance from us.
    D. its frequency.
    E. None of the above

34. Which of the following events represents an example of nuclear fission?

    A. The merging of a proton and an electron to form a neutron.
    B. The production of energy resulting from a collision between an electron and a positron.
    C. The merging of two atomic nuclei to get a new nucleus having greater mass than either of the original two.
    D. The decay of an isolated neutron.
    E. The splitting of an atomic nucleus into two new nuclei, both of which have less mass than the original.

35. Suppose that a sample of matter masses 0.00025 g. That's the equivalent of

    A. 0.025 mg.
    B. 0.25 mg.
    C. 2.5 mg.
    D. 25 mg.
    E. 250 mg.

36. Imagine that 10 kg of matter comes into contact with 10 kg of antimatter, so that the samples completely annihilate each other to yield pure energy. How much energy will the reaction produce?
    A. 9.0×10^{17} J
    B. 1.1×10^{18} J
    C. 2.2×10^{18} J
    D. 1.8×10^{18} J
    E. 2.2×10^{18} J

37. In theory, we can express the magnitude of a momentum vector in
    A. gram-centimeters per second squared ($g \cdot cm/s^2$).
    B. gram-centimeters per second ($g \cdot cm/s$).
    C. gram-seconds per centimeter ($g \cdot s/cm$).
    D. gram-seconds per centimeter squared ($g \cdot s/cm^2$).
    E. centimeter-seconds per gram ($cm \cdot s/g$).

38. The term *heat death of the universe* means
    A. the eventual destruction of the earth by the sun.
    B. the eventual collapse of the universe into a fireball.
    C. the ultimate end product of heat entropy in the cosmos.
    D. the eventual cooling of the solar system after the sun dies.
    E. the eventual collapse of our galaxy into a black hole.

39. Technically, when we talk about the weight of an object, we refer to
    A. a speed vector.
    B. a velocity vector.
    C. an acceleration vector.
    D. a force vector.
    E. a displacement vector.

40. The individual atoms of different isotopes of a given element always have the same number of
    A. protons.
    B. neutrons.
    C. electrons.
    D. negatrons.
    E. ions.

41. Suppose that a 50-ml sample of liquid masses 80 g. What's the mass density of this sample?
    A. 1.6×10^3 kg/m^3
    B. 160 g/cm^3
    C. 16 mol/cm^3
    D. 1.6×10^5 mol/m^3
    E. We need more information to answer this question.

42. **If we allow an object to fall freely on the moon, the object will**
    A. travel downward at a constant speed.
    B. accelerate downward at a constant rate.
    C. accelerate downward at a rate that increases with time.
    D. decelerate downward at a constant rate.
    E. decelerate downward at a rate that increases with time.

43. **The fundamental SI of material quantity is the**
    A. kilogram.
    B. newton.
    C. pound.
    D. mole.
    E. joule.

44. **If we could place a proton and an electron side by side and compare them, we would find that**
    A. The proton masses more than the electron; the two particles carry the same quantity of electric charge having the same polarity.
    B. The proton and the electron have equal mass; the two particles carry the same quantity of electric charge having opposite polarity.
    C. The proton masses less than the electron; the two particles carry the same quantity of electric charge having the same polarity.
    D. The proton and the electron have equal mass; the electron carries an electrical charge while the proton does not.
    E. The proton masses more than the electron; the two particles carry equal quantities of electric charge having opposite polarities.

45. **Suppose that an atom has 17 protons in the nucleus and 18 electrons "orbiting" the nucleus. What, approximately, is the atomic mass?**
    A. We need more information to answer this question.
    B. 17 AMU
    C. 18 AMU
    D. 35 AMU
    E. 306 AMU

46. **Imagine that we sit on the surface of Mars. We pump all of the martian "air" out of a sealed container, leaving a perfect vacuum inside. Then we introduce 10 g of matter into the container. The matter distributes itself evenly throughout the chamber within a few seconds. We repeat this experiment with 20 g, 30 g, and 40 g of the same substance, always with the same result: uniform distribution of matter inside the container. We can conclude that the substance is**
    A. a negative isotope.
    B. a cation.
    C. a gas.
    D. an anion.
    E. a positive isotope.

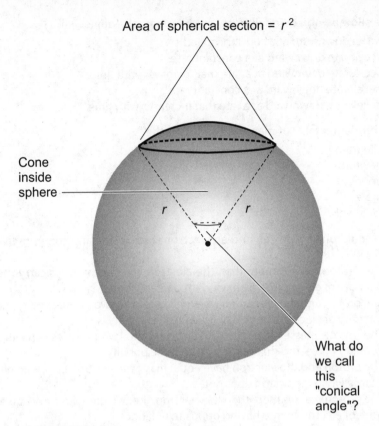

Area of spherical section = r^2

Cone inside sphere

r r

What do we call this "conical angle"?

FIGURE TEST I-3 · Illustration for Part I Test Question 47.

47. **Figure Test I-3 illustrates a sphere of radius r with a cone inside. The vertex (or tip) of the cone coincides with the center of the sphere. The area of the sphere's surface enclosed by the cone's base perimeter equals the square of the sphere's radius. We call the solid angle (or "conical angle") at the cone's vertex**

   A. a radian.
   B. an angular degree.
   C. a conical degree.
   D. a steradian.
   E. a solid degree.

48. **Suppose that a sample of substance has a weight density of 5.4×10^4 N/m^3 at the earth's surface. If we take the sample to the surface of Mars, where the acceleration of gravity is 37 percent as strong as on earth, we will find that**

   A. the weight density has remained the same.
   B. the weight density has decreased to 2.0×10^4 N/m^3.
   C. the weight density has decreased to 7.4×10^3 N/m^3.
   D. the weight density has increased to 1.5×10^5 N/m^3.
   E. the weight density has increased to 3.9×10^5 N/m^3.

49. Which of the following nonstandard "units" could we theoretically use to express *accumulated power consumption* over time?

    A. The joule per hour
    B. The gram-centimeter
    C. The gram-centimeter per second
    D. The watt per minute
    E. The gram-centimeter squared per second squared

50. Imagine that we have access to a machine that collects energy over a period of time, and then converts all the energy into matter. We employ that machine to collect sunlight for a few weeks, and then push a button marked "Make Matter!" How can we calculate the mass of the resulting material sample in kilograms?

    A. Divide the total collected energy in joules by the square of the speed of light in meters per second.
    B. Multiply the total collected energy in joules by the square of the speed of light in meters per second.
    C. Divide the square of the speed of light in meters per second by the total collected energy in joules.
    D. Divide speed of light in meters per second by the total collected energy in joules.
    E. We can't!

Part II

Electricity, Magnetism, and Electronics

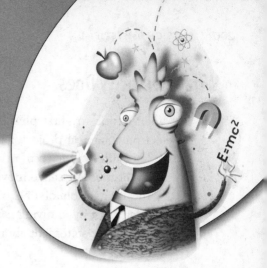

Direct Current

Now that you have a solid grasp of classical physics, let's learn about the particles and forces that allow us to light our homes, communicate with people on the other side of the world, and build machines that people would have called "magical" or even "impossible" a few decades ago.

CHAPTER OBJECTIVES

In this chapter, you will

- Learn how electrical current flows.
- Explore the nature of static electricity.
- See how voltage, current, and resistance relate to each other.
- Learn the symbols for circuit diagrams.
- Discover how resistances combine.
- Calculate currents and voltages with Kirchhoff's laws.

What Electricity Does

When I took my first physics course in middle school, the teacher used an electromechanical projector that rendered audiovisual data on a large screen, while celluloid film 16 mm wide clattered through its gear box. We watched several of these so-called movies narrated by an articulate old college professor. At the conclusion of one of his lectures, the professor said, "We evaluate electricity not by knowing *what it is*, but by codifying *what it does*." I've never heard a better statement about electricity, either before then or since.

Conductors

In some materials, the electrons can move easily from atom to atom. In others, the electrons can move only with difficulty. In still other substances, the electrons rarely move among the atoms at all, unless we apply an overwhelming external force of exactly the right sort. We call a substance of the first type, in which we can easily get the electrons to wander among the atoms, an *electrical conductor*, a *conducting medium*, or simply a *conductor*.

Elemental silver conducts electricity better than any other "everyday" substance at room temperature. Copper and aluminum also constitute excellent electrical conductors. Iron, steel, and various other metals conduct electricity fairly well. Some liquids, such as elemental mercury at room temperature, make good conductors. Salt water conducts fairly well, especially with high concentrations of salt. Gases do not conduct well under most circumstances, because the relatively large distances between atoms make the free exchange of electrons difficult. Nevertheless, if a sample of gas becomes ionized, it can conduct to some extent.

Electrons in a conductor do not move in a steady stream, like the molecules of water as they pass through a garden hose. Instead, the electrons "jump" from the outer shells of atoms to the outer shells of other atoms nearby. Figure 7-1 illustrates how such a "jump" takes place (in simplistic form, of course). In a typical electrical circuit, trillions of electrons make "leaps" like this every second.

Imagine a long line of people, each one constantly passing a baton to the neighbor on the right. If there are plenty of batons all along the line, and if everyone keeps passing batons along as they come, we see a steady stream of batons moving along the line. This scenario represents what happens in a good conductor as electricity flows through it. If the people become tired and pass fewer batons along, the rate of flow decreases, representing what happens in a

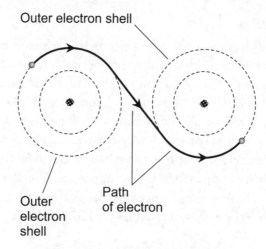

Outer electron shell

Outer electron shell

Path of electron

FIGURE 7-1 · In an electrical conductor, electrons can pass easily from atom to atom.

poor electrical conductor. In the extreme, if the people refuse to pass the batons, we have the analog of conditions in a material that does not conduct electricity at all.

Insulators

If the people refuse to pass the batons in the above-described situation, the line represents an *electrical insulator*. Such substances prevent electrical currents from flowing except under certain circumstances. Most gases constitute good electrical insulators (because they don't conduct well). Glass, dry wood, paper, and plastics also make effective electrical insulators. Pure water insulates as well, although it will conduct to some extent when we dissolve certain minerals, particularly salts, in it. Some metal oxides form good insulators despite the fact that the metal in its pure form conducts.

Engineers use the term *dielectric* to describe insulating materials or objects deliberately placed into a system to prevent any flow of electrons that could equalize an *electric charge* difference between two points. Dielectrics constitute an important ingredient in the manufacture of *capacitors*, electrical components that can store electron surpluses and deficiencies. When we have two electric charge "zones" of opposite *polarity* (called *plus* and *minus*, *positive* and *negative*, or + and −) in close proximity but kept apart by an insulating barrier, we call the individual charge "zones" *electric poles*, and we call the pair of charge "zones" an *electric dipole*.

Resistors

Some substances, such as carbon, conduct electricity fairly well, but not very well. We can alter the conductivity by adding impurities such as clay to a paste of finely granulated carbon. We call an electrical component, deliberately manufactured for the purpose of controlling the amount of current that flows in an electrical circuit or system, a *resistor*. A component with high resistance exhibits low conductance, and a component with low resistance has high conductance.

Engineers and technicians express and measure electrical resistance in units called *ohms*, sometimes symbolized by the uppercase Greek letter omega (Ω). In practical circuits, engineers often need to specify larger units such as kilohms (symbolized kΩ or k) or megohms (symbolized MΩ or M), where

$$1 \text{ k} = 1000 \text{ ohms} = 10^3 \text{ ohms}$$

and

$$1 \text{ M} = 1000 \text{ k} = 1{,}000{,}000 \text{ ohms} = 10^6 \text{ ohms}$$

As the ohmic value of a component increases, so does the resistance, in direct proportion. In an electrical utility system such as we find in our homes, schools, and businesses, we want to keep the resistance as low as possible. Resistance converts electrical energy into heat energy. We call the resulting phenomenon *resistance loss* or *ohmic loss*.

We can minimize the ohmic losses in electrical systems by using thick wires and high voltages to carry the currents. (We'll learn more about the relationship among current, voltage, and resistance shortly.) This fact explains why utility companies employ gigantic towers, high voltages, and massive cables to transport electrical energy from generating plants to distant end users.

Current

Whenever electricity flows within a substance, we have an *electric current*. We can express the magnitude of an electric current in terms of the number of *charge carriers*, or particles containing a unit electric charge, passing a single point in one second.

Charge carriers come in two main forms: electrons, which have a *unit negative charge*, and *holes*, a rather curious term that describes, in an abstract way, single-electron absences or shortages within atoms, and which carry a *unit positive charge*. Ionized atoms or particles can act as charge carriers, and in some cases, atomic nuclei can as well. These types of particles carry whole-number multiples of a unit electric charge. Ions can have positive or negative polarity. Atomic nuclei always carry positive charge.

In most situations where we see an electric current, a great many charge carriers pass any given point in 1 second. Even a current that we would consider "tiny" comprises a huge charge-carrier transfer rate in terms of sheer numbers. For example, in a household electric circuit, a small "night light" carries a current of roughly 6 *hundred quadrillion* (6×10^{17}) charge carriers per second! Engineers and scientists rarely write or speak about electric current directly in terms of charge carriers per second. Instead, they use units of *coulombs per second*. A *coulomb* (symbolized C) represents approximately 6.24×10^{18} unit negative or positive charges. A current of 1 coulomb per second (1 C/s) represents 1 *ampere* (symbolized A). The ampere constitutes the standard unit of electric current throughout the world.

In practical circuits, we'll often need smaller units than the ampere, such as the milliampere (symbolized mA) or the microampere (symbolized μA), where

$$1 \text{ mA} = 0.001 \text{ A} = 10^{-3} \text{ A}$$

and

$$1 \text{ μA} = 0.001 \text{ mA} = 0.000001 \text{ A} = 10^{-6} \text{ A}$$

When a current flows through a component or medium having finite, non-zero resistance—and even the best conductors exhibit some resistance—that component or medium grows hot. In an old-fashioned *incandescent lamp*, the resistance of a coiled *filament* gives rise to heat and light emissions when it carries significant current. Even the best incandescent lamp creates more heat than light. *Fluorescent lamps* perform better in this respect; they produce more light and less heat than incandescent lamps for a given amount of current. Stated another way, fluorescent lamps require less current than incandescent lamps to give off a certain amount of light. *Light-emitting-diode* (LED) *lamps* perform best of all, at least at the time of this writing.

If you connect a lamp to a battery, a theoretical physicist, true to her discipline, will tell you that the current flows out of the positive battery terminal, through the wire and the lamp, and back into the negative battery terminal. When we talk about current flow in this manner, we talk about the so-called *conventional current*. The individual electrons, which constitute most of the charge carriers in the wire and the lamp, actually move in the opposite direction. They go from the negative battery terminal, through the wire and the lamp, to the positive battery terminal.

Still Struggling

"Why," you might ask, "does conventional current go against the movement of electrons? That's counter-intuitive!" That idea may seem strange to us today, but when scientists first began to experiment with electricity, they didn't know about electrons, holes, ions, or charged particles. Early experimenters had no clue as to what the strange phenomenon called electricity *actually was*. They only saw *what it did*. They found it more reasonable to say that current goes from plus to minus, than to say that it goes from minus to plus. That old method of expression never died, so it has survived to this day.

Static Electricity

Charge carriers, particularly electrons, can build up or become deficient on physical objects, creating a peculiar sort of "electrical tension" that lay people call "static." You've experienced this phenomenon while walking on a carpeted floor during the winter, especially in a place with low atmospheric humidity. An excess or shortage of electrons develops in your body. You acquire a charge of *static electricity*. It doesn't go anywhere—until you touch some metallic object that's connected to an electrical ground or to some large fixture. Then a discharge occurs, accompanied by a spark and a distinct physical sensation! The current surge, which accompanies this sudden electrical discharge, causes a "little pop" that some people find quite annoying.

If you managed to acquire a much greater static-electric charge than you normally do when you shuffle around on your carpet in the winter, your hair would stand on end, because every hair would repel every other. Objects that carry the same electric charge, caused either by an excess or a deficiency of electrons, experience a mutual force of repulsion. If you were massively charged, the discharge spark—which would inevitably occur sooner or later—might jump several centimeters, and you'd feel more than a "little pop"! Such a large charge quantity can present a genuine physical danger. Static-electric (also called *electrostatic*) charge buildup of this magnitude does not happen with ordinary carpet and shoes, fortunately. But a device called a *Van de Graaff generator*, found in some high-school physics labs, can cause a spark this large. You must use great care when operating this type of device for physics experiments.

On the grand scale of the earth's atmosphere, *lightning* occurs between clouds, and between clouds and the surface. This "spark" constitutes a greatly magnified version of the little spark you get after shuffling around on the carpet. Until the discharge occurs, an electrostatic charge difference exists within individual clouds, between different clouds, or between clouds and the ground. Figure 7-2 shows four types of lightning. The discharge can occur within a single cloud (*intracloud lightning*, at A), between two different clouds (*intercloud lightning*, at B), from a cloud to the surface (*cloud-to-ground lightning*, at C), or from the surface to a cloud (*ground-to-cloud lightning*, at D). In a lightning

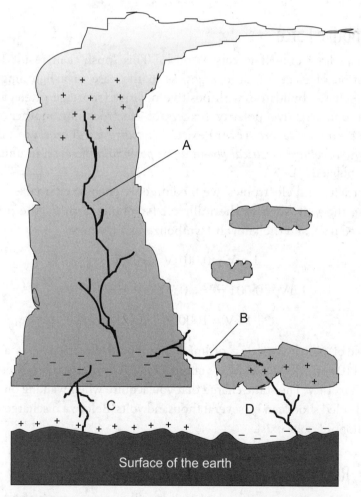

FIGURE 7-2 • Lightning can occur within a single cloud (A), between clouds (B), or between a cloud and the surface (C and D).

discharge, technically called a *stroke*, we consider the direction of the discharge to coincide with the direction in which the electrons move (contrary to the conventional current). In cloud-to-ground or ground-to-cloud lightning, the concentration of charge carriers on the earth's surface follows the thunderstorm like a shadow as the prevailing winds blow the storm along.

The current in a lightning stroke can approach a million amperes (1,000,000 A). The current actually flows for only a fraction of a second, but a single stroke of lightning can displace many coulombs of charge. Such an extreme current surge can start fires, cause containers holding volatile liquids or gases (such as gasoline or methane) to explode, and electrocute people and animals.

Electromotive Force

Current can flow only if it gets a "push." This "push" can result from an accumulation of electrostatic charges, as in the case of a lightning stroke. When the charge builds up, with positive polarity (shortage of electrons) in one place and negative polarity (excess of electrons) in another place, a powerful *electromotive force* (EMF) exists. We express and measure this force, also known as *voltage, electrical potential*, or *potential difference*, in units called *volts* (symbolized V).

In electricity and electronics, we'll commonly encounter smaller or larger units than the volt, such as the millivolt (symbolized mV), the microvolt (symbolized μV), and the kilovolt (symbolized kV), where

$$1 \text{ mV} = 0.001 \text{ V} = 10^{-3} \text{ V}$$

$$1 \text{ μV} = 0.001 \text{ mV} = 0.000001 \text{ V} = 10^{-6} \text{ V}$$

$$1 \text{ kV} = 1000 \text{ V} = 10^{3} \text{ V}$$

In most of North America, ordinary household electricity has a voltage between 110 and 130 V; usually it's about 117 V. An automotive battery has an EMF of 12 to 14 V. The static charge that you acquire when walking on a carpet with hard-soled shoes can be several thousand volts. Before a discharge of lightning, millions of volts exist.

Voltage, Resistance, and Current Relate!

An EMF of 1 V, across a resistance of 1 ohm, will cause a current of 1 A to flow. This relation forms the basis for a classic principle in electricity theory called

Ohm's law, which we'll explore in some detail shortly. For the moment, we can sketch this principle with the following simplistic threefold example:

- If we double the voltage across a component but leave the component's resistance constant, we double the current through that component.

- If we double a component's resistance but keep the voltage across it constant, we halve the current through that component.

- If we double the current through a component having a fixed resistance, we double the potential difference across that component.

We can have a large potential difference across a component, or between two points, without any current flow at all. A situation of this sort arises immediately before a lightning stroke occurs, and immediately before you touch a metallic object after walking on a carpet in dry weather. A similar state of affairs exists between the two "live" slots of an electrical outlet. It's true of a lantern battery when nothing is connected to it. No current flows in any of these scenarios, but you can cause current to flow if you provide a *conductive path* between two points having a relative difference of potential.

Even a large EMF might not drive much current through a conductor or resistor. Once again, think about your body after you have spent some time shuffling around on a carpet. Although the EMF seems deadly in terms of sheer numerical magnitude (thousands of volts), not many coulombs of charge accumulate on your body. Therefore, not many electrons flow through your finger, in relative terms, when you touch a metallic object. You don't get a severe shock, even though the voltage far exceeds that which appears at the electrical outlets in your house. Don't get the idea that "wall outlet voltage" presents no danger, however. If plenty of coulombs are available to generate current, even a moderate EMF such as 117 V can deliver a lethal electrical wallop. This fact explains why you should never repair an electrical device with the power on. The utility power source can pump an unlimited number of charge carriers through your body—more than enough to kill you.

Electrical Diagrams

If you want to understand how electrical circuits work, you must know how to read electrical wiring diagrams, called *schematic diagrams*. These diagrams use so-called *schematic symbols*. You can think of these symbols as something like an alphabet in a pictographic language such as Chinese. Before you grow intimidated by this comparison, rest assured that you'll find it easier to learn schematic symbology than to learn a foreign language!

The simplest schematic symbol represents a wire or electrical conductor: a straight, solid line. Sometimes dashed lines represent conductors, but usually, we draw broken (dashed) lines to partition diagrams into constituent circuits, or to indicate that certain components interact with each other or operate in step with each other. Conductor lines almost always go either horizontally across, or vertically up and down the page, so the imaginary charge carriers seem to march in formation like soldiers. This convention keeps the diagram neat and easy to read.

When two conductor lines cross, you should assume that they *do not* join at the crossing point unless you see a heavy, black dot where the two lines meet. The dot should always show up clearly wherever conductors are meant to be connected, no matter how many of them meet at the junction.

We symbolize a resistor by drawing a "zig-zaggy" line (Fig. 7-3A). We symbolize a *variable resistor*, such as a *rheostat* or *potentiometer*, with a "zig-zaggy" line slashed by an arrow (Fig. 7-3B), or with a "zig-zaggy" line with an arrow pointing at it (Fig. 7-3C).

We portray an *electrochemical cell* by drawing two parallel lines, one longer than the other. The longer line represents the positive terminal (Fig. 7-4A). We symbolize a *battery*—a set of two or more cells connected end-to-end with the positive end of one cell going to the negative end of the next—with an alternating sequence of parallel lines, long-short-long-short (Fig. 7-4B).

We can denote a metering device as a circle. Sometimes the circle has an arrow inside it, and the meter type, such as mA (milliammeter) or V (voltmeter), appears alongside the circle, as shown in Fig. 7-5A. Sometimes the meter type designator appears inside the circle, and we have no arrow (Fig. 7-5B). It doesn't

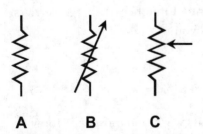

FIGURE 7-3 · At A, a fixed resistor. At B, a two-terminal variable resistor. At C, a three-terminal potentiometer.

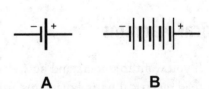

FIGURE 7-4 · At A, an electrochemical cell. At B, a battery.

matter which of these two schemes we use, as long as we remain symbolically consistent everywhere within a given diagram or presentation.

Other common schematic symbols include the *lamp*, the *capacitor*, the *air-core coil*, the *iron-core coil*, the *chassis ground*, the *earth ground*, the *alternating-current* (AC) *source*, the set of *terminals*, and the *black box* (which can stand for almost anything). Figure 7-6 shows all of these.

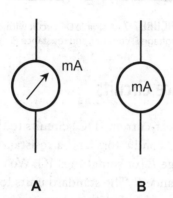

A   **B**

FIGURE 7-5 · Meter symbols can have their designators outside (A) or inside (B).

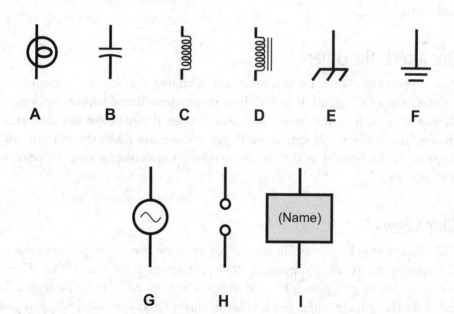

FIGURE 7-6 · Schematic symbols for an incandescent lamp (A), a fixed-value capacitor (B), an air-core coil (C), an iron-core coil (D), a chassis ground (E), an earth ground (F), an AC source (G), connecting terminals (H), and a black box (I).

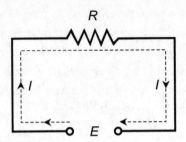

FIGURE 7-7 · A simple DC circuit with voltage E, current I, and resistance R.

Voltage/Current/Resistance Circuits

We can simplify most direct-current (DC) circuits to three major components: a voltage source, a set of conductors, and a resistance, as shown in Fig. 7-7. We call the source voltage E (or sometimes V). We call the circuit current I. We call the circuit resistance R. The standard units for these components are the volt (V), the ampere (A), and the ohm, respectively. Italicized letters represent mathematical variables (voltage, current, and resistance in this case). Nonitalicized characters represent symbols for units such as volts, amperes, and ohms.

One Affects the Other

You already know that a relationship exists among the voltage, current, and resistance in a DC circuit. If one of these parameters should happen to change, then one or both of the others will also change. If you make the resistance smaller, the current will get larger. If you reduce the EMF, the current will decrease. If the current in the circuit increases, the voltage across the resistor will increase.

Ohm's Law

The classical rule known as Ohm's law got its name from *Georg Simon Ohm*, a German physicist who supposedly first expressed it in the 1800s. Three formulas denote this principle in its entirety as applicable to DC circuits. To calculate the voltage when you know the current and the resistance, you can use the formula

$$E = IR$$

To calculate the current when you know the voltage and the resistance, you can use the formula

$$I = E/R$$

To calculate the resistance when you know the voltage and the current, you can use the formula

$$R = E/I$$

You need only memorize (yes, memorize!) one of these formulas in order to derive the other two. Think of the following abbreviations:

- The letter E stands for "electromotive force."
- The letter I stands for "imposed current."
- The letter R stands for "resistance."

In Ohm's law, these three variables appear in alphabetical order with the equals sign after the E, telling you that

$$E = IR$$

If you want Ohm's law to give you the correct results, you must use the proper units. Under most circumstances, you'll want to use the *standard units* of volts, amperes, and ohms. If you use volts, milliamperes (mA), and ohms, or if you use kilovolts (kV), microamperes (µA), and megohms, you can't expect to get the right answers. If you see initial quantities in units other than volts, amperes, and ohms, you should convert to these standard units before you begin your calculations. After you've done all the arithmetic, you can convert the individual units to whatever you like. For example, if you get 13,500,000 ohms as a calculated resistance, you might prefer to call it 13.5 megohms. But in the calculation, you should use the number 13,500,000 (or 1.35×10^7) and stick to ohms for the units.

Current Calculations

In order to determine the current in a circuit, you must know the voltage and the resistance, or else figure out a way to deduce them. Figure 7-8 illustrates a circuit consisting of a variable DC generator, a voltmeter, some wire, an ammeter, and a calibrated, wide-range potentiometer (variable resistor). The actual component values do not appear in this diagram, but you can assign specific values to generate sample Ohm's law problems. You can assume that the wire conducts perfectly, so it doesn't contribute any resistance to the circuit.

PROBLEM 7-1

Suppose that you set the DC generator (Fig. 7-8) to produce 10 V, and you set the potentiometer to 10 ohms of resistance. How much current flows?

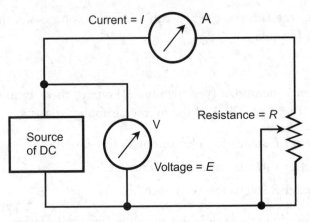

FIGURE 7-8 • Circuit for working Ohm's law problems.

SOLUTION

Use the formula $I = E/R$. Plug in the values for E and R; they both equal 10, expressed in volts and ohms. Simple arithmetic tells you that

$$I = E/R$$
$$= 10/10$$
$$= 1.0 \text{ A}$$

PROBLEM 7-2

Suppose that the DC generator (Fig. 7-8) produces 100 V and the potentiometer has a resistance of 10.0 k. What's the current?

SOLUTION

First, convert the resistance to ohms; 10.0 k equals 10,000 ohms. Then plug the values into the formula, getting

$$I = E/R$$
$$= 100/10,000$$
$$= 0.0100 \text{ A}$$

You might prefer to express this current as 10.0 mA (that is, 10.0 milliamperes, representing 10/1000 of an ampere).

Voltage Calculations

You can employ Ohm's law to calculate the potential difference across a component when you know the current through it, and when you also know its resistance.

 **PROBLEM 7-3**

Suppose that you set the potentiometer (Fig. 7-8) to 100 ohms, and meter A indicates a current flow of 10.0 mA. What is the DC source voltage?

SOLUTION

Use the formula $E = IR$. First, convert the current to amperes; 10.0 mA equals 0.0100 A. Then multiply to get

$$E = IR$$
$$= 0.0100 \times 100$$
$$= 1.00 \text{ V}$$

Resistance Calculations

You can use Ohm's law to determine the resistance between two points in a DC circuit when you know the voltage between those points and the current that flows in the circuit.

PROBLEM 7-4

In the circuit of Fig. 7-8, suppose that the voltmeter V reads 24 V and the ammeter A shows 3.0 A. What's the resistance of the potentiometer?

SOLUTION

You can use the formula $R = E/I$ and plug in the values directly, because you already have them in volts and amperes. Calculating, you obtain

$$R = E/I$$
$$= 24/3.0$$
$$= 8.0 \text{ ohms}$$

Power Calculations

You can calculate the power P (in watts, symbolized W) in a DC circuit such as that shown in Fig. 7-8 using the formula

$$P = EI$$

where E represents the voltage (in volts) and I represents the current (in amperes). If you don't know the voltage straightaway, you can calculate it from the current and the resistance. Recall the Ohm's law formula for obtaining voltage:

$$E = IR$$

If you know I and R but not E, you can get the power P as follows:

$$P = (IR)I$$
$$= I^2R$$

You can calculate the power even if you don't know the current. Suppose that you know only the voltage and the resistance. Remember the Ohm's law formula for obtaining current:

$$I = E/R$$

You can calculate the power P as follows:

$$P = E(E/R)$$
$$= E^2/R$$

PROBLEM 7-5

Refer again to Fig. 7-8. Suppose that the voltmeter reads 12 V and the ammeter shows 50 mA. How much power does the potentiometer dissipate?

✓ SOLUTION

Use the formula $P = EI$. First, convert the current to amperes, getting $I = 0.050$ A. Then you can calculate

$$P = EI$$
$$= 12 \times 0.050$$
$$= 0.60 \text{ W}$$

How Resistances Combine

When you interconnect electrical components or devices having DC resistance, their ohmic values combine according to specific rules. Sometimes, the combined resistance exceeds that of any one of the components or devices. In other cases, the combined resistance amounts to something less than that of any one of the components or devices.

Resistances in Series

When two or more components having DC resistance appear in a *series* arrangement (you connect them end-to-end to form a "string" or "chain" of components), their ohmic values add up to get the total resistance.

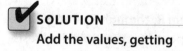 **PROBLEM 7-6**

Suppose that you connect three components in series. They have resistances of 112 ohms, 470 ohms, and 680 ohms, as shown in Fig. 7-9. What is the total resistance R of the series combination?

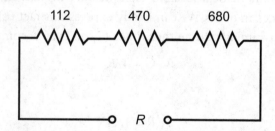

FIGURE 7-9 · Three specific resistances connected in series.

✓ SOLUTION

Add the values, getting

$$R = 112 + 470 + 680$$

$$= 1262 \text{ ohms}$$

You can round this off to 1260 ohms. The extent of the precision depends on the *tolerances* of the components—how much their actual resistances can vary, as a result of manufacturing processes, from the values specified by the vendor.

Resistances in Parallel

When you wire up two or more components having DC resistance in a *parallel* arrangement (you connect them "across" each other so that all the left-hand terminals join and all the right-hand terminals join), they behave differently than they do in a series configuration. In general, if you have a resistor of a certain value and you place other resistors in parallel with it, the overall resistance goes down.

You can evaluate resistances in parallel by considering them as *conductances* instead. We express conductance in units called *siemens*, symbolized S. (The word "siemens" serves both in the singular and the plural sense). When you read some older physics or engineering documents, you might see *mho* ("ohm" spelled backwards) used as the fundamental unit of conductance; the mho and the siemens quantify the same thing. When we connect conductances in parallel, their values add up the way resistances add up in series. If we change all the ohmic values to siemens, we can add these figures up, and convert the final answer back to ohms.

Engineers use the uppercase, italic letter G to symbolize conductance as a parameter or mathematical variable. Conductance in siemens equals the reciprocal of resistance in ohms. We can neatly express this fact using the following two formulas, assuming that neither R nor G can ever become zero:

$$G = 1/R$$

and

$$R = 1/G$$

PROBLEM 7-7

Consider five resistances in parallel. Call them R_1 through R_5, and call the total resistance R as shown in Fig. 7-10. Suppose that the individual resistors have values as follows:

$$R_1 = 100 \text{ ohms}$$
$$R_2 = 200 \text{ ohms}$$
$$R_3 = 300 \text{ ohms}$$
$$R_4 = 400 \text{ ohms}$$
$$R_5 = 500 \text{ ohms}$$

What's the total resistance R of this parallel combination?

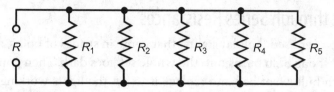

FIGURE 7-10 · Five general resistances connected in parallel.

SOLUTION

Converting the resistances to conductances, you get the following values:

$$G_1 = 1/100 = 0.0100 \text{ S}$$
$$G_2 = 1/200 = 0.00500 \text{ S}$$
$$G_3 = 1/300 = 0.00333 \text{ S}$$
$$G_4 = 1/400 = 0.00250 \text{ S}$$
$$G_5 = 1/500 = 0.00200 \text{ S}$$

When you add these conductances arithmetically, you obtain

$$G = 0.0100 + 0.00500 + 0.00333 + 0.00250 + 0.00200$$
$$= 0.02283 \text{ S}$$

The total resistance is therefore

$$R = 1/G$$
$$= 1/0.02283$$
$$= 43.80 \text{ ohms}$$

Because you know the input numbers to only three significant figures, you should round this result off to 43.8 ohms.

You can calculate a net parallel resistance without converting to conductances. Refer again to Fig. 7-10. The resistances combine according to the formula

$$R = 1/(1/R_1 + 1/R_2 + 1/R_3 + 1/R_4 + 1/R_5)$$

Once in awhile, you'll encounter a situation in which multiple resistances of equal value appear in parallel. In a case of that sort, the total resistance equals the resistance of any one component divided by the number of components. For example, two 100-ohm resistors combine in parallel to yield a *net resistance* of 100/2 = 50 ohms; four of the same resistors combine in parallel to yield 100/4 = 25 ohms; five of them combine in parallel to yield 100/5 = 20 ohms.

Current through Series Resistances

Have you ever used "holiday lights" that come in strings of bulbs, all connected in series? If one bulb burns out, the whole set goes dark. Then you have to find out which bulb went bad and replace it to get the lights working again. If the string has 12 bulbs, each one works with about 10 V. You plug in the whole bunch, and the utility mains, supplying 120 V, drive exactly the proper amount of current through each bulb.

In a series circuit such as a string of light bulbs, the current at any given point equals the current at any other point. You can insert an ammeter in series with the circuit, and the meter will show the same reading no matter where you put it. This rule holds true in any series DC circuit, regardless of the component values, and regardless of whether or not they all have the same resistance.

If the individual bulbs in a series-connected string have different resistances, some of them will consume more power than others. If you "short out" one of the bulbs by placing a wire directly across its terminals, that bulb will go dark because the potential difference across it will go down to zero. However, the current through the whole chain will increase, because the overall resistance of the string will go down. As a result, each of the remaining bulbs will carry a little bit too much current; as a result, another bulb will likely burn out after a little time passes. If you "short out" that bulb too, the current through the remaining bulbs will increase even more; all of the bulbs will have to carry considerably more current than they should, and a third bulb will blow out almost immediately.

Occasionally, you'll encounter a situation in which multiple resistances of equal value appear in series. Then the total resistance equals the resistance of any one component multiplied by the number of components. For example, two 100-ohm resistors combine in series to produce a net resistance of $100 \times 2 = 200$ ohms; six of the same resistors combine in series to give you $100 \times 6 = 600$ ohms; 14 of them combine in series to yield $100 \times 14 = 1400$ ohms.

Voltages across Series Resistances

In a series circuit, the voltage gets split up among the components, although not necessarily into equal potential differences. In any case, the sum of the potential differences across each component always equals the DC power-supply or battery voltage. This fact holds true regardless of the component resistances, and regardless of whether or not those resistances all have the same value.

Examine the schematic diagram of Fig. 7-11. Each individual resistor carries the same current as any of the others. Each resistor R_n has a potential difference E_n across it, equal to the product of the current and the resistance of that particular resistor. These E_n's appear in series just like the cells in a battery, or the bulbs in a string of "holiday lights" of the sort described a moment ago.

How do you know that the sum of the E_n's in the circuit of Fig. 7-11 adds up to the battery voltage? Well, imagine for a moment that it didn't—that the E_n's across all the resistors added up to something more or less than the supply voltage, E. You'd observe a "phantom EMF" someplace in the circuit, and that "phantom EMF" would add or take away voltage. But an EMF can't come out of nowhere.

Look at this another way. The voltmeter V in Fig. 7-11 *must* indicate the voltage E that the battery supplies, because the meter appears directly across the battery. The meter V also shows the sum of the E_n's across the set of resistors, because the meter appears directly across the whole set of R_n's. Obviously, the meter must display the same voltage whether you think of it as measuring the battery voltage or the sum of the voltages across the series combination of resistors. Therefore, E equals the sum of the E_n's.

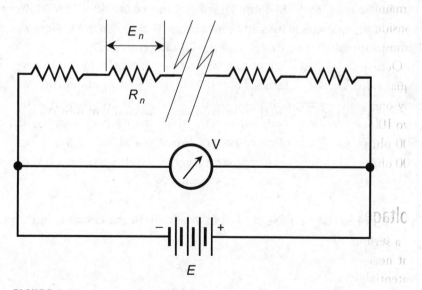

FIGURE 7-11 · Analysis of voltage in a series DC circuit.

Still Struggling

How do you find the voltage across any particular resistor R_n in a circuit like the one in Fig. 7-11? Remember Ohm's law for finding voltage:

$$E = IR$$

Remember, too, that you must use volts, ohms, and amperes when making calculations. If you want to calculate the current I in the circuit, you need to know the total resistance and the supply voltage. Then you can use the formula

$$I = E/R$$

First find the current in the whole circuit. Then find the voltage across any particular resistor.

 PROBLEM 7-8

Suppose that 10 series-connected resistors exist in a series circuit of the sort shown in Fig. 7-11. Five resistors have values of 10 ohms, and the other five have values of 20 ohms. The power source provides 15 V DC. How much voltage appears across any one of the 10-ohm resistors? How much voltage appears across any one of the 20-ohm resistors?

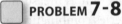 **SOLUTION**

First, you can find the total resistance of the circuit as follows:

$$R = (10 \times 5) + (20 \times 5)$$
$$= 50 + 100$$
$$= 150 \text{ ohms}$$

Then you can find the current at any point in the circuit using Ohm's law, getting

$$I = E/R$$
$$= 15/150$$
$$= 0.10 \text{ A}$$

Now you know the current through each individual resistor in the circuit, as well as through the battery itself: 0.10 A. If a particular resistor has a value of, say, $R_* = 10$ ohms, then you can calculate the voltage E_* across that resistor as

$$E_* = IR_*$$
$$= 0.10 \times 10$$
$$= 1.0 \, V$$

If another of the resistors has a value of, say, $R_\# = 20$ ohms, then the voltage $E_\#$ across that component equals

$$E_\# = IR_\#$$
$$= 0.10 \times 20$$
$$= 2.0 \, V$$

You have five resistors with 1.0 V across each, for a total of 5.0 V. You have five more resistors with 2.0 V across each, for a total of 10 V. Therefore, the sum of the voltages E across the whole set of resistors is

$$E = 5.0 + 10$$
$$= 15 \, V$$

Voltage across Parallel Resistances

Imagine a set of ornamental light bulbs connected in parallel, rather than in series as described earlier. You'll find it easier to repair a parallel-wired string of holiday lights, when one bulb burns out, than to repair a series-wired string. In a parallel circuit, only the bad bulb will go dark, allowing you to identify it straightaway. Also, the failure of one bulb in a parallel circuit does not cause *catastrophic system failure*. (That term explains itself pretty well, doesn't it?)

In a parallel circuit, the voltage across each component always equals the supply or battery voltage. The current drawn by each component depends only on its individual resistance, not on the resistances of other components in the circuit. In this sense, the components in a parallel-wired circuit work independently, as opposed to the series-wired circuit in which they all interact.

If you suddenly remove any single component or *branch* of a parallel circuit, the conditions in the other components or branches do not change. If you add a new component or branch in parallel with all the rest, assuming that the power supply can handle the additional current load, conditions in the previously existing components or branches remain exactly as they were.

Currents through Parallel Resistances

Imagine that R represents the net resistance of a set of resistors, all connected in parallel across a single battery. Let E represent the battery voltage. Suppose that we use an ammeter called A to measure the current I_n carried by the circuit branch containing the resistance R_n, as shown in Fig. 7-12. Then we measure the I_n's (that is, the currents) carried by each and every branch of the circuit, one by one, and add those currents up. We will invariably find that all the I_n's add up to the total current, I, drawn from the battery. Current splits up among parallel-connected components, in the same way as voltage splits up among series-connected components.

PROBLEM 7-9

Suppose that the battery in the circuit of Fig. 7-12 delivers 12 V. Further suppose that you have 12 resistors, each with a value of 120 ohms, in the parallel circuit. What's the total current, I, drawn from the battery?

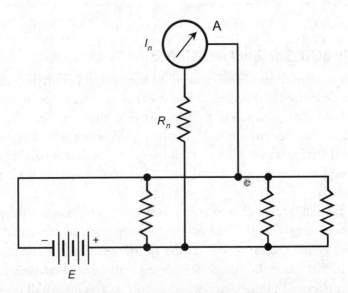

FIGURE 7-12 · Analysis of current in a parallel DC circuit.

SOLUTION

First, you should calculate the total resistance. All the resistors have the same value, making the task easy. You can simply divide $R_n = 120$ by 12, getting a net resistance of $R = 10$ ohms. Then you can calculate the current I using Ohm's law as follows:

$$I = E/R$$
$$= 12/10$$
$$= 1.2 \text{ A}$$

PROBLEM 7-10

In the circuit of Fig. 7-12, what does the ammeter A indicate, given the component values described in Problem 7-9?

SOLUTION

To solve this problem, you must determine the current that flows in any particular branch of the circuit—that is, through any one of the resistors. Because you have a parallel connection, the potential difference equals 12 V across every branch. You already know that $R_n = 120$ for every single one of the resistors. Therefore, you can calculate I_n, the current in any particular branch as follows:

$$I_n = E/R_n$$
$$= 12/120$$
$$= 0.10 \text{ A}$$

You can (and should) verify that all the I_n's add up to the total current, I. Let's do that here. The circuit contains 12 identical branches, each carrying 0.10 A. Therefore, the total current is

$$I = 0.10 \times 12$$
$$= 1.2 \text{ A}$$

Power Distribution in Series Circuits

Return your attention to series circuits. You can calculate the power in a circuit containing resistors in series by determining the current I, in amperes, that the circuit carries. Then you can easily calculate the power P_n (in watts) dissipated by any particular resistor of value R_n (in ohms) using the formula

$$P_n = I^2 R_n$$

Still Struggling

The total *wattage* (power in watts) dissipated in a series circuit always equals the sum of the wattages dissipated in each resistor. In this sense, the distribution of the power in a series circuit mimics the distribution of the voltage.

 PROBLEM 7-11

Suppose we have a series circuit with a 150 V battery and three resistors having the following values:

$$R_1 = 330 \text{ ohms}$$
$$R_2 = 680 \text{ ohms}$$
$$R_3 = 910 \text{ ohms}$$

How much power does R_2 dissipate?

 SOLUTION

First, let's determine the current that the entire circuit demands. Because the resistors appear in series, the total resistance is

$$R = 330 + 680 + 910$$
$$= 1920 \text{ ohms}$$

Therefore, the total current is

$$I = 150/1920$$
$$= 0.078125 \text{ A}$$
$$= 78.1 \text{ mA}$$

The power dissipated by R_2 is therefore

$$P_2 = I^2 R_2$$
$$= 0.078125 \times 0.078125 \times 680$$
$$= 4.15 \text{ W}$$

Power Distribution in Parallel Circuits

When you connect resistances in parallel, each particular resistance R_n dissipates a certain amount of power P_n according to the formula

$$P_n = I_n{}^2 R_n$$

where I_n represents the current through R_n. Unless all the R_n's happen to equal each other, the individual I_n's will differ. You can also find the power $P_{n'}$ dissipated by a particular resistor of value $R_{n'}$ using the formula

$$P_n = E^2/R_n$$

where E represents the voltage of the power supply or battery. This same voltage appears across each individual resistor in the circuit.

 Still Struggling

In a parallel circuit, the total dissipated wattage equals the sum of the wattages dissipated by the individual resistances. A parallel circuit therefore splits up the power just as a series circuit does.

 PROBLEM 7-12

Suppose that a circuit contains three resistances of

$$R_1 = 22 \text{ ohms}$$
$$R_2 = 47 \text{ ohms}$$
$$R_3 = 68 \text{ ohms}$$

in parallel across a battery that supplies $E = 3.0$ V. Calculate the power dissipated by each resistor.

 **SOLUTION**

To begin, you can find E^2, the square of the supply voltage, as follows:

$$E^2 = 3.0 \times 3.0$$
$$= 9.0$$

Then you can calculate the individual wattages as follows:

$$P_1 = E^2/R_1 = 9.0/22 = 0.4091 \text{ W}$$
$$P_2 = E^2/R_2 = 9.0/47 = 0.1915 \text{ W}$$
$$P_3 = E^2/R_3 = 9.0/68 = 0.1324 \text{ W}$$

You should round these results off to $P_1 = 0.41$ W, $P_2 = 0.19$ W, and $P_3 = 0.13$ W, respectively.

Kirchhoff's Laws

The physicist *Gustav Robert Kirchhoff* (1824–1887) performed research and experimentation in electricity many years before wireless communication, the Internet, and other "modern marvels" existed. Nevertheless, we can give him credit for two fundamental laws governing the behavior of DC circuits.

Kirchhoff's Current Law

Kirchhoff reasoned that in any DC circuit, the electric current going into any particular point equals the current going out of that point, no matter how many branches lead into or out of the point. Figure 7-13 illustrates an example

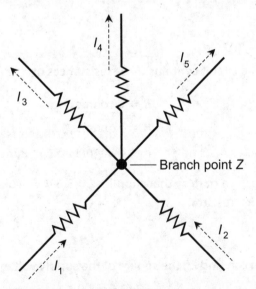

FIGURE 7-13 · Kirchhoff's current law. In this example, $I_1 + I_2 = I_3 + I_4 + I_5$.

where two currents I_1 and I_2 enter a *branch point* Z, while three currents I_3, I_4, and I_5 leave the same point Z. *Kirchhoff's current law* tells us that

$$I_1 + I_2 = I_3 + I_4 + I_5$$

Kirchhoff's current law derives its truth from an almost trivial principle. For a moment, think of electric current as the movement of water through a network of pipes. In a closed system—one in which the pipes don't leak and no water comes in between the "start" and the "finish"—the volume of water flowing toward any point in a given time period must equal the volume of water flowing away from that point in the same time period. Charge carriers in a DC circuit follow the same rule.

Still Struggling

In a closed system, electric charge carriers, like water molecules, cannot come out of nowhere, nor can they vanish into nothingness. If the number of charge carriers per unit time entering a particular point weren't the same as the number of charge carriers per unit time leaving that point, we'd have to somehow create charge carriers from nothing, or else make charge carriers vanish into nothing. Because charge carriers comprise physical objects, that means we'd have to create or destroy matter—an impossible task.

 **PROBLEM 7-13**

Refer to Fig. 7-13. Suppose that each of the two resistors below point Z has a value of 100 ohms, and all three resistors above point Z have values of 10.0 ohms. Suppose that each 100-ohm resistor carries 500 mA (0.500 A). How much current flows through any single one of the 10.0-ohm resistors? How much voltage appears across any single one of the 10.0-ohm resistors?

SOLUTION

The total current into Z is 500 mA + 500 mA = 1.00 A. This current splits equally three ways among the 10.0-ohm resistors, because all three resistors happen to have equal values. Therefore, the current through any one of them must equal 1.00/3 A, which is 0.333 A. You might want to call it 333 mA. You can use Ohm's law to calculate the voltage across any one of the 10.0-ohm resistors as follows:

$$E = IR$$

$$= 0.333 \times 10.0$$

$$= 3.33 \text{ V}$$

Kirchhoff's Voltage Law

According to *Kirchhoff's voltage law*, the sum of all the potential differences, as you go around a circuit from some fixed point and return there from the opposite direction, and taking polarity into account, always works out to zero.

Consider the rule that you've learned concerning series circuits: The voltages across all of the resistors add up to the supply voltage. However, the *polarities* of the potential differences across the resistors *oppose* the battery polarity. This concept should become clear when you draw a schematic diagram of a series circuit with all the components, including the battery or other power source. Figure 7-14 illustrates an example with a single battery and four resistors.

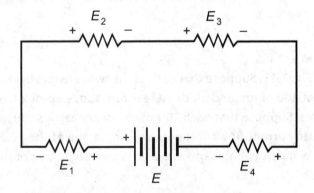

FIGURE 7-14 · Kirchhoff's voltage law. In this example, $E + E_1 + E_2 + E_3 + E_4 = 0$ when we take each component's voltage polarity into account.

Still Struggling

The key to understanding Kirchhoff's voltage law involves the fact that it refers to *single points*. Of course, a potential difference can exist between different points in a circuit. But no point can harbor an electrical potential difference *with respect to itself!* This concept, like the one that gives rise to Kirchhoff's current law, constitutes a near-triviality. When he wrote his voltage law, Kirchhoff must have realized that voltage can't appear out of nowhere, nor can it vanish into nothingness.

PROBLEM 7-14

Refer to Fig. 7-14. Suppose that the four resistors have values as follows:

$$R_1 = 50 \text{ ohms}$$
$$R_2 = 60 \text{ ohms}$$
$$R_3 = 70 \text{ ohms}$$
$$R_4 = 80 \text{ ohms}$$

Further suppose that the current I through each resistor equals 500 mA (0.500 A). What's the battery voltage E?

SOLUTION

Using Ohm's law, you can determine the voltage across each of the four resistors, one by one:

$$E_1 = IR_1 = 0.500 \times 50 = 25 \text{ V}$$
$$E_2 = IR_2 = 0.500 \times 60 = 30 \text{ V}$$
$$E_3 = IR_3 = 0.500 \times 70 = 35 \text{ V}$$
$$E_4 = IR_4 = 0.500 \times 80 = 40 \text{ V}$$

The battery voltage E equals the sum of the voltages across the individual resistors as follows:

$$E = E_1 + E_2 + E_3 + E_4$$
$$= 25 + 30 + 35 + 40$$
$$= 130 \text{ V}$$

QUIZ

Refer to the text in this chapter if necessary. A good score is eight correct. Answers are at the back of the book.

1. Imagine that we connect a battery to a 100-ohm resistor, so that a certain constant number of charge carriers pass through the resistor every millisecond. Then we increase the size of the battery, multiplying the charge-carrier flow rate through the 100-ohm resistor by a factor of 4. What happens to the power dissipated by that resistor?

    A. Nothing. It stays the same.
    B. It doubles.
    C. It quadruples.
    D. It increases by a factor of 16.

2. Suppose that we continue the experiment described in Question 1. We double the value of the resistance from 100 to 200 ohms, but make no changes to the battery or to any other part of the circuit. What happens to the rate of charge-carrier flow through the resistor?

    A. Nothing. It stays the same.
    B. It decreases to 1/2 of its former rate.
    C. It decreases to 1/4 of its former rate.
    D. It decreases to 1/16 of its former rate.

3. Imagine two resistors connected in series. One of the resistors has a value of 50 ohms and carries 200 mA of current. The other resistor has an unknown value, but we know that the two resistors, combined, dissipate 6.0 W of power. The voltage across the unknown resistor is

    A. 4.0 V.
    B. 10 V.
    C. 20 V.
    D. 40 V.

4. Suppose that a component carries 200 mA of DC when we connect a 4.50-V battery directly across it. The component has a resistance of

    A. 113 ohms.
    B. 101 ohms.
    C. 22.5 ohms.
    D. 0.900 ohms.

5. Suppose that a component carries 3.0 A of DC when we connect a 12-V battery directly across it. The component has a conductance of

    A. 48 S.
    B. 4.0 S.
    C. 0.75 S.
    D. 0.25 S.

6. If we enter an "alternative frame of mind," we can mathematically express the potential difference between two points in terms of

   A. ampere-ohms.
   B. amperes per ohm.
   C. ohms per ampere.
   D. watts per ohm.

7. Suppose that 3.12×10^{17} charge carriers flow at a steady rate past a particular point in a DC circuit every 100 ms (0.100 s). How much current flows through that point?

   A. 5.00 mA
   B. 50.0 mA
   C. 500 mA
   D. We need more information to calculate the current.

8. If the potential difference across a resistor equals 20 V and the resistor carries 10 mA of DC, then the resistor dissipates

   A. 200 mW.
   B. 400 mW.
   C. 2 W.
   D. 4 W.

9. Imagine that we have six 2-W lamps connected in series with a battery that supplies 12 V. If we remove one of the lamps and don't replace it with anything at all, what will happen to the current through any of the remaining five lamps?

   A. It will remain the same.
   B. It will increase slightly.
   C. It will decrease slightly.
   D. It will drop to zero.

10. Suppose that we connect a 7.5-V battery directly across a 10,000-ohm resistor. How much power will that resistor dissipate?

    A. 0.75 mW
    B. 5.6 mW
    C. 7.5 mW
    D. 56 mW

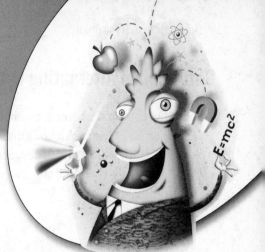

chapter **8**

Alternating Current

We can express DC in terms of two variables: the *polarity* (or direction) and the *amplitude*. Alternating current (AC) presents us with a more complicated scenario because more variables exist: the *period* (and its reciprocal, the *frequency*), the *waveform*, and the *phase*.

In this chapter, you will

- Learn how period and frequency relate.
- Explore various waveforms.
- Analyze an AC cycle in small parts.
- Express and compare waveform amplitudes.
- Define and calculate AC phase angles.

Definition of Alternating Current

Direct current has a polarity, or direction, that stays the same over a long period of time. Although the amplitude can vary—the number of amperes, volts, or watts can fluctuate—the charge carriers always flow in the same direction through the circuit. In AC, the polarity reverses regularly and repeatedly.

Period

In a *periodic AC wave*, the type discussed in this chapter, the mathematical function of amplitude versus time recurs indefinitely. We define the period as the length of time between one repetition of the pattern, or one wave *cycle*, and the next. Figure 8-1 illustrates an example for a simple AC wave.

In theory, a wave period can range from a fraction of a second to millions of years. The charged particles held captive by the sun's magnetic field reverse direction over periods measured in years. Some radio waves exhibit periods of less than 10^{-12} of a second. When we express or measure a wave period in seconds, we symbolize it as T.

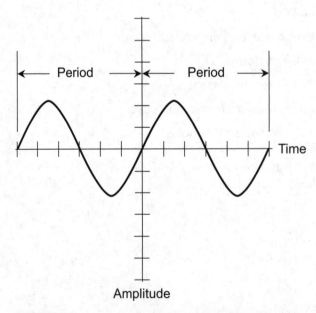

FIGURE 8-1 · An AC wave repeats precisely once per period.

Frequency

The frequency of a wave, denoted f, equals the reciprocal of the period. We can state this fact as the two formulas

$$f = 1/T$$

and

$$T = 1/f$$

In the "olden days" (prior to the 1970s), engineers specified low and medium frequencies in *cycles per second*, abbreviated cps. They specified high frequencies in terms of *kilocycles*, *megacycles*, or *gigacycles*, representing thousands, millions, or billions (thousand-millions) of cycles per second. Nowadays, we use the *hertz*, abbreviated Hz, as the standard unit of frequency.

Still Struggling

We can interchange the word "hertz" with the phrase "cycles per second." Therefore, 1 Hz = 1 cps, 10 Hz = 10 cps, and so on. We express high frequencies in *kilohertz* (kHz), *megahertz* (MHz), *gigahertz* (GHz), and *terahertz* (THz). These units relate as follows:

$$1 \text{ kHz} = 1000 \text{ Hz}$$
$$1 \text{ MHz} = 1000 \text{ kHz} = 10^6 \text{ Hz}$$
$$1 \text{ GHz} = 1000 \text{ MHz} = 10^9 \text{ Hz}$$
$$1 \text{ THz} = 1000 \text{ GHz} = 10^{12} \text{ Hz}$$

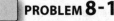

 PROBLEM 8-1

An AC wave has a period of 5.000×10^{-6} s. What's the frequency in hertz? In kilohertz? In megahertz?

SOLUTION

First, we find the frequency f_{Hz} in hertz by taking the reciprocal of the period in seconds, obtaining

$$f_{Hz} = 1 / (5.000 \times 10^{-6})$$
$$= 2.000 \times 10^5 \text{ Hz}$$

Next, we divide f_{Hz} by 1000 or 10^3 to get the frequency f_{kHz} in kilohertz:

$$f_{kHz} = f_{Hz} / 10^3$$
$$= 2.000 \times 10^5 / 10^3$$
$$= 200.0 \text{ kHz}$$

Finally, we divide f_{kHz} by 1000 or 10^3 to get the frequency f_{MHz} in megahertz:

$$f_{MHz} = f_{kHz} / 10^3$$
$$= 200.0 / 10^3$$
$$= 0.2000 \text{ MHz}$$

Waveforms

If we graph the current or voltage in an AC system as a function of time, we obtain a *waveform*. Alternating currents can have an infinite variety of waveforms. Following are the simplest types.

Sine Wave

In its purest manifestation, AC has a *sine-wave*, or *sinusoidal*, waveform. Figure 8-1 illustrates an example of a sine wave. Any AC wave that concentrates all of its energy at a single frequency will invariably exhibit a perfect sinusoidal waveform. Conversely, any perfect sine-wave AC signal contains energy at one, and only one, frequency, called its *fundamental frequency*.

In practice, a waveform can look like a *sinusoid* when we view it on a laboratory instrument such as an *oscilloscope* even if, in reality, small amounts of energy exist at frequencies other than the fundamental frequency. An oscilloscope shows amplitude as a function of time; engineers say that it provides a *time-domain display*. Common utility AC in the United States graphs as a 60-Hz sinusoid if we display it on an oscilloscope. But some energy exists at many other frequencies. These components show up if we scrutinize utility AC on a more sophisticated instrument called a *spectrum analyzer*, which portrays amplitude as a function of frequency. Engineers call this sort of picture a *frequency-domain display*.

In my book *Electricity Experiments You Can Do at Home*, I offer a simple way for you to connect a "pickup antenna" to a personal computer, download a freeware program from the Internet, and actually see some of the "impurities" in AC utility energy. You can make your computer function as a low-frequency spectrum analyzer. Yet, if you connect the same "pickup antenna" to a conventional oscilloscope, you'll see a waveform that looks like a "pure" sinusoid. Appearances can deceive!

Square Wave

If you look at a perfect *square wave* on an oscilloscope, you'll see a pair of parallel, dashed lines, one having positive polarity and the other having negative polarity (Fig. 8-2A). The polarity transitions sometimes show up as thin vertical lines. Figure 8-2B portrays the traditional graphic rendition of a square wave with these vertical lines drawn in. Note that the waveform does not really consist of perfect squares, but it maintains its negative and positive polarity states for equal lengths of time.

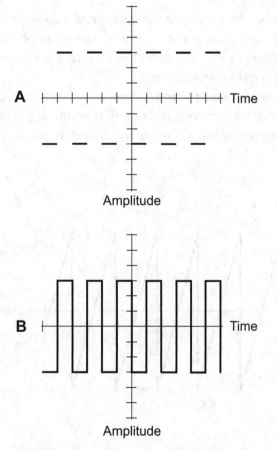

FIGURE 8-2 · At A, a theoretically perfect square wave. At B, the conventional rendition of the same wave.

Some square waves have negative and positive peaks of equal magnitude; for example, half of the time the amplitude equals +*k*, and the other half of the time the amplitude equals −*k* volts, amperes, or watts (where *k* remains constant). Some square waves have positive and negative magnitudes that differ. You might even see a square wave in which the polarity never actually reverses, but the magnitude changes periodically. However, you can call the wave "square" as long as it maintains both amplitude states for equal lengths of time.

If the length of time for which the amplitude holds at one state differs from the length of time the amplitude holds at the other state, you cannot legitimately call it a square wave; you must describe it with the more general term *rectangular wave*.

Sawtooth Waves

Some AC waves reverse polarity at constant, finite, measurable rates. The slope of the amplitude-versus-time line on a time-domain graph indicates how fast the magnitude changes. We call signals having such waveforms *sawtooth waves* because of their serrated appearance.

Figure 8-3 illustrates a sawtooth wave in which the positive-going slope (*rise*) goes up so fast that we can consider it vertical on the display as with a square wave, but the negative-going slope (*decay*) descends at a finite, nonzero rate.

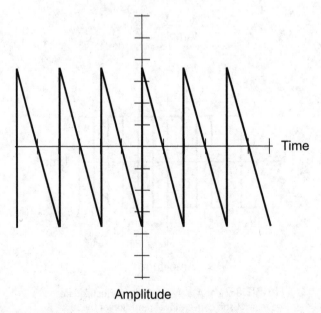

FIGURE 8-3 · A fast-rise, slow-decay sawtooth wave.

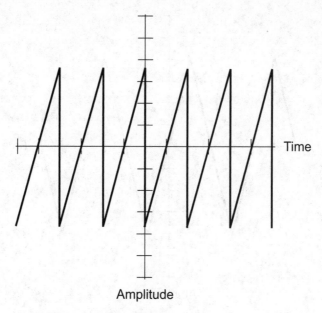

Time

Amplitude

FIGURE 8-4 · A slow-rise, fast-decay sawtooth wave, also called a ramp wave.

The period of the wave equals the time between points at identical positions on two successive pulses.

Another form of sawtooth wave exhibits a gradual positive-going slope and an instantaneous negative-going transition. We call this type of wave a *ramp* (Fig. 8-4). You will encounter ramp waves in the scanning signal generators for old-fashioned *cathode-ray-tube* (CRT) television sets and oscilloscopes.

Still Struggling

The term "sawtooth" applies rather loosely to a wide variety of waveforms. Sawtooth waves can have rise and decay slopes in an infinite number of different combinations. There's only one binding criterion: In a true sawtooth wave, the rise and decay events must take place in a *linear* fashion, meaning that the rise and decay "curves" must both constitute *straight lines*. Figure 8-5 illustrates a special example in which the positive-going slope equals the negative-going slope, producing a *triangular wave*.

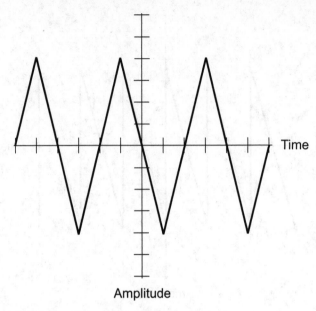

Time

Amplitude

FIGURE 8-5 · A triangular wave.

PROBLEM 8-2

Suppose that each horizontal division in Fig. 8-5 represents 1.00 microsecond (1.00 μs or 1.00×10^{-6} s). What's the period of this triangular wave? What's the frequency?

✔ SOLUTION

Let's evaluate the wave from a point where it crosses the time axis going upward, and then find an immediately adjacent point (to the right or left) where the wave crosses the time axis going upward again. A close look at the graph tells us that any two such points always occur four horizontal divisions apart. The period, T, therefore equals 4.00 μs or 4.00×10^{-6} s. To calculate the frequency, we take the reciprocal of the period, obtaining

$$f = 1/T$$
$$= 1/(4.00 \times 10^{-6})$$
$$= 2.50 \times 10^5 \text{ Hz}$$

We might call this frequency 250 kHz or 0.250 MHz. If you're astute, you might say that we base this level of accuracy on the assumption that we can make an exact visual interpolation of Fig. 8-5—perhaps an arrogant assumption!

Fractions of a Cycle

Scientists and engineers break an AC cycle, particularly a sine-wave cycle, down into small parts for analysis and reference. We can think of a sinusoidal-wave cycle as a single revolution of a point in a circular path. In that context, the full cycle equals 360 angular degrees of circular motion, a half cycle represents 180 degrees, a quarter of a cycle represents 90 degrees, and so on.

Around and Around

Imagine that you swing a glowing ball around your head at the end of a string a couple of meters long, at a rate of one revolution per second. The ball describes a circle in space as shown in Fig. 8-6A. Suppose that as you swing the ball around, it always stays at the same level; that is, its trajectory lies in a horizontal plane. Further suppose that you carry out this exercise in a pitch-dark gymnasium.

If your friend stands at a good distance (maybe 25 m) from you with her eyes in the plane of the ball's path, she sees only the glowing ball, a phosphorescent dot in the darkness, oscillating back and forth. She sees the dot move toward the right, slow down, and then reverse its direction, going back toward the left (Fig. 8-6B). Then it moves faster, then slower again, reaching its left-most point,

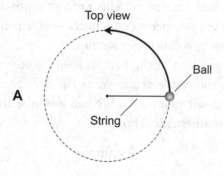

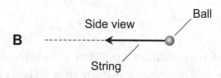

FIGURE 8-6 · At A, a swinging ball and string as seen from above. At B, the same situation as viewed from a distance in the plane of the ball's path.

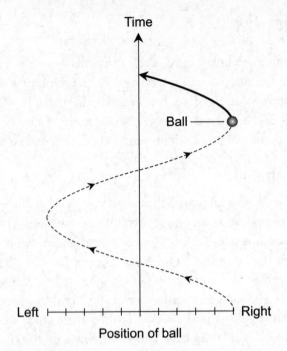

FIGURE 8-7 · Position of ball as seen edge-on, as a function of time.

at which it turns around once more. This process continues, with a frequency of 1 Hz, or a complete cycle per second, because you swing the ball around at the rate of exactly one revolution per second.

If you graph the position of the ball, as seen by your friend, with respect to time, you'll get a sine wave as shown in Fig. 8-7. This wave has the same characteristic shape as all sine waves. We can describe the standard, or basic, sine wave using the mathematical function

$$y = \sin x$$

in the Cartesian, or rectangular, (x, y) coordinate plane. The more general form for the equation of a sine wave is

$$y = a \sin bx + c$$

where a represents a real-number constant that tells us the wave "height," b represents another real-number constant that tells us how "stretched-out" the wave appears, and c represents a third real-number constant that describes how far, either positively or negatively, from the time axis the wave's average value lies.

Degrees

If we want to specify small fractions of an AC cycle, we can divide it into 360 equal degrees, symbolized ° or deg (but we can also write out the whole word). We assign 0° to the point in the cycle where the magnitude equals zero and the amplitude trends upward (in the positive direction). We assign the same point on the next cycle the value 360°. We can break down the wave fractionally as follows:

- The point 1/4 of the way through the cycle lies at 1/4 × 360°, or 90°.
- The point halfway through the cycle lies at 1/2 × 360°, or 180°.
- The point 3/4 of the way through the cycle lies at 3/4 × 360°, or 270°.

Figure 8-8 illustrates how this scheme works. We can slice the cycle into portions as small as we want. For example:

- The point 1/8 of the way through the cycle lies at 1/8 × 360°, or 45°.
- The point 1/10 of the way through the cycle lies at 1/10 × 360°, or 36°.
- The point 1/100 of the way through the cycle lies at 1/100 × 360°, or 3.6°.

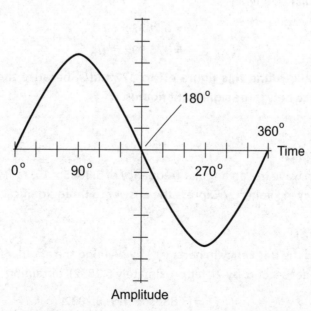

FIGURE 8-8 · Engineers divide a wave cycle into 360° of phase.

Radians

While engineers commonly use degrees to fractionalize AC wave cycles, physicists do it a little differently. We can divide the complete cycle into exactly 2π, or approximately 6.2832, equal parts: the ratio of the circumference of a circle to its radius. One *radian*, symbolized rad (although we can write out the whole word), therefore equals 360° divided by 2π, which works out to approximately 57.296°.

We can express the frequency of an AC wave in radians per second (rad/s) rather than in hertz (cycles per second). Because a complete 360° wave cycle contains 2π rad, the *angular frequency* of a wave, in radians per second (rad/s), equals 2π times the "ordinary" frequency in hertz. Physicists symbolize angular frequency as a variable in equations by writing a lowercase, italicized Greek letter omega (ω).

PROBLEM 8-3

What's the angular frequency of most AC utility electricity in the United States? Assume the "ordinary" frequency is 60.0 Hz.

SOLUTION

Multiply the frequency in hertz by 2π. If you take this value as 6.2832, then the angular frequency is

$$\omega = 6.2832 \times 60.0$$
$$= 376.992 \text{ rad/s}$$

You should round this figure off to 377 rad/s, because the input data extends to only three significant figures.

PROBLEM 8-4

A certain wave has an angular frequency of 3.8865×10^5 rad/s. What is the frequency in kilohertz? Express the answer to three significant figures.

SOLUTION

First find the frequency in hertz (f_{Hz}) by dividing the angular frequency, in radians per second, by 2π (approximately 6.2832), obtaining

$$f_{Hz} = (3.8865 \times 10^5) / 6.2832$$
$$= 6.1855 \times 10^4 \text{ Hz}$$

To obtain the frequency in kilohertz (f_{kHz}), divide by 10^3 and then round off to three significant figures, as follows:

$$f_{kHz} = 6.1855 \times 10^4 / 10^3$$

$$= 61.855 \text{ kHz} \approx 61.9 \text{ kHz}$$

Amplitude

The term *amplitude* also goes by the names *magnitude, level, strength,* or *intensity.* Depending on the quantity that we want to express or measure, we can specify the amplitude of an AC wave in amperes (for current), volts (for voltage), or watts (for power).

Instantaneous Amplitude

We define the *instantaneous amplitude* of an AC wave as the voltage, current, or power at some precise moment (or instant) in time. In an AC wave, the instantaneous value constantly changes. The manner in which it varies depends on the waveform. We can graphically portray instantaneous wave amplitude by drawing or locating a specific point on the curve that we get when we graph the wave function.

Average Amplitude

The *average amplitude* of an AC wave represents the mathematical average (that is, the *arithmetic mean*) of all the instantaneous voltage, current, or power levels evaluated over exactly one wave cycle, or over any exact whole number of wave cycles.

A pure AC sine wave always exhibits an average amplitude of zero. The same holds true for a pure AC square wave or triangular wave. However, the average amplitude of a sawtooth wave does not necessarily equal zero. You can get an idea of why these propositions hold true by scrutinizing the waveforms graphed in Figs. 8-1 through 8-5.

TIP *If you've taken any calculus courses, you can guess that in order to precisely calculate the average amplitude of an AC wave, you must find the integral of the waveform, evaluate that integral over a full cycle, and then divide the resulting quantity by the period of the wave. That technique goes beyond the level of this book, but you'll encounter it if you take any courses in electrical engineering.*

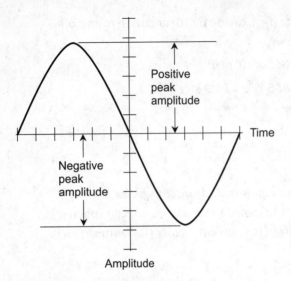

FIGURE 8-9 · Positive and negative peak amplitudes. In this case, the positive peak amplitude equals the negative peak amplitude.

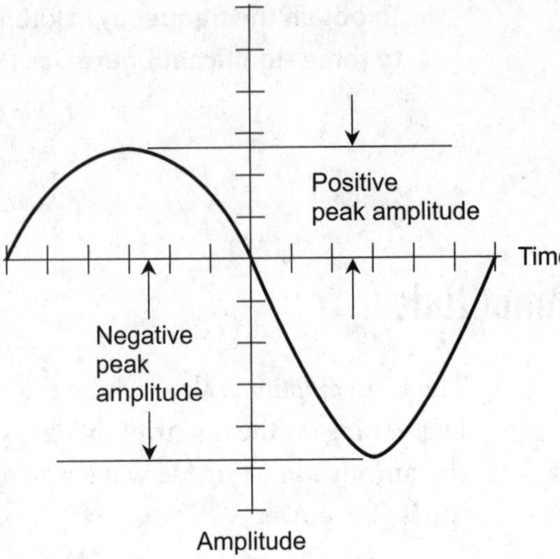

FIGURE 8-10 · A wave in which the positive and negative peak amplitudes differ.

Positive and Negative Peak Amplitude

The *peak amplitude* of an AC wave defines the maximum extent, either positive or negative, that the instantaneous amplitude attains. In many waves, the positive peak amplitude equals the negative peak amplitude. But sometimes they differ. Figure 8-9 illustrates a sine wave in which the positive peak amplitude equals the negative peak amplitude. Figure 8-10 shows a distorted wave that has different positive and negative peak amplitudes.

Peak-to-Peak Amplitude

We define the *peak-to-peak* (pk-pk) *amplitude* of a wave as the difference between the positive peak amplitude and the negative peak amplitude. Figure 8-11 shows graphically how this principle works for a pure sine wave. The peak-to-peak amplitude of any wave equals the positive peak amplitude plus the absolute value of the negative peak amplitude. The peak-to-peak amplitude therefore expresses how much the wave level "swings" from one extreme to the other during its cycle.

In many waves, the peak-to-peak amplitude equals exactly twice the peak amplitude. This happens when the positive and negative peak amplitudes are the same. If the positive and negative peak amplitudes differ, however, we can't define the peak-to-peak amplitude in such a simple way.

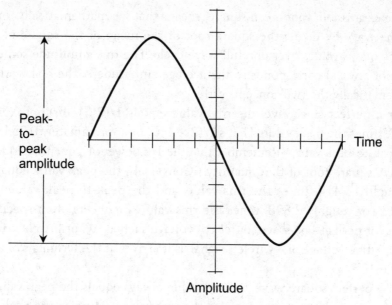

Peak-
to-
peak
amplitude

Time

Amplitude

FIGURE 8-11 · Peak-to-peak wave amplitude.

Still Struggling

Once in awhile, if we encounter a truly complicated waveform, we'll have trouble figuring out any of these values; we might even find the task impossible. This sort of thing happens, for example, in "waveforms" that represent *broadband signals* and *electrical noise*.

Root-Mean-Square Amplitude

Occasionally, we will want to express the *effective amplitude* of an AC wave. We define effective amplitude as the voltage or current that a DC source would have to produce in order to cause the same general effect as our AC wave in a real-world circuit or system. When we say that a wall outlet has 117 V between its terminals, we mean to say that it has 117 *effective volts*. The most common figure for effective AC levels is called the *root-mean-square*, or rms, value.

The expression "root mean square" means that we mathematically "operate on" the wave by taking the square root of the mean of the square of all its instantaneous values over one full wave cycle. The rms amplitude sometimes (but not always) corresponds to the average amplitude in the real world, but mathematically the two concepts differ.

For a perfect sine wave, the rms value equals $1/(2^{1/2})$, or approximately 0.707, times the peak value. That's $1/(2 \times 2^{1/2})$, or approximately 0.354, times the peak-to-peak value. (Remember that the 1/2 power of a quantity equals the positive square root of that quantity.) Conversely, the peak value equals $2^{1/2}$, or roughly 1.414, times the rms value, and the peak-to-peak value equals $2 \times 2^{1/2}$, or roughly 2.828, times the rms value. Engineers often specify rms values for perfect sine-wave sources of voltage such as AC utility electricity or the effective voltage of a wireless signal as it arrives at the terminals of a radio receiver.

For a perfect square wave, the rms value always equals the peak value, and the peak-to-peak value equals twice the rms value and twice the peak value. For sawtooth and irregular waves, the relationship between the rms value and the peak value depends on the exact shape of the wave. The rms value, however, never exceeds the peak value, regardless of the waveform.

Superimposed DC

Once in awhile, we'll encounter a wave that exists as a "hybrid" of AC and DC put together. We can get an AC/DC combination by connecting a source of DC, such as a battery, in series with a source of AC, such as the utility power line. Any AC wave can have a *DC component* superimposed on it. If the DC component exceeds the peak value of the AC wave, then fluctuating, or pulsating, DC results.

Suppose, for example, that we connect a 200 V DC battery in series with the 117-V rms utility power-line output. Pulsating DC will appear, with an average value of 200 V but with instantaneous values much higher and lower than that. Figure 8-12 illustrates the waveform that we will get in this scenario.

PROBLEM 8-5

Suppose that an AC sine wave measures 60 V pk-pk. The wave contains no DC component. What's the positive peak voltage? What's the negative peak voltage?

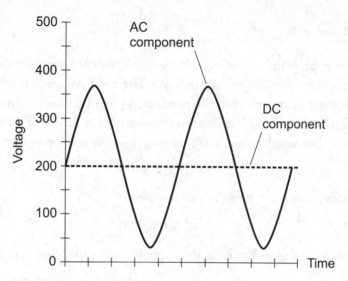

FIGURE 8-12 · Composite AC/DC wave resulting from 117-V rms AC in series with a source of +200 V DC.

✓ SOLUTION

In this case, the positive peak voltage equals half the peak-to-peak voltage, or +30 V pk. The negative peak voltage equals −1 times the positive peak voltage, or −30 V pk. Half of the wave's peaks attain instantaneous voltages of +30 V, and the other half of the peaks attain instantaneous voltages of −30 V.

PROBLEM 8-6

Suppose that we superimpose a DC component of +10 V on the sine wave described in Problem 8-5. What are the positive and negative peak voltages?

✓ SOLUTION

In the situation of Problem 8-5, the positive peak voltage equals +30 V and the negative peak voltage equals −30 V. When we superimpose a DC component of +10 V on the wave, both the positive peak and the negative peak voltages increase by 10 V. The positive peak voltage therefore becomes +30 + 10 V or +40 V, and the negative peak voltage becomes −30 V + 10 V or −20 V.

Phase Angle

Engineers express the extent of the time displacement between two waves having identical frequencies in various ways. The most common scheme involves determining a quantity called the *phase angle*, denoted by the lowercase italic Greek letter phi (ϕ). We can express phase angles as values of ϕ larger than or equal to 0° but smaller than 360°, ranging over the half-open interval

$$0° \leq \phi < 360°$$

If we express the same range in radians, we get

$$0 \text{ rad} \leq \phi < 2\pi \text{ rad}$$

Occasionally, we'll read or hear about phase angles specified over a range of

$$-180° < \phi \leq +180°$$

In radians, we represent that range as

$$-\pi \text{ rad} < \phi \leq +\pi \text{ rad}$$

We can define phase angles only for pairs of waves that have identical frequencies. If the frequencies differ, the phase angle between the two waves changes from moment to moment, so we cannot define a fixed value for it.

Phase Coincidence

The term *phase coincidence* means that two waves having identical frequencies begin at exactly the same moment in time. When portrayed on a graph, they appear to "line up" or coincide. Figure 8-13 illustrates an example of phase coincidence for two waves having different amplitudes. (If the amplitudes were the same, we would see only one wave on the coordinate grid!) The two waves differ in phase by 0°—in other words, not at all.

If two sine waves of identical frequency exist in phase coincidence, then the positive peak amplitude of the *resultant wave*, which constitutes another sine wave having the same frequency as the original two, equals the sum of the positive peak amplitudes of the two individual waves (called *composite waves*, *component waves*, or *constituent waves*). The negative peak amplitude of the resultant wave equals the sum of the negative peak amplitudes of the composite waves. The phase of the resultant wave coincides with the phase of either composite wave.

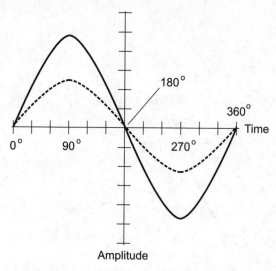

FIGURE 8-13 • Two sine waves in phase coincidence.

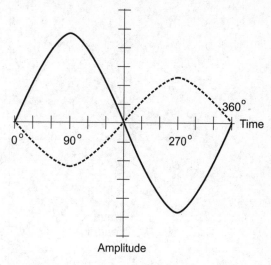

FIGURE 8-14 • Two sine waves in phase opposition.

Phase Opposition

When two sine waves having identical frequency begin exactly 1/2 cycle (180°) apart, we say that they're in *phase opposition* or that they have *opposing phase*. Figure 8-14 illustrates an example. If two sine waves have the same peak-to-peak amplitude and opposing phase, and if neither of them has a DC component, they completely cancel each other out, because they have equal and opposite instantaneous amplitudes at every moment in time. In that case, we get a resultant wave equivalent to no voltage or current at all.

In a situation such as that shown in Fig. 8-14, where two sine waves have the same frequency, different amplitudes, and opposing phase, the resultant constitutes a sine wave whose peak-to-peak value equals to the difference between the peak-to-peak values of the two composite waves. The phase of the resultant wave corresponds to the phase of the stronger of the two composite waves. The frequency of the resultant wave equals the frequency of the two composite waves.

Leading Phase

Imagine two sine waves, which we call wave X and wave Y, having the same frequency. If wave X begins a fraction of a cycle earlier than wave Y, then we say that wave X *leads* wave Y in phase, or that wave X has the *leading phase*. For this situation to hold true, wave X must begin its cycle less than 180° before wave Y.

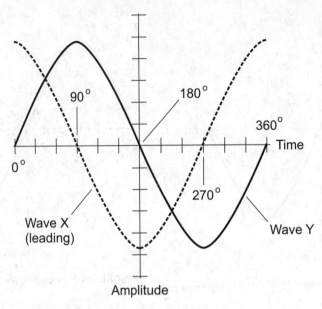

FIGURE 8-15 • Wave X leads wave Y by 90°.

Figure 8-15 shows wave X leading wave Y by 90°, which equals 1/4 of a cycle. The phase angle can range from 0° to 180°, noninclusive. If we call the phase angle ϕ, then

$$0° < \phi < +180°$$

If we want to express that range in radians, we write

$$0 \text{ rad} < \phi < +\pi \text{ rad}$$

When we say that wave X has a phase of $+\phi$ degrees or radials relative to wave Y, we mean that wave X leads wave Y by ϕ degrees or radians.

Lagging Phase

Suppose that wave X begins its cycle more than 180°, but less than 360°, ahead of wave Y. In this situation, we do better to imagine that wave X starts its cycle later than wave Y, by some value between, but not including, 0° and 180°. Then we say that wave X *lags* wave Y, or that wave X has the *lagging phase*.

Figure 8-16 shows wave X lagging wave Y by 90°. The phase angle can amount to anything between, but not including, −180° and 0°. We express lagging phase as a negative angle ϕ such that

$$-180° < \phi < 0°$$

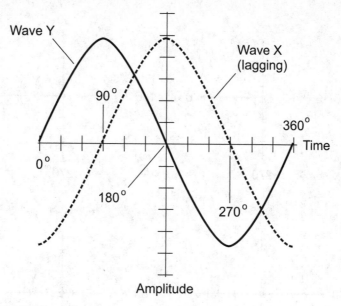

FIGURE 8-16 • Wave X lags wave Y by 90°.

Expressed in radians, we have the equivalent range

$$-\pi \text{ rad} < \phi < 0 \text{ rad}$$

When we say that wave X has a phase of $-\phi$ degrees or radians relative to wave Y, we mean that wave X lags wave Y by ϕ radians or degrees.

Vector Representations of Phase

If a sine wave X leads another sine wave Y of the same frequency by ϕ degrees, then we can draw the waves as vectors on a coordinate plane. We orient vector X precisely ϕ degrees *counterclockwise* from vector Y. If wave X lags Y by ϕ degrees, then we orient wave X exactly ϕ degrees *clockwise* from Y. If the two waves precisely coincide with each other in phase, then their vectors line up. If the two waves precisely oppose each other in phase, then their vectors point in opposite directions.

The drawings of Fig. 8-17 show four phase relationships between waves X and Y. In this particular set of examples, we assume that neither wave has a DC component. We also assume that wave X has twice the peak-to-peak amplitude of wave Y, so that vector X always measures twice the length of vector Y.

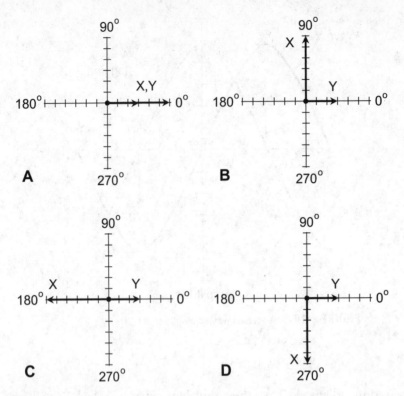

FIGURE 8-17 • Vector representations of phase. At A, waves X and Y coincide in phase. At B, wave X leads wave Y by 90°. At C, waves X and Y exist in phase opposition. At D, wave X lags wave Y by 90°.

When we examine these vector graphs, we can notice the following facts about the waves:

- At A, waves X and Y exist in phase coincidence.
- At B, wave X leads wave Y by 90° or $\pi/2$ rad.
- At C, waves X and Y exist in phase opposition.
- At D, wave X lags wave Y by 90° or $\pi/2$ rad.

In all cases, we can imagine (as time passes) that the vectors both turn counterclockwise at a constant rate of one complete rotation per wave cycle. They always stay the same angular distance apart from each other as they move.

Mathematically, we can represent a perfect sine wave, having no DC component and constant frequency, as a vector whose magnitude expresses its peak-to-peak value, and whose angle on the coordinate plane expresses its instantaneous phase. The vector constantly rotates, just as the ball described earlier in this chapter orbits your head if you put it on the end of a string and whirl the whole assembly around.

Still Struggling

In a perfect sine wave having constant, zero average amplitude and constant frequency, the vector magnitude never varies. If the waveform does not constitute a sinusoid, the vector magnitude varies in time. As you can guess, there exist infinitely many possible functions for the magnitude versus instantaneous phase angle. Some such functions can get complicated indeed!

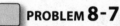

PROBLEM 8-7

Consider three perfect sine waves X, Y, and Z, none of which has a DC component, and all of which have identical frequency. Suppose that wave X leads wave Y by 0.5000 rad, while wave Y leads wave Z by precisely 1/8 cycle. By how many degrees does wave X lead or lag wave Z?

SOLUTION

Let's convert all of the phase angles to degrees. One radian equals approximately 57.30°. Therefore

$$0.5000 \text{ rad} = 57.30° \times 0.5000$$

$$= 28.65°$$

A phase difference of 1/8 cycle equals 360°/8, or 45.00°, where we consider the values 360° and 8 as mathematically exact and we express the resultant to four significant figures. The phase angles add directly. Therefore, we can conclude that wave X leads wave Y by 28.65° + 45.00°, or 73.65°.

PROBLEM 8-8

Consider three perfect sine waves X, Y, and Z, none of which has a DC component, and all of which have identical frequency. Suppose that wave X leads wave Y by 0.5000 rad, while wave Y *lags* wave Z by precisely 1/8 cycle. By how many degrees does wave X lead or lag wave Z?

SOLUTION

In this situation, wave X leads wave Y to the same extent as it does in Problem 8-7 (that is, by 28.65°). The phase difference between Y and Z has the same extent as before but the opposite sense: wave Y lags wave Z by 45.00°. We can also say that wave Y leads wave Z by −45.00°. In this scenario, wave X leads wave Z by 28.65° + (−45.00°), which equals 28.65° − 45.00° or −16.35°. We conclude that wave X lags wave Z by 16.35°. We might also express the situation by saying that wave Z leads wave X by 16.35°.

QUIZ

Refer to the text in this chapter if necessary. A good score is eight correct. Answers are at the back of the book.

1. A ramp constitutes a special form of
    A. sine wave.
    B. sawtooth wave.
    C. square wave.
    D. rectangular wave.

2. Imagine two sine waves having the same frequency. Their vector representations point at an angle of $\pi/4$ radians with respect to each other, representing a displacement of
    A. 1/32 cycle.
    B. 1/16 cycle.
    C. 1/8 cycle.
    D. 1/4 cycle.

3. Suppose that an AC sawtooth wave has a peak-to-peak voltage of 3.00 V. The wave has a superimposed DC component of −2.00 V. What's the rms voltage?
    A. We need more information to answer this question.
    B. 1.50 V rms
    C. 1.00 V rms
    D. 0.50 V rms

4. What's the peak-to-peak voltage of AC utility electricity at 117 V rms?
    A. 41.4 V
    B. 82.7 V
    C. 165 V
    D. 331 V

5. Consider two sine waves having the same frequency, but differing in phase by 30°. That's equivalent to a phase angle of
    A. $\pi/12$ rad.
    B. $\pi/10$ rad.
    C. $\pi/8$ rad.
    D. $\pi/6$ rad.

6. If we double the frequency of an AC wave, then its period
    A. does not change.
    B. doubles.
    C. quadruples.
    D. decreases to half its former value.

7. Suppose that in Fig. 8-18, each horizontal division represents 1.00 ms, and each vertical division represents 10 V. What's the frequency of the wave?

    A. We can't define it directly, because it's ambiguous.
    B. 2.00 kHz
    C. 1.00 kHz
    D. 500 Hz

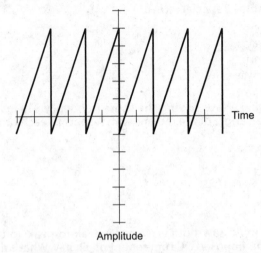

FIGURE 8-18 • Illustration for Quiz Questions 7 and 8.

8. Suppose that in Fig. 8-18, each horizontal division represents 1.00 ms, and each vertical division represents 10 V. What's the approximate peak-to-peak voltage of the wave?

    A. We can't define it, because it's ambiguous.
    B. 60 V
    C. 50 V
    D. 10 V

9. Imagine that the "sunspot cycle" for a distant star has a period of 25.000 earth years. What's the frequency of this cycle in hertz? Consider one earth year to comprise 365.24 earth days, and one earth day to comprise 24.000 hours.

    A. 3.1689×10^{-8} Hz
    B. 1.2676×10^{-9} Hz
    C. 7.9222×10^{-7} Hz
    D. We cannot answer this question, because no unit of frequency exists smaller than the hertz.

10. Suppose that an irregular AC wave attains a maximum instantaneous voltage of +1.35 V and a minimum instantaneous voltage of −0.35 V. What's the peak-to-peak voltage?

    A. 1.70 V
    B. 1.00 V
    C. 0.85 V
    D. We can't define it, because it's ambiguous.

chapter **9**

Magnetism

Electric and magnetic phenomena interact. A *magnetic field* arises when electric charge carriers move relative to an observer. Conversely, when an electrical conductor moves in a magnetic field, an electric current flows in that conductor.

CHAPTER OBJECTIVES

In this chapter, you will

- Observe the earth's magnetic field.
- Learn what causes magnetic force.
- Analyze magnetic flux and poles.
- Determine magnetic field quantity and flux density.
- Compare permanent magnets with electromagnets.
- See some applications of magnetism.

Geomagnetism

The earth has a core consisting largely of iron, heated to the extent that some of it liquefies. As the earth rotates on its axis, the iron in the core flows in complicated subterranean rivers, eddies, and convection patterns. This flow gives rise to a huge *geomagnetic field* that surrounds our planet and extends thousands of kilometers into space.

Earth's Magnetic Poles and Axis

The geomagnetic field has poles, just as an old-fashioned bar magnet does. On the earth's surface, these magnetic poles exist in the arctic and antarctic regions, but not at the *geographic poles* (the points where the earth's axis intersect the surface). The *geomagnetic axis* that connects the geomagnetic poles tilts somewhat with respect to the *geographic axis* on which the earth rotates.

Charged subatomic particles from the sun, constantly streaming outward through the solar system, distort the geomagnetic field. This so-called *solar wind* "blows" the geomagnetic field out of symmetry. On the side of the earth facing the sun, the *geomagnetic lines of flux* compress. On the side of the earth opposite the sun, the geomagnetic lines of flux dilate. Similar "blown-by-the-wind" flux distortion occurs in the magnetic fields surrounding the other planets in our solar system, notably Jupiter.

The Magnetic Compass

Thousands of years ago, science-minded people noticed the presence of the geomagnetic field, even though no one knew exactly "how it worked" or "what made it." When hung from thin threads or cords, certain rocks called *lodestones* oriented themselves in a generally north-south direction. Old-world seafarers and explorers correctly attributed this effect to the presence of a "force" in the air. The reasons for this phenomenon remained unknown for centuries, but adventurers put it to good use. Even today, a *magnetic compass* makes a valuable navigation aid. It can work when more sophisticated navigational devices, such as *Global Positioning System* (GPS) equipment, fail.

The geomagnetic field interacts with the magnetic field around a compass needle, which comprises a small bar magnet. The interaction produces a *magnetic force* on the compass needle, causing the needle to align itself parallel to the geomagnetic lines of flux. In most locations, this force operates not only in a horizontal plane (parallel to the earth's surface), but vertically. The vertical force component vanishes at the *geomagnetic equator*, a line running around the globe equidistant from the two geomagnetic poles. As the *geomagnetic latitude* increases, either toward the north or the south geomagnetic pole, the magnetic force pulls up and down on the compass needle more and more. We call the extent of the vertical force component at any particular place the *geomagnetic inclination*. Have you ever noticed this effect when using a magnetic compass? One end of the needle dips slightly toward the compass face, while the other end tilts upward toward the glass.

TIP *A magnetic compass doesn't always "tell the truth." Because the earth's geo-magnetic axis and geographic axis don't coincide, the needle of a magnetic com-pass usually points somewhat to the east or west of geographic north, which is the direction going toward the north geographic pole along line of longitude over the surface. The extent of the discrepancy in the compass reading depends on our sur-face location. At any particular location on the earth's surface, we call the angular difference between geomagnetic north (that is, north according to a compass) and geographic north (or true north)* the geomagnetic declination *for that location.*

Magnetic Force

As children, most of us discovered that magnets "stick" to some metals. Iron, nickel, a few other elements, and alloys or solid mixtures containing any of them constitute *ferromagnetic materials*. Magnets exert an attractive force on samples of these metals. Magnets do not exert force on other metals, however, unless those metals carry electric currents. Electrically insulating substances never "attract magnets" under normal conditions.

Cause and Strength

When we bring a *permanent magnet* near a sample of ferromagnetic material, the atoms in the material line up to a certain extent, temporarily magnetizing the sample. This atomic alignment produces a magnetic force between the atoms of the sample and the atoms in the magnet. Every single atom acts as a tiny magnet; when these "nanomagnets" act in concert with one another, the whole sample behaves as a magnet. Permanent magnets always attract samples of ferromagnetic material.

If we bring two permanent magnets into close proximity, we observe a stron-ger magnetic force than we do when we bring either of the permanent magnets near a sample of ferromagnetic material. The mutual force between two per-manent magnets can manifest as attraction (the two magnets pull toward each other) or repulsion (the two magnets push away from each other), depending on the way we align them. Either way, the intensity of the force increases as we bring the magnets closer together.

Some magnets produce fields and forces so powerful that no human being can pull them apart if they get "stuck" together, and no person can bring them all the way together against their mutual repulsion. We can build *electromagnets* having fields stronger than any known permanent magnet. (We'll explore how these devices work later in this chapter.) Industrial workers use huge electromagnets to carry heavy pieces of scrap iron or steel from place to place. Other electromagnets can provide sufficient repulsion to suspend one object above another, an effect known as *magnetic levitation*.

Electric Charge Carriers in Motion

Whenever the atoms in a sample of ferromagnetic material align to some extent rather than existing in random orientations, a magnetic field surrounds the sample. A magnetic field can also result from the motion of electric *charge carriers*. In an electrical conductor such as a wire, electrons move in incremental "hops" from atom to atom in a well-defined overall direction. In a permanent magnet, the movement of orbiting electrons occurs in such a manner that an *effective current* arises.

Magnetic fields can result from the motion of charged particles through space, as well as through an electrical conductor. The sun constantly ejects protons and helium nuclei, both of which carry positive electric charges. These particles produce effective currents as they travel through space. These effective currents in turn generate magnetic fields. When the magnetic fields produced by the charged particles interact with the geomagnetic field, the charged particles accelerate toward the geomagnetic poles.

When an eruption on the sun called a *solar flare* occurs, the sun ejects far more charged subatomic particles than usual. As these particles approach and accelerate toward the geomagnetic poles, their magnetic fields, working together, disrupt the geomagnetic field, spawning a *geomagnetic storm*. Such an event upsets ionized atoms in the earth's upper atmosphere, affecting "shortwave radio" communications and producing the *aurora borealis* ("northern lights") and *aurora australis* ("southern lights"), phenomena familiar to people who dwell at high latitudes. If a geomagnetic storm reaches sufficient intensity, it can interfere with wire communications and electric power transmission at the surface.

Lines of Flux

Physicists consider magnetic fields to consist of *flux lines*, or *lines of flux*. The intensity of a particular magnetic field depends on the number of flux lines passing at right angles through a region having a certain cross-sectional area, such as a centimeter squared (cm²) or a meter squared (m²). The flux lines don't comprise material fibers or threads, of course, but we can witness their effects by doing simple experiments.

Have you seen the classical school-lab demonstration in which iron filings lie on a sheet of paper, and then the experimenter holds a permanent magnet underneath the sheet? The filings arrange themselves in a pattern that shows, roughly, the "shape" of the magnetic field in the vicinity of the magnet. A bar magnet has a field whose lines of flux exhibit a characteristic symmetrical pattern (Fig. 9-1).

Another experiment involves passing a current-carrying wire at a right angle through a horizontally oriented sheet of paper. The iron filings bunch up in

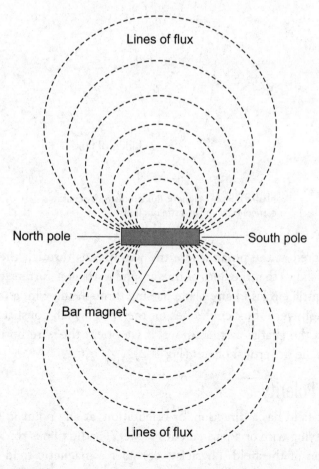

FIGURE 9-1 · Magnetic flux around a bar magnet.

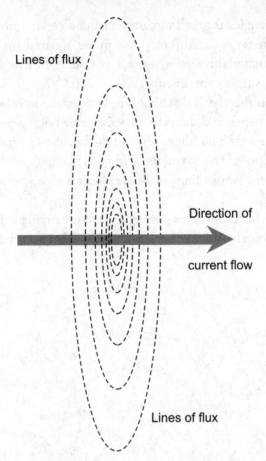

Lines of flux

Direction of

current flow

Lines of flux

FIGURE 9-2 · Magnetic flux produced by charge carriers traveling in a straight line.

circles centered at the point where the wire passes through the paper. This experiment shows that the lines of flux around a straight, current-carrying wire form concentric circles in any plane passing through the wire at a right angle. The center of every "flux circle" lies on the wire, which constitutes the path along which the charge carriers move. If we orient the wire horizontally, the "flux circles" lie in vertical planes (Fig. 9-2).

Magnetic Polarity

A magnetic field has a direction, or orientation, at any point in space near a current-carrying wire or a permanent magnet. The flux lines run parallel with the direction of the field. Physicists consider a magnetic field to begin, or originate, at a *north pole*, and to end, or terminate, at a *south pole*. In the case of a permanent magnet, we can usually (but not always) determine the locations of the magnetic poles on the object. With a current-carrying wire, the magnetic field goes around and around endlessly, like a dog chasing its own tail.

A charged electric particle, such as a proton hovering in space, forms an *electric monopole*, and the electric flux lines around it aren't closed. A positive or negative electric charge pole does not have to mate with a charge pole of the opposite sense; it can exist all alone. The electric flux lines around any stationary, charged particle run outward from that particle in straight radial lines for a theoretically infinite distance. But magnetic fields behave according to stricter laws. Under normal circumstances, all magnetic flux lines form closed loops. In the vicinity of a magnet, we can always find a starting point (the north pole) and an ending point (the south pole). Around a current-carrying wire, the loops all form closed circles.

Still Struggling

The poles of a permanent magnet do not match the geomagnetic poles. In fact, they precisely oppose! The north geomagnetic pole actually constitutes a magnetic south pole (therefore attracting the north magnetic pole of a hiker's compass). Similarly, the south geomagnetic pole constitutes a magnetic north pole.

Magnetic Dipoles

You might at first suppose that the magnetic field around a current-carrying wire arises from a monopole, or that no poles exist at all. When you look at Fig. 9-2, for example, the concentric flux circles don't appear to originate or terminate anywhere, do they? The flux circles go around and around with no beginning or ending points. But you can assign originating and terminating points to those circles, thereby defining a *magnetic dipole*—a pair of opposite magnetic poles in close proximity.

Imagine that you hold a flat piece of paper next to a current-carrying wire, so that the wire runs along one edge of the sheet. The magnetic circles of flux surrounding the wire pass through the sheet of paper, entering one side and emerging from the other side. You have a "virtual magnet" whose north pole coincides with the face of the paper sheet from which the flux circles emerge. The south pole coincides with the opposite face of the sheet, into which the flux circles plunge.

The flux lines in the vicinity of a magnetic dipole always connect the two poles. Some flux lines appear straight in a local sense, but in the larger sense they always form curves. The greatest magnetic field strength around a bar magnet occurs near the poles, where the flux lines converge or diverge. Around a current-carrying wire, the greatest field strength occurs near the wire.

Magnetic Field Strength

Physicists and engineers express the overall magnitude or quantity of a magnetic field in units called *webers*, symbolized Wb. We can employ a smaller unit, the *maxwell* (Mx), for weak fields. One weber equals 100,000,000 (10^8) maxwells. Therefore

$$1 \text{ Wb} = 10^8 \text{ Mx}$$

and

$$1 \text{ Mx} = 10^{-8} \text{ Wb}$$

The Tesla and the Gauss

If you have a permanent magnet or electromagnet, you might see its "strength" expressed in terms of webers or maxwells. But more often, you'll hear or read about units called *teslas* (T) or *gauss* (G). These units define the concentration, or intensity, of the magnetic field as its flux lines pass at right angles through flat regions having specific cross-sectional areas.

The *flux density*, or number of "flux lines per unit cross-sectional area," often yields a more useful expression for magnetic effects than the overall quantity of magnetism. In equations, we denote flux density with the letter B. A flux density of 1 tesla equals 1 weber per meter squared (1 Wb/m^2). A flux density of 1 gauss equals 1 maxwell per centimeter squared (1 Mx/cm^2). As things work out, 1 gauss equals 0.0001 (10^{-4}) tesla, so we have the relations

$$1 \text{ G} = 10^{-4} \text{ T}$$

and

$$1 \text{ T} = 10^4 \text{ G}$$

If you want to convert from teslas to gauss (not gausses!), multiply by 10^4. If you want to convert from gauss to teslas, multiply by 10^{-4}.

Still Struggling

If the distinctions between webers and teslas, or between maxwells and gauss, confuse you, think of an ordinary light bulb. Suppose a bulb emits 15 W of visible-light power. If you enclose the bulb completely inside a chamber, then 15 W of visible light strike the interior walls of the chamber, regardless of the size of the chamber (or, for that matter, its geometric shape). But this notion doesn't give you a useful notion of the brightness of the light. You know that a single bulb produces plenty of light if you want to illuminate a small closet, but nowhere near enough light to illuminate a gymnasium or auditorium.

To get an idea of how well a bulb will illuminate a particular surface, you must know the number of visible-light watts *per unit area* on that surface. When you say that a bulb gives off so-many watts of light overall, it's like saying that a magnet has an overall magnetic-field quantity of so-many webers or maxwells. When you say that the bulb produces so-many watts of light per unit area, it's like saying that a magnetic field has a flux density of so-many teslas or gauss.

Magnetomotive Force

When we work with magnetic fields produced by electric currents, we can quantify a phenomenon called *magnetomotive force* with a unit called the *ampere-turn* (At). This unit describes itself well: the number of amperes flowing in a loop or coil of wire, times the number of turns that the loop or coil contains.

If we bend a length of wire into a loop and drive 1 A of current through it, we get 1 At of magnetomotive force in the vicinity of the loop. If we wind the same length of wire (or any other length) into a 50-turn coil and keep driving 1 A of current through it, the resulting magnetomotive force increases by a factor of 50, to 50 At. If we then reduce the current in the 50-turn loop to 1/50 A or 20 mA, the magnetomotive force goes back down to 1 At.

Sometimes, engineers employ a unit called the *gilbert* to express magnetomotive force. One gilbert (1 Gb) equals exactly $5/(2\pi)$ At, or approximately 0.7958 At. This relation yields conversion algorithms as follows:

- If we want to determine the number of ampere-turns when we know the number of gilberts, we should multiply by $5/(2\pi)$ or approximately 0.7958.

- If we want to determine the number of gilberts when we know the number of ampere-turns, we should multiply by $2\pi/5$ or approximately 1.257.

Flux Density versus Current

In a straight wire carrying a steady, direct electric current and surrounded by *free space* (air or a vacuum), we observe the greatest flux density near the wire, and diminishing flux density as we get farther away from the wire. We can use a simple formula to express magnetic flux density as a function of the current in a straight wire and the distance from the wire. Like many formulas in physics, it works perfectly under ideal circumstances, but discrepancies commonly occur in the flawed "real world."

Imagine an infinitely thin, absolutely straight, infinitely long length of wire (the ideal case). Suppose that the wire carries a current of I amperes. Let's represent the flux density (in teslas) as B. Now let's consider a point P that lies at a distance r (in meters) from the wire as measured in a plane perpendicular to the wire, as shown in Fig. 9-3. We can find the flux density at the point P using the formula

$$B = 2 \times 10^{-7} \, I/r$$

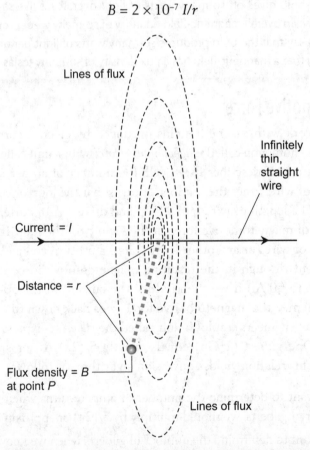

FIGURE 9-3 • The magnetic flux density varies inversely with the distance from a wire carrying a constant direct current.

We can consider the value of the constant, 2×10^{-7}, mathematically exact to any desired number of significant figures.

Of course, we'll never encounter a wire with zero thickness or infinite length. But as long as the wire thickness constitutes a small fraction of r, and as long as the wire lies along a reasonably straight line near point P, this formula works quite well in most practical scenarios.

PROBLEM 9-1

What is the flux density B_t in teslas at a distance of 200 mm from a straight, thin wire carrying 400 mA of DC?

SOLUTION

First, we must convert all quantities to units in the International System (SI). This means that we have $r = 0.200$ m and $I = 0.400$ A. We can input these values directly into the formula for flux density to obtain

$$B_t = 2 \times 10^{-7} \, I/r$$
$$= 2.00 \times 10^{-7} \times 0.400/0.200$$
$$= 4.00 \times 10^{-7} \, T$$

PROBLEM 9-2

In the above-described scenario, what is the flux density B_g (in gauss) at point P?

SOLUTION

To figure this out, we must convert from teslas to gauss. This task involves merely multiplying the result in Solution 9-1 by 10^4 to get

$$B_g = 4.00 \times 10^{-7} \times 10^4$$
$$= 4.00 \times 10^{-3} \, G$$

Electromagnets

The motion of electric charge carriers always produces a magnetic field. This field can reach considerable intensity in a tightly coiled wire having many turns and carrying a large current. When we place a ferromagnetic rod called a *core* inside a coil as shown in Fig. 9-4, the magnetic lines of flux concentrate in the core, and we obtain an *electromagnet*. Most electromagnets have cylindrical

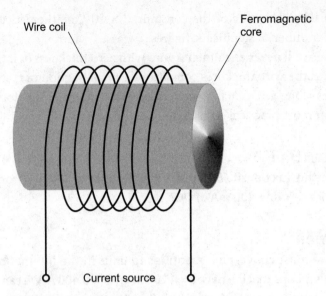

Wire coil

Ferromagnetic core

Current source

FIGURE 9-4 · A simple electromagnet.

cores. The length-to-radius ratio can vary from extremely low (a fat pellet) to extremely high (a thin rod). Regardless of the length-to-radius ratio, the flux produced by current in the wire temporarily magnetizes the core.

Direct-Current Types

You can build a DC electromagnet by wrapping a couple of hundred turns of insulated wire around a large iron bolt or nail. You can find these items in any good hardware store. You should test the bolt for ferromagnetic properties before you leave the store, if possible. (If a permanent magnet "sticks" to the bolt, then the bolt comprises a ferromagnetic material.) Ideally, the bolt should measure at least 3/8 inch in diameter and several inches long. You must use insulated or enameled wire (not bare wire!), preferably made of solid, soft copper.

Wind the wire at least several dozen (if not 100 or more) times around the bolt in the same direction. Secure the wire in place with electrical tape or masking tape. A large 6-V "lantern battery" can provide plenty of DC to operate the electromagnet. If you like, you can connect two or more such batteries in parallel to increase the maximum deliverable current. Never leave the coil connected to the battery for more than a few seconds at a time.

TIP *Do not use a lead-acid automotive battery for this experiment. The near short-circuit produced by an electromagnet can cause the acid from such a battery to boil out, seriously injuring you.*

All DC electromagnets have defined north and south poles, just as permanent magnets do. However, an electromagnet can get much stronger than any permanent magnet. The magnetic field exists only as long as the coil carries current. When you remove the power source, the magnetic field nearly vanishes. A small amount of *residual magnetism* remains in the core after current stops flowing in the coil, but this field has minimal intensity.

Alternating-Current Types

Do you suspect that you can make an electromagnet extremely powerful if, rather than using a lantern battery for the current source, you plug the ends of the coil directly into an AC utility outlet? In theory, you can—but in practice, you'll expose yourself to the danger of electrocution, expose your house to the risk of electrical fire, and most likely cause a fuse or circuit breaker to open, "killing" the power to the device. In any event, don't try it. Some buildings (such as college dorms) may lack adequate overcurrent protection. If you want to build and test a safe AC electromagnet, my book *Electricity Experiments You Can Do at Home* (McGraw-Hill, 2010) offers instructions for doing it.

Some commercially manufactured electromagnets operate from 60-Hz utility AC. These magnets "stick" to ferromagnetic objects. The polarity of the magnetic field reverses every time the direction of the current reverses, producing 120 fluctuations, or 60 complete north-to-south-to-north polarity changes, every second, as shown graphically in Fig. 9-5.

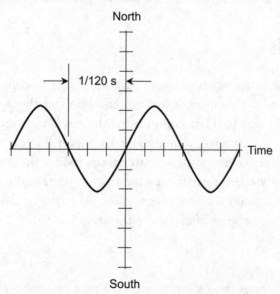

FIGURE 9-5 • The polarity and field intensity in a 60-Hz AC electromagnet vary as a function of time.

If you bring a permanent magnet or DC electromagnet near either "pole" of an AC electromagnet, no net force results from the AC electromagnetism itself, because equal and opposite attractive and repulsive forces occur between the alternating magnetic field and any steady external field. But the permanent magnet or DC electromagnet attracts the core of the AC electromagnet, whether the AC device carries current or not.

PROBLEM 9-3

Suppose that we apply 80-Hz AC to an electromagnet instead of the standard 60 Hz. What will happen to the interaction between the alternating magnetic field and a nearby permanent magnet or DC electromagnet?

✔ SOLUTION

Assuming that the behavior of the core material remains the same, the situation at 80 Hz will not change from the 60-Hz case. In theory, the AC frequency makes no difference in the behavior of an AC electromagnet. In practice, however, the magnetic field weakens at high AC frequencies because the AC electromagnet's *inductance* tends to impede the flow of current. This so-called *inductive reactance* depends on the number of coil turns, and also on the characteristics of the ferromagnetic core. We'll learn about inductive reactance later in this book.

Magnetic Materials

Ferromagnetic substances cause magnetic lines of flux to compress relative to their distribution in a vacuum. A few materials do exactly the opposite; they cause the lines of flux to dilate compared with their distribution in a vacuum. We call the latter substances *diamagnetic*. Examples include wax, dry wood, bismuth, and silver. No diamagnetic material can reduce the strength of a magnetic field by anywhere near the factor that ferromagnetic substances can increase it. We quantify the ferromagnetic or diamagnetic characteristics of a substance in terms of *permeability* and *retentivity*.

Permeability

Permeability expresses the extent to which a medium concentrates magnetic lines of flux relative to the flux density in a vacuum. By convention, scientists assign a permeability value of precisely 1 to a vacuum. Imagine that we drive constant, identical direct currents through two identical wire coils, one with an air core and surrounded by air, and the other with a vacuum core and surrounded by a vacuum. In that situation, the flux density inside the air coil will exceed the flux density inside the vacuum coil by a factor so small that, in most circumstances, we can ignore it. Therefore, we can consider air to have a permeability value of 1 for practical purposes.

If we place a ferromagnetic core inside the coil, the flux density increases, sometimes by a large factor. By definition, that factor equals the permeability. If a certain material causes the flux density inside a coil to increase by a factor of 60 compared with the flux density in a vacuum, then that material has a permeability of 60. The permeability of iron can range from approximately 60 (impure) to 8000 (highly refined). If we use a special metallic alloy called a *permalloy* as the core material in an electromagnet, we can increase the flux density inside the coil by as much as 1,000,000 (10^6) times. Table 9-1 lists permeability values for some common substances.

If, for some reason, we feel compelled to make an electromagnet as weak as possible, we can use a sample of diamagnetic material such as dry wood or wax

TABLE 9-1   Permeability values for some common materials.

Substance	Permeability (Approx.)
Air, dry, at sea level	1
Alloys, ferromagnetic	3000–1,000,000
Aluminum	Slightly more than 1
Bismuth	Slightly less than 1
Cobalt	60–70
Iron, powdered and pressed	100–3000
Iron, solid, refined	3000–8000
Iron, solid, unrefined	60–100
Nickel	50–60
Silver	Slightly less than 1
Steel	300–600
Vacuum	1
Wax	Slightly less than 1
Wood, dry	Slightly less than 1

for the core material. Diamagnetic materials always have permeability values smaller than 1 (but never very much smaller). Usually, engineers use diamagnetic objects to keep magnets physically separated while minimizing the interaction between them.

Retentivity

Certain ferromagnetic materials stay magnetized better than others. When we subject a substance such as iron to a magnetic field as intense as it can handle—say by enclosing it in a wire coil carrying high current—some residual magnetism will remain when the current stops flowing in the coil. Retentivity, also known as *remanence*, quantifies the extent to which a substance can "memorize" a magnetic field imposed on it.

Suppose that we wind a wire coil around a sample of ferromagnetic material and then drive so much DC through the coil that the magnetic flux inside the core reaches its maximum possible density. We call this condition *core saturation*. We measure the flux density in this situation, and get a figure of B_{max} (in teslas or gauss). Now suppose that we remove the current from the coil, and then we measure the flux density inside the core again, obtaining a figure of B_{rem} (in teslas or gauss, as before). We can express the retentivity, B_r, of the core material as a ratio according to the formula

$$B_r = B_{rem}/B_{max}$$

or as a percentage using the formula

$$B_{r\%} = 100 \, B_{rem}/B_{max}$$

As an example, suppose that a metal rod can attain a flux density of 135 G when enclosed by a coil carrying DC. Imagine that 135 G represents the maximum possible flux density for that material. (For any substance, such a maximum always exists, unique to that substance; further increasing the coil current or number of turns will not magnetize it any further.) Now suppose that we remove the current from the coil, and 19 G remain in the rod. Then the retentivity, B_r, equals

$$B_r = 19/135$$

$$= 0.14$$

As a percentage, we have

$$B_{r\%} = 100 \times 19/135$$

$$= 14\%$$

Certain ferromagnetic substances exhibit high retentivity, and therefore make excellent permanent magnets. Other ferromagnetic materials have poor retentivity. They might work as the cores of electromagnets, but they do not make good permanent magnets.

If a ferromagnetic substance has low retentivity, it can function as the core for an AC electromagnet, because the polarity of the magnetic field in the core follows along closely as the current in the coil alternates. If the material has high retentivity, however, the material acts "magnetically sluggish" and has trouble following the current reversals in the coil. Substances of this sort don't work well in AC electromagnets. Engineers call such "magnetic sluggishness" *hysteresis*.

 PROBLEM 9-4

Suppose that we wind a coil of wire around a metal core to make an electromagnet. We find that by connecting a variable DC source to the coil, we can drive the magnetic flux density in the core up to 0.500 T but no higher. When we shut down the current source, the flux density inside the core drops to 500 G. What's the retentivity of this metal?

 SOLUTION

First, let's convert both flux density figures to the same units. We recall that $1 \text{ T} = 10^4 \text{ G}$. Therefore, the maximum possible core flux density in gauss equals $0.500 \times 10^4 = 5000 \text{ G}$ when current flows in the coil; we already know that it's 500 G after we remove the current. "Plugging in" these numbers gives us the ratio

$$B_r = 500/5000$$
$$= 0.100$$

or the percentage

$$B_{r\%} = 100 \times 500/5000$$
$$= 100 \times 0.100$$
$$= 10.0\%$$

Permanent Magnets

Industrial engineers can make any suitably shaped sample of ferromagnetic material into a permanent magnet of the sort you played with as a child (and maybe still use to stick notes to your refrigerator door). The strength of a permanent magnet depends on two factors:

- The retentivity of the material that we use to make it
- The amount of effort that we put into magnetizing it

The manufacture of powerful permanent magnets requires the use of an alloy with high retentivity. The most "magnetizable" alloys derive from specially formulated mixtures of aluminum, nickel, and cobalt, occasionally including trace amounts of copper and titanium. Industrial engineers place samples of the selected alloy inside heavy wire coils carrying high, continuous DC for long periods of time.

You can magnetize any piece of iron or steel to some extent. Some technicians use magnetized tools when installing or removing screws from hard-to-reach places in computers, wireless transceivers, and other devices. If you want to magnetize a tool, stroke its metal shaft with the end of a powerful bar magnet several dozen times. But beware: Once you've imposed residual magnetism in a tool, that tool will remain at least slightly magnetized for good.

Flux Density Inside a Long Coil

Consider a long coil of wire, commonly known as a *solenoid*, having n turns in a single layer, and whose length measures s meters. Suppose that this coil carries a steady direct current of I amperes, and has a ferromagnetic core of permeability μ. Assuming that the core has not reached a state of saturation, we can calculate the flux density B_t (in teslas) inside the material using the formula

$$B_t = 4\pi \times 10^{-7} \, \mu n I/s$$
$$\approx 1.2566 \times 10^{-6} \, \mu n I/s$$

If we want to calculate the flux density B_g (in gauss), we can use the formula

$$B_g = 4\pi \times 10^{-3} \, \mu n I/s$$
$$\approx 0.012566 \, \mu n I/s$$

PROBLEM 9-5

Imagine that we've constructed a DC electromagnet, and we drive a certain constant DC through its coil. The coil measures 20 cm long, and has 100 turns of wire. The core, which has permeability $\mu = 100$, has not reached a state of saturation. The core and the coil have the same length. We measure the flux density inside the core as $B_g = 20$ G. How much current flows in the coil?

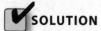

SOLUTION

Let's start by ensuring that we'll use the proper units in our calculation. We're told that the electromagnet measures 20 cm in length, so we can set $s = 0.20$ m. The flux density equals 20 G. Using algebra, we can rearrange the second formula above so that it solves for I. We start with

$$B_g = 0.012566\ \mu n I/s$$

Dividing through by I, we get

$$B_g/I = 0.012566\ \mu n/s$$

When we divide both sides by B_g, we obtain

$$I^{-1} = 0.012566\ \mu n/(sB_g)$$

Finally, we can take the reciprocal of both sides to get

$$I = 79.580\ sB_g/(\mu n)$$

Derivations of this sort require that we never divide an equation through by any quantity that can become zero in a practical situation. That's not a problem here. (We have no interest in situations involving the absence of current, no turns of wire, permeability values of zero, or coils having zero length!) Now let's plug in the numbers from the original statement. We calculate

$$I = 79.580\ sB_g/(\mu n)$$
$$= 79.580 \times 0.20 \times 20/(100 \times 100)$$
$$= 79.580 \times 4.0 \times 10^{-4}$$
$$= 0.031832\ \text{A}$$
$$= 31.832\ \text{mA}$$

We should round this result off to 32 mA, because our input data accuracy only extends to two significant figures.

Magnetic Machines

A solenoid with a movable ferromagnetic core can do various things. Electrical relays, bell ringers, meters, electric "hammers," and other mechanical devices make use of the principle of the solenoid. More sophisticated electromagnets, sometimes in conjunction with permanent magnets, allow us to construct motors and generators.

Chime

Figure 9-6 illustrates a *bell ringer*, also known as a *chime*. Its solenoid comprises an electromagnet. The ferromagnetic core has a hollow region in the center, along its axis, through which a steel rod called the *hammer* passes. The coil contains many turns of wire, so the electromagnet produces high flux density if a substantial current passes through the coil.

When no current flows in the coil, gravity holds the hammer down so that it rests on the base plate. When a pulse of current passes through the coil, the rod moves upward at high speed. The magnetic field "wants" the ends of the rod, which has the same length as the core, to align with the ends of the core. But the rod's upward momentum causes it to pass through the core and strike the ringer. Then the steel rod falls back to its resting position, allowing the ringer to reverberate.

Relay

We can't always locate electrical switches near the devices they control. For example, suppose that we want to switch a communications system between

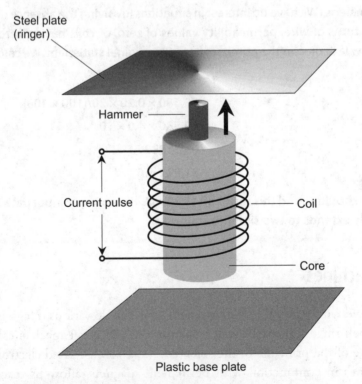

FIGURE 9-6 · A bell ringer, also known as a chime.

two different antennas from a station control point 50 m away. Wireless antenna systems carry high-frequency AC (the radio signals) that must remain within certain parts of the circuit. A *relay* makes use of a solenoid to allow remote-control switching.

Figure 9-7A illustrates a simple relay, and Fig. 9-7B shows the schematic diagram for the same device. A "springy strip" holds a movable lever, called the

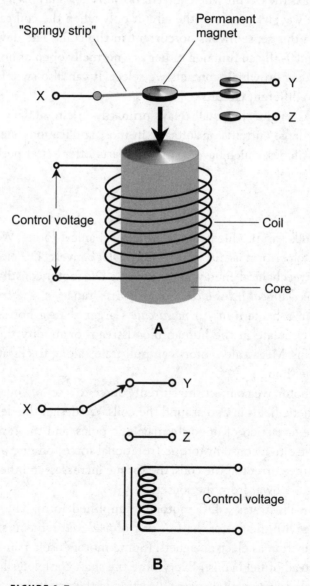

FIGURE 9-7 • A simple relay (at A) and the schematic symbol for the same relay (at B).

armature, to one side (upward in this diagram) when no current flows through the coil. Under these conditions, terminal X contacts terminal Y, but X does not contact Z. When a sufficient current flows in the coil, the armature moves to the other side (downward in this illustration), disconnecting terminal X from terminal Y, and connecting X to Z.

A *normally closed relay* completes the circuit when no current flows in the coil, and breaks the circuit when current flows. A *normally open relay* works in the opposite way; it completes the circuit only when the coil carries current. ("Normal" in this sense means no current in the coil.) The device shown in Figs. 9-7A and 9-7B can function either as a normally open or normally closed relay, depending on which contacts we select. It can also switch a single line between two different circuits.

These days, engineers install relays primarily in circuits and systems that must handle large currents or voltages. In most applications, electronic semiconductor switches, which have no moving parts, offer better performance and reliability than relays.

DC Motor

Magnetic fields can produce considerable mechanical forces. We can harness these forces to perform useful work. A *DC motor* converts DC electrical energy into rotating mechanical energy. In this sense, a DC motor constitutes a specialized *electromechanical transducer*. Such devices range in size from *nanoscale* (smaller than a bacterium) to *megascale* (larger than a house). Nanoscale motors can circulate in the human bloodstream or modify the behavior of internal organs. Megascale motors can pull trains along tracks at hundreds of kilometers per hour.

In a DC motor, we connect the current source to a set of coils, thereby producing magnetic fields in and around the coils. We design the device in such a way that the attraction of opposite magnetic poles, and the repulsion of like poles, gives rise to a constant torque (rotational force) on one of the coils. As we increase the current in the coils, the torque increases, and the device draws more electrical power from the source.

Figure 9-8 illustrates a DC motor in simplified form. The *armature coil* rotates along with the motor shaft. A pair of *field coils* remains stationary. The field coils function as electromagnets. (Some motors use a pair of permanent magnets instead of field coils.) Every time the shaft completes half a rotation, the *commutator* reverses the current direction in the armature coil, so the shaft torque continues in the same sense (clockwise or counterclockwise).

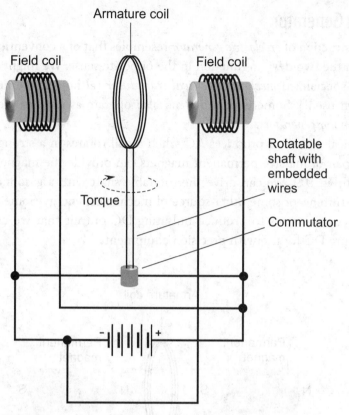

FIGURE 9-8 · A simplified drawing of a DC motor. Straight lines represent wires. Intersecting lines indicate electrical connections only when a dot appears at the point where the lines cross.

 Still Struggling

You might wonder, "Why doesn't a DC motor turn until the armature and field coils attract to their maximum extent, and then stop there? Why does the armature coil keep on going?" The answer lies in the fact that the armature coil has mass, so its rotational momentum carries it beyond the point of maximum attraction to a "point of no return." We should note, however, that the armature coil must possess a certain minimum mass. Otherwise, it will indeed "freeze up" at the point of maximum attraction—and the motor won't work.

Electric Generator

The construction of an *electric generator* resembles that of a conventional motor, although the two devices function in the opposite sense. We might call a generator a specialized *mechano-electrical transducer* (although I've never heard that term used!). Some generators can also operate as motors; we call such devices *motor-generators*.

A typical generator produces AC when a coil rotates in a strong magnetic field. A pair of massive permanent magnets can provide the magnetic field as shown in Fig. 9-9. We can drive the rotatable shaft with a gasoline-powered motor, a turbine, or some other source of mechanical energy. Some generators employ commutators to produce pulsating DC output that we can *filter* to obtain pure DC for use with precision equipment.

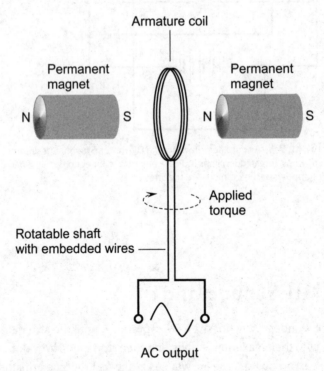

FIGURE 9-9 • A simple AC generator.

QUIZ

Refer to the text in this chapter if necessary. A good score is eight correct. Answers are at the back of the book.

1. If a wire coil has 100 turns and carries 200 mA of DC, then it produces a magnetomotive force of

    A. 2.00×10^4 At.
    B. 500 At.
    C. 20.0 At.
    D. 2.00×10^{-3} At.

2. The magnetic field around an electromagnet driven by sine-wave AC

    A. fluctuates in intensity, and periodically reverses polarity.
    B. fluctuates in intensity, but maintains the same polarity at all times.
    C. maintains constant intensity, but periodically reverses polarity.
    D. maintains constant intensity and the same polarity at all times.

3. A sample of diamagnetic material

    A. works well as the core for an AC electromagnet.
    B. maintains residual magnetism for a long time, making it ideal for manufacturing permanent magnets.
    C. causes magnetic lines of flux to "bunch up" more than they do in a vacuum.
    D. exhibits a permeability value of less than 1.

4. The magnetic flux near a straight wire carrying constant DC takes the form of

    A. straight lines parallel to the wire.
    B. straight lines perpendicular to the wire.
    C. circles whose centers lie on the wire.
    D. spirals that spin out from the wire.

5. If we want to make a powerful electromagnet that will produce an alternating magnetic field when we drive AC through its coil, we should choose a core material that has

    A. high permeability and high retentivity.
    B. high permeability and low retentivity.
    C. low permeability and high retentivity.
    D. low permeability and low retentivity.

6. Suppose that a rod made of a certain ferromagnetic material can support a maximum flux density of 800 G when a large DC flows in a wire coil surrounding it. When we remove the rod from inside the coil, the flux density in the rod goes down to 20 G. What's the permeability of this material?

    A. We must have more information to answer this question.
    B. 40

C. 1.6×10^3
D. 1.6×10^4

7. Consider a ferromagnetic rod measuring 10 cm long, surrounded by 400 turns of wire that carries steady DC. The rod, which has permeability of 1.5×10^5, has not reached a state of saturation. We measure the flux density inside the core material as $B_t = 0.30$ T. Based on this information, we know that the coil must carry

A. 0.40 mA.
B. 40 μA.
C. 2.5 A.
D. 25 A.

8. Which of the following units can we use to denote the overall quantity of a magnetic field?

A. The gauss
B. The maxwell
C. The tesla
D. Any of the above

9. What's the magnetic flux density at a point 2.50 m away from a straight, thin wire carrying 500 mA of DC?

A. 4.00×10^{-6} T
B. 2.50×10^{-4} T
C. 2.00×10^{-3} G
D. 4.00×10^{-4} G

10. Imagine that we wind 100 turns of wire in a coil around a rod-shaped ferromagnetic core. We drive 20 mA of DC through the coil. Then we wind some more wire around the core in additional layers, obtaining a 400-turn coil. Once again we drive 20 mA of DC through the wire. If the core never reaches a state of saturation at any time, and if the overall length and diameter of the coil both remain essentially the same, the increase in the number of coil turns causes the flux density inside the core to

A. increase by a factor of 16.
B. increase by a factor of 4.
C. decrease by a factor of 4.
D. decrease by a factor of 16.

chapter 10

More about Alternating Current

In DC electrical circuits, the current, voltage, resistance, and power relate according to simple equations. The same equations work for AC circuits, provided that the components merely dissipate energy, and never store or release it. If a component stores or releases energy in an AC system, we say that it exhibits *reactance*. When we combine a component's reactance and resistance in a certain mathematical way, we get an expression of the component's *impedance*, which completely defines and quantifies how it opposes, or *impedes*, the flow of AC.

CHAPTER OBJECTIVES

In this chapter, you will

- Learn how inductors work and behave.
- Analyze inductive reactance.
- Learn how capacitors work and behave.
- Analyze capacitive reactance.
- See how reactance combines with resistance.
- Graph impedance points and vectors.

Inductance

When a component stores AC electrical energy in the form of a magnetic field, then that component has *inductance*, and we call it an *inductor*. Inductors often (but not always) comprise coiled-up lengths of wire.

The Property of Inductance

Imagine a perfectly conducting wire measuring 1,000,000 (10^6) km long. Suppose that we bend it into a gigantic, single-turn loop somewhere in the vacuum of outer space, and then connect its ends to the terminals of a battery. Current immediately begins to flow in the wire, producing a magnetic field. In the first few milliseconds, the magnetic field has small extent because the current has gotten only partway around the loop. Over the next little while, as the charge carriers (mainly electrons) make their way around the whole circumference of the loop, the magnetic field grows.

Once the current has established itself throughout the loop, a certain amount of energy will exist in the magnetic field, and we will have created an inductor. The amount of energy *in joules* stored *per ampere* of current flowing in the loop—the loop's inductance—will depend on how long it took for the current to get established all the way around. That time will depend, in turn, on the length of the loop, assuming that we create it entirely in a vacuum.

Practical Inductors

In practice, of course, we can't make wire loops anywhere near as large as the one described above, but we can wind long lengths of wire into small coils. When we do this, we increase the magnetic flux density by a large factor, for a given length of wire and a given amount of current, compared with the flux density that we would get if we made the same length of wire into a single-turn loop. If we place a high-permeability ferromagnetic rod inside a coil of wire, we increase the flux density still more, as we've learned from our study of magnetism.

Whenever we do anything to increase the flux density inside a coil, given a constant current flowing through the wire, we increase the inductance of the coil. In a real-world wire coil, the inductance depends on the number of turns, the radius, the length-to-radius ratio, and the permeability of the material inside it. In general, the inductance increases if we do any of the following things while holding all other factors constant:

- Increase the number of turns.
- Increase the coil radius.

- Decrease the coil length.
- Increase the permeability of the core material.

The Unit of Inductance

Whenever we connect a source of DC across an inductor, it takes awhile for the current to establish itself uniformly throughout the inductor and eventually level off at a certain maximum. The time it takes for the current to build up to its maximum depends on the inductance. As the inductance increases, the current takes longer to build up.

The standard unit of inductance expresses the ratio between the rate of current change and the DC voltage that we connect across an inductor. An inductance of 1 *henry* (1 H) represents a *current increase* of 1 ampere per second (1 A/s) in the first few instants of time after we connect it to a DC source of 1 volt (1 V), or a *current decrease* of 1 A/s in the first few instants after we disconnect it from a DC source of 1 V. Conversely, 1 H represents an *induced voltage* of 1 V DC across an inductor through which we force a current that increases or decreases at the rate of 1 A/s. Therefore, one henry equals 1 volt per (ampere per second), which translates to 1 volt-second per ampere ($V \cdot s/A$ or $V \cdot s \cdot A^{-1}$).

The henry constitutes a large unit of inductance. Rarely will you see an inductor with a value that high. Usually, we find inductances expressed in *millihenrys* (mH), *microhenrys* (μH) or *nanohenrys* (nH), defined as follows:

$$1 \text{ mH} = 0.001 \text{ H} = 10^{-3} \text{ H}$$

$$1 \text{ } \mu\text{H} = 0.001 \text{ mH} = 10^{-6} \text{ H}$$

$$1 \text{ nH} = 0.001 \text{ } \mu\text{H} = 10^{-9} \text{ H}$$

Small coils with few turns of wire have small inductances, in which the current changes quickly and the induced voltage generally remains small. Huge coils with ferromagnetic cores, and having many turns of wire, have large inductances in which the current changes slowly and the induced voltage can become large. The terms *induct*or and *induct*ance arise from the fact that applied currents *induce* potential differences, and applied voltages induce changing currents.

Inductors in Series

Imagine that we place two or more current-carrying inductors in close proximity and connect them in series. As long as the magnetic fields around those inductors don't interact, their inductances add up arithmetically, just as

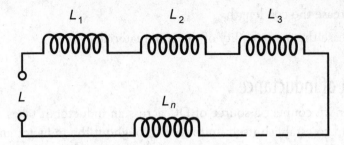

FIGURE 10-1 · Inductances in series add like resistances in series.

resistances in series combine. The total or net inductance equals the sum of the individual inductances. We must, of course, use the same size unit for all the inductors if we want this simple rule to work.

Suppose that we have inductances L_1, L_2, L_3, ..., L_n connected in series as shown in Fig. 10-1. Provided that the magnetic fields of the inductors do not interact—that is, as long as no *mutual inductance* exists between or among the components—we can calculate the total inductance L using the formula

$$L = L_1 + L_2 + L_3 + \ldots + L_n$$

Inductors in Parallel

If no mutual inductance exists among two or more parallel-connected inductors, their values add up like resistances in parallel. Consider several inductances L_1, L_2, L_3, ..., L_n connected in parallel as shown in Fig. 10-2. If we want to calculate the net inductance L, we can use the formula

$$L = 1/(1/L_1 + 1/L_2 + 1/L_3 + \ldots + 1/L_n)$$

$$= (1/L_1 + 1/L_2 + 1/L_3 + \ldots + 1/L_n)^{-1}$$

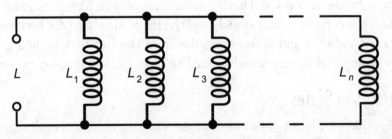

FIGURE 10-2 · Inductances in parallel add like resistances in parallel.

Again, as with inductances in series, we must make sure that all the units agree. We must not mix microhenrys with millihenrys, or henrys with nanohenrys. The unit that we choose to define the individual component values will correspond with the unit that we get for the final answer.

Still Struggling

What, you may ask, happens if mutual inductance does exist among inductors connected together? If the magnetic fields of multiple inductors work together, then the resulting mutual inductance *increases* the net inductance to a value larger than the above-described formulas would indicate. If the magnetic fields of multiple inductors work against each other, then the mutual inductance *reduces* the net inductance to a value smaller than the above formulas would indicate.

PROBLEM 10-1

Consider three inductors connected in series with no mutual inductance. Suppose that they have inductances of 1.50 mH, 150 μH, and 120 μH. What is the net inductance of the combination?

✔ SOLUTION

We must start by converting all of the inductances to the same unit; then we can add them up arithmetically. Let's use millihenrys (mH). We must multiply the second and third values by 0.001 (10^{-3}) to convert from microhenrys to millihenrys. Therefore, the net series inductance equals

$$L = (1.50 + 0.150 + 0.120) \text{ mH}$$
$$= 1.77 \text{ mH}$$

PROBLEM 10-2

What is the total inductance of the same three inductors connected in parallel, still assuming no mutual inductance?

✔SOLUTION

First, we convert all the inductances to the same units. Let's use millihenrys again. Then we take the reciprocals of these numbers. The value of the first inductance is 1.50 mH, so the reciprocal of this equals 0.667. Similarly, the reciprocals of the second and third inductances are 6.667 and 8.333. We can add up these "reciprocal inductances" to get the reciprocal of the net parallel inductance

$$L^{-1} = 0.667 + 6.667 + 8.333$$
$$= 15.667$$

Finally, we take the reciprocal of L^{-1} to obtain the net parallel inductance as

$$L = 1/15.667$$
$$= 0.0638 \text{ mH}$$

If you don't like values expressed in small fractions of units, you can convert this result to 63.8 μH.

Inductive Reactance

We can express DC resistance as a number ranging from zero (a perfect conductor) to extremely large values, increasing without limit through thousands, millions, and billions of ohms. We call resistance a *scalar quantity*, because we can render it on a one-dimensional *scale* like a number line. In fact, we can represent all possible values of DC resistance along a single straight *half-line* (also called a *ray*).

Given a certain DC voltage, the current decreases as the resistance increases, in accordance with Ohm's law. The same rule holds for pure AC through a pure resistance, as long as we specify both the voltage and the current in terms of the peak values, the peak-to-peak values, or the root-mean-square values. Once we place an inductance in an AC circuit, however, the DC version of Ohm's law no longer works.

Inductors and DC

Suppose that you have some wire that conducts electricity very well. If you wind a length of the wire into a coil and connect it to a battery, the wire draws a small amount of current at first, but the current quickly becomes large, no matter how you configure the wire. You might wind it into a single-turn loop, let it lie haphazardly on the floor, or wrap it around a wooden stick. In any case, you'll get a large current equal to

$$I = E/R$$

where I represents the current (in amperes), E represents the DC battery voltage (in volts), and R represents the DC resistance of the wire (in ohms).

You can make an electromagnet by passing DC through a coil wound around an iron or steel rod. You'll still observe a large, constant current in the coil. In a practical electromagnet, the coil heats up as some of the electrical energy dissipates in the wire; not all of the electrical energy contributes to the magnetic field. If you increase the battery voltage and also increase the ability of the battery to produce large currents, the wire in the coil will heat up more. Ultimately, if you increase the battery voltage enough, and if the battery can deliver unlimited current, the wire will heat to the melting point.

Inductors and AC

Now suppose that you change the voltage source across the coil from DC to pure AC (that is, AC having no DC component). Imagine that you can vary the frequency of the AC from a few hertz to hundreds of hertz, then kilohertz, then megahertz.

At low AC frequencies, you'll get a lot of current in the coil, just as you did with the battery. But the coil exhibits a certain amount of inductance, and it takes a little time for current to establish itself in the coil. Depending on how many turns the coil has, and on whether the core consists of air or a ferromagnetic material, you'll reach a point, as you steadily increase the AC frequency, when the coil starts to get "sluggish." The current won't have time to fully establish itself in the coil before the AC polarity reverses.

At sufficiently high AC frequencies, the current through a coil has difficulty following the changes in the instantaneous voltage across it. Just as the coil starts to "think" that it can act a short circuit, the AC voltage wave passes its peak, goes back to zero, and then "tries" to pull the current the other way. In effect, this "sluggishness" behaves a lot like DC resistance. As you raise the AC frequency, the coil's opposition to current increases. Eventually, if you keep on increasing the frequency, the coil will fail to acquire a significant current flow before the voltage polarity reverses. The coil will then act like a large-value

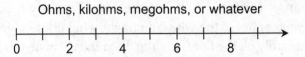

FIGURE 10-3 • We can portray inductive reactance on a half-line starting at zero and going in the positive direction.

resistor. With respect to AC, an inductor functions as a sort of frequency-dependent resistor.

The opposition that the coil offers to AC goes by the rather arcane name of *inductive reactance*. We express or measure inductive reactance in ohms, just as we do with resistance. Inductive reactance can vary as resistance does, from near zero (a short piece of wire) to a few ohms (a small coil) to kilohms or megohms (bigger and bigger coils, or coils with ferromagnetic cores operating at high AC frequencies). We can portray inductive reactance values along a half-line (Fig. 10-3), just as we can do with resistance. The numerical values on the half-line start at zero and increase positively without limit.

Inductive Reactance and Frequency

Inductive reactance constitutes one of two forms of reactance. (We'll examine the other form later in this chapter.) In mathematical expressions, we symbolize reactance as X, and we symbolize inductive reactance as X_L.

If the frequency of an AC source equals f (in hertz) and the inductance of a coil equals L (in henrys), then we can calculate the inductive reactance X_L (in ohms) using the formula

$$X_L = 2\pi f L$$

$$\approx 6.2832\, fL$$

This same formula applies if we specify the frequency f in kilohertz and the inductance L in millihenrys. It also applies if we express f in megahertz and L in microhenrys. If we quantify frequency in thousands, we must quantify inductance in thousandths; if we quantify frequency in millions, we must quantify inductance in millionths.

? Still Struggling

Inductive reactance increases linearly with increasing AC frequency, so the function of X_L versus f shows up as a straight line when we plot its graph on a rectangular coordinate plane. Inductive reactance also increases linearly with inductance, so the function of X_L versus L also appears as a straight line on a rectangular graph. Summarizing:

- If we hold L constant, then X_L varies in direct proportion to f.
- If we hold f constant, then X_L varies in direct proportion to L.

Figure 10-4 illustrates these relations graphically.

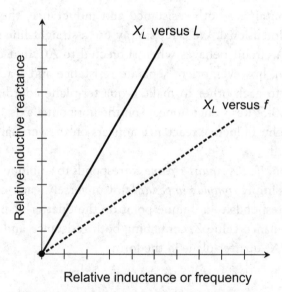

FIGURE 10-4 • Inductive reactance X_L varies in direct proportion to the inductance L, and also in direct proportion to the AC frequency f.

PROBLEM **10-3**

An inductor has a value of 10.00 mH. What's its inductive reactance at an AC frequency of 100.0 kHz?

✓ SOLUTION

We have units of millihenrys (thousandths of henrys) and kilohertz (thousands of hertz), so we can apply the above formula directly. Using 6.2832 to represent an approximation of 2π, we obtain

$$X_L = 6.2832\ fL$$
$$= 6.2832 \times 100.0 \times 10.00$$
$$= 6283.2\ \text{ohms}$$

Because our input data extends to only four significant figures, we should round this result off to 6283 ohms.

The Resistance versus Inductive Reactance (RX_L) Quarter-Plane

In a circuit containing both resistance and inductance, the characteristics become two-dimensional. We can't simply use a straight-line scale to portray the way such a circuit behaves when subjected to AC that can vary in frequency. We can, however, orient separate resistance and reactance half lines perpendicular to each other to make a quarter-plane coordinate system, as shown in Fig. 10-5. Resistance appears on the horizontal axis, increasing as we move to the right. Inductive reactance appears on the vertical axis, increasing as we go upward.

Each point on the RX_L *quarter-plane* corresponds to a unique *complex-number impedance* (or simply *complex impedance*). Conversely, each complex impedance value corresponds to a unique point on the quarter-plane. We express a complete impedance value Z, containing both resistance and inductive reactance, on the RX_L quarter-plane in the form

$$Z = R + jX_L$$

where R represents the resistance (in ohms), X_L represents the inductive reactance (also in ohms), and j equals the unit imaginary number, that is, the positive square root of -1. Electrical engineers sometimes call the positive square root of -1 the *j operator*. (If you're uncomfortable with imaginary and complex numbers, you might need to review your pre-calculus or intermediate algebra text.)

Suppose that we have a pure resistance, say $R = 5$ ohms. In this case the complex impedance equals $Z = 5 + j0$. We can plot it at the point $(5, j0)$ on the RX_L quarter-plane. If we have a pure inductive reactance such as $X_L = 3$ ohms,

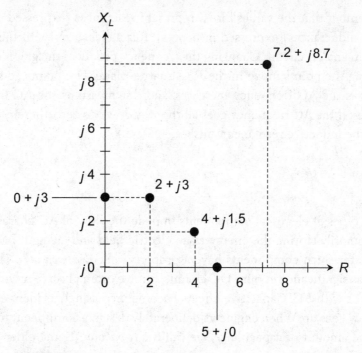

FIGURE 10-5 • The resistance versus inductive reactance (RX_L) quarter-plane, showing some specific impedances.

then the complex impedance equals $Z = 0 + j3$, and its point belongs at $(0, j3)$ on the RX_L quarter-plane. Engineers sometimes incorporate both resistance and inductive reactance into electronic circuit designs. Then we encounter complex impedance values such as the following:

$$Z = 2 + j3$$
$$Z = 4 + j1.5$$
$$Z = 7.2 + j8.7$$

Figure 10-5 shows graphical representations of the five complex impedances mentioned in this paragraph.

All practical coils have some resistance, because no real-world wire conducts current perfectly. All resistors have a tiny bit of inductive reactance; all electrical components have wire leads at each end, and any length of wire (even a straight one) exhibits some inductance. Therefore, in an AC circuit, we will never encounter a *mathematically perfect* pure resistance such as $5 + j0$, or a *mathematically perfect* pure reactance such as $0 + j3$. We can approach these ideals, but we can never actually attain them (except maybe in quiz and test problems!).

Remember that the values for X_L represent reactances (expressed in ohms), and not inductances (expressed in henrys). Reactances vary with the AC frequency in an RX_L circuit. Changing the frequency produces the graphical effect of making the points move in the RX_L quarter-plane. The points go vertically upward as the AC frequency increases, and downward as the AC frequency decreases. If the AC frequency goes all the way down to zero, thereby resulting in DC, the inductive reactance vanishes.

Capacitance

Certain types of electrical components impede the flow of AC charge carriers by temporarily storing the energy as an electric field and giving it back to the circuit later. Such components have a property called *capacitance*. This effect rarely has significance in pure DC circuits, but we often observe it with pulsating DC or with AC. Capacitive effects become increasingly evident as the AC frequency goes up. When engineers deliberately design a component to provide a specific amount of capacitance, we call it a *capacitor*. (Some older texts use the term *condenser*.)

The Property of Capacitance

Imagine two huge, flat sheets of metal, both excellent electrical conductors. Suppose that these sheets have surface areas equivalent to the land area of the state of Nebraska. We place one sheet exactly over the other, and separate them uniformly by a few centimeters. Now suppose that we connect these metal sheets to a battery, the upper sheet going to the positive terminal and the lower sheet going to the negative terminal. The sheets will acquire electric charges: the top sheet positive and the bottom sheet negative.

The charging-up process won't occur all at once. It will take a little while because the sheets span gigantic surface areas. If the plates were small, they would both charge up almost instantly, attaining a potential difference equal to the voltage of the battery. But because the plates cover so much area, it takes some time for the negative plate to "fill up" with electrons from the battery, and it takes roughly the same amount of time for the positive plate to get electrons "pulled off" into the battery.

Over time, the potential difference between the two plates rises until it equals the battery voltage. An electric field forms in the space between the plates. This field remains small at first; the plates don't charge right away. But the size of the field increases over a period of time, depending on the actual sizes of the sheets, and also depending on the spacing between them. This electric field stores energy. We express capacitance in terms of the ability of the plates, and of the space between them, to store energy in the form of an electric field. In formulas and equations, we symbolize capacitance by writing the uppercase italic letter C.

Practical Capacitors

Obviously, we can't construct a capacitor from metal plates the size of Nebraska. However, we can place two foil sheets or strips in close proximity and separate them with a thin, nonconducting sheet of material such as paper. Then we can roll the whole assembly up to obtain a large effective surface area. Such a device exhibits significant capacitance. We can increase the capacitance further by stacking up several plates, separating them with thin nonconducting layers, and connecting alternate pairs of plates together to make a two-terminal "meshed-sheet" component.

In a capacitor, the electric flux concentration increases when we place a *dielectric material* of a certain type between the plates. The dielectric increases the effective surface area of the plates, so we can make a physically small component exhibit a lot of capacitance. In general, the capacitance between two conducting plates or sheets increases if we do any of the following things while holding all other factors constant:

- Increase the total surface area of the sheets, as long as they overlap precisely.
- Decrease the separation distance between the sheets.
- Increase the *dielectric constant* of the material separating the sheets.

Dielectric constant constitutes the electrostatic equivalent of magnetic permeability. A vacuum has a dielectric constant of 1. Dry air has a dielectric constant slightly higher than that of a vacuum, but the difference rarely amounts to anything worth bothering about in practical situations. Some substances have high dielectric constants that multiply the effective capacitance many times. Others have low dielectric constants, serving primarily to hold the plates of a capacitor at a uniform physical distance from each other.

In theory, if the dielectric constant of a material equals x, then placing that material between the plates of a capacitor will increase the capacitance by a factor of x compared with the capacitance when a vacuum exists between the plates. In practice, this rule holds true only for *ideal dielectrics*. An ideal dielectric does not turn any of the energy contained in the electric field into heat. In addition, the dielectric rule holds true only if all the electric lines of flux between the plates pass through the dielectric material alone, and not through anything else. While we can never attain such perfect scenarios, we can come close.

The Unit of Capacitance

When we connect a source of DC such as a battery between the plates of a capacitor, it takes some time before the electric field reaches its full intensity. The voltage builds up at a rate that depends on the capacitance. The greater the capacitance, the slower the rate of change of voltage in the plates.

The unit of capacitance expresses the ratio between the amount of current flowing and the rate of voltage change across the plates of a capacitor. A capacitance of 1 *farad*, abbreviated F, represents a current flow of 1 ampere (1 A) while the potential difference increases or decreases at the rate of 1 volt per second (1 V/s). In other words, 1 farad equals 1 ampere per (volt per second), which translates to 1 ampere-second per volt (A · s/V or A · s · V^{-1}). A capacitance of 1 F also accumulates 1 C of charge quantity when we apply 1 V to the plates. Therefore, in effect, one farad equals 1 coulomb per volt (C/V or C · V^{-1}).

The farad represents a huge unit of capacitance. You'll almost never see a real-world capacitor with a value of 1 F. Engineers commonly express capacitance values in terms of the *microfarad* (μF) and the *picofarad* (pF), such that:

$$1 \ \mu F = 0.000001 \ F = 10^{-6} \ F$$
$$1 \ pF = 0.000001 \ \mu F = 10^{-12} \ F$$

Interelectrode Capacitance

We rarely observe significant mutual interaction between well-designed capacitors. However, at very high AC frequencies, *interelectrode capacitance* can sometimes pose a problem for engineers. This effect shows up as a tiny capacitance between wires or other electrical conductors in close proximity.

Capacitors in Series

When we connect two or more capacitors in series, their values combine just as resistances combine in parallel, assuming that no mutual capacitance exists among the components. If we connect two capacitors of the same value in series, the net capacitance equals half the capacitance of either component alone. In general, if we have several capacitors connected in series, we observe a net capacitance smaller than that of any of the individual components. As with resistances and inductances, we should always use the same size units when we calculate the net capacitance of any combination.

Consider several capacitors with values C_1, C_2, C_3, ..., C_n connected in series as shown in Fig. 10-6. We can find the net capacitance C using the formula

$$C = 1/(1/C_1 + 1/C_2 + 1/C_3 + \ldots + 1/C_n)$$

$$= (1/C_1 + 1/C_2 + 1/C_3 + \ldots + 1/C_n)^{-1}$$

Still Struggling

Do you wonder why the capacitor symbols in Fig. 10-6 (and everywhere else where they appear in this book) consist of one straight line and one curved line? Engineers commonly use this notation because, in many situations, one end of the capacitor (with the curved line) either connects directly to, or faces toward, a *common ground* point (a neutral point, with a reference voltage of zero). In the circuit of Fig. 10-6, no common ground exists, so it doesn't matter which way the capacitors go. In some older texts, you'll see capacitor symbols with two parallel, straight lines, rather than one straight line and one curved line. You need not pay any attention to those little curved lines in this course.

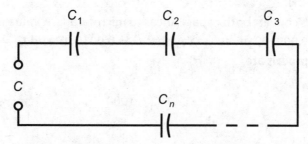

FIGURE 10-6 • Capacitances in series add like resistances in parallel.

PROBLEM 10-4

Suppose that two capacitances, $C_1 = 0.10\ \mu F$ and $C_2 = 0.050\ \mu F$, appear in series. What is the net capacitance?

✔ SOLUTION

Let's use microfarads as the unit for our calculations. Using the above formula, we first find the reciprocals of the individual capacitances, getting

$$1/C_1 = 10$$

and

$$1/C_2 = 20$$

We add these numbers to obtain the reciprocal of the net series capacitance

$$C^{-1} = 10 + 20$$
$$= 30$$

Finally, we take the reciprocal of C^{-1} to obtain

$$C = 1/30$$
$$= 0.033\ \mu F$$

PROBLEM 10-5

Imagine that we connect two capacitors with values of $0.0010\ \mu F$ and $100\ pF$ in series. What's the net capacitance?

✔ SOLUTION

First, let's convert both capacitances to microfarads. A value of 100 pF represents $0.000100\ \mu F$, so we say that $C_1 = 0.0010\ \mu F$ and $C_2 = 0.000100\ \mu F$. The reciprocals are

$$1/C_1 = 1000$$

and

$$1/C_2 = 10,000$$

Now we can calculate the reciprocal of the series capacitance as

$$C^{-1} = 1000 + 10,000$$
$$= 11,000$$

Therefore

$$C = 1/11,000$$
$$= 0.000091 \ \mu F$$

We do better to state this capacitance as 91 pF.

TIP   *You can solve Problem 10-5 by choosing picofarads to work with, rather than microfarads. In either case, you must use caution when you place the decimal points. Always double-check your arithmetic when things get "messy" like this. Calculators will take care of the decimal placement problem, sometimes using exponent notation and sometimes not, but a calculator can only work with what you put into it. If you enter a wrong number, you'll get a wrong answer, and if you miss a digit, you'll be off by a factor of 10 (an order of magnitude).*

Capacitors in Parallel

Capacitances in parallel add like resistances in series. The net capacitance equals the sum of the individual component values, as long as we stick to the same units all the way through our calculations.

Suppose that we connect capacitors C_1, C_2, C_3, ..., C_n in parallel, as shown in Fig. 10-7. As long as we observe no mutual capacitance between the components, we can calculate the net capacitance C with the formula

$$C = C_1 + C_2 + C_3 + ... + C_n$$

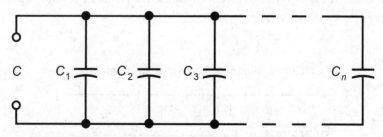

FIGURE 10-7 • Capacitances in parallel add like resistances in series.

PROBLEM 10-6

Imagine three capacitors connected in parallel, having values of $C_1 = 0.100$ μF, $C_2 = 0.0100$ μF, and $C_3 = 0.00100$ μF. What's the net parallel capacitance?

✔ SOLUTION

We simply add the values up, because all of the capacitances are expressed in the same size units (microfarads). We have

$$C = 0.100 + 0.0100 + 0.00100$$

$$= 0.11100 \text{ μF}$$

We can round off this result to $C = 0.111$ μF.

Capacitive Reactance

Inductive reactance "meets" its counterpart in the form of *capacitive reactance*. We can portray capacitive reactance graphically as a half-line or ray, starting at the same zero point as inductive reactance, but running off in the opposite direction, having negative ohmic values as shown in Fig. 10-8. When we combine the ray for capacitive reactance with the ray for inductive reactance, we get a complete real-number line with ohmic values ranging from huge negative numbers, through zero, to huge positive numbers.

Capacitors and DC

Imagine two large, parallel metal plates, as described earlier. If we connect them to a source of DC, they draw a large amount of current at first, as they become electrically charged. But as the plates reach equilibrium, this current diminishes, and when the two plates attain the same potential difference throughout, the current goes down to zero.

Ohms, kilohms, megohms, or whatever

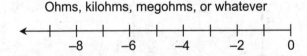

FIGURE 10-8 • We can portray capacitive reactance on a half-line starting at zero and going in the negative direction.

If we increase the voltage of the battery or power supply, we eventually reach a point at which sparks jump between the plates of our capacitor. Ultimately, if the power supply can deliver the necessary voltage, this sparking, or *arcing*, becomes continuous. Under these conditions, the pair of plates no longer acts like a capacitor at all. When we place excessive voltage across a capacitor, the dielectric no longer provides electrical separation between the plates. We call this undesirable condition *dielectric breakdown*.

In an air-dielectric or vacuum-dielectric capacitor, dielectric breakdown manifests itself as a temporary affair, rarely causing permanent damage to the component. The device operates normally after we reduce the voltage so that the arcing stops. However, in capacitors made with solid dielectric materials such as mica, paper, polystyrene, or tantalum, dielectric breakdown can burn or crack the dielectric, causing the component to conduct current even after we reduce the voltage back down below the arcing threshold. If a capacitor suffers this sort of damage, we must remove it from the circuit, discard it, and replace it with a new one.

Capacitors and AC

Suppose that we change the power source connected to a capacitor from DC to AC. Imagine that we can adjust the frequency of this AC from a low initial value of a few hertz, up to hundreds of hertz, then to many kilohertz, and finally to many megahertz or gigahertz.

At first, the voltage between the plates follows along with the voltage of the power source as the AC polarity alternates. But the set of plates has a certain amount of capacitance. They can charge up quickly if they have small surface areas and/or if a lot of space exists between them, but they can't charge instantaneously. As we increase the frequency of the applied AC, we eventually reach a point at which the plates can't charge up very much before the AC polarity reverses. Just as the plates begin to get a good charge, the AC passes its peak and starts to discharge them. As we raise the frequency still further, the set of plates acts increasingly like a short circuit. Eventually, if we keep raising the AC frequency, the period of the wave becomes much shorter than the charge/discharge time, and current flows in and out of the plates just as fast as it would if the plates were directly shorted, or removed altogether and replaced with a piece of wire.

Capacitive reactance quantifies the opposition that a capacitor offers to AC. We express and measure capacitive reactance in ohms, just as we do with inductive reactance or pure resistance. But capacitive reactance, by convention, has negative values rather than positive ones. Capacitive reactance, denoted X_C in mathematical formulas, can vary from near zero (when the plates are huge and close together, and/or the frequency is very high) to a few negative ohms, to many negative kilohms or megohms.

Capacitive reactance varies with frequency. It gets *larger negatively* as the frequency goes down, and smaller negatively as the applied AC frequency goes up. This behavior runs contrary to what happens with inductive reactance, which gets *larger positively* as the frequency goes up. Sometimes, nontechnical people talk about capacitive reactance in terms of its *absolute value*, with the minus sign removed. Then we might say that X_C increases as the frequency goes down, or that X_C decreases as the frequency goes up. However, you should learn to work with negative X_C values and stick with that convention. That way, the mathematics will give you the most accurate representation of how AC circuits behave when they contain capacitance.

Capacitive Reactance and Frequency

In a purely mathematical sense, capacitive reactance "mirrors" inductive reactance. In another sense, X_C constitutes a continuation of X_L into negative values, something like the extensions of the Celsius or Fahrenheit temperature scales to values "below zero."

If we specify the frequency of an AC source (in hertz) as f, and if we specify the capacitance of a component (in farads) as C, then we can calculate the capacitive reactance using the formula

$$X_C = -1/(2\pi f C)$$

$$= -(2\pi f C)^{-1}$$

$$\approx -(6.2832 f C)^{-1}$$

This formula will also work if we input f in megahertz and C in microfarads (μF). It would even apply for values of f in kilohertz (kHz) and values of C in *millifarads* (mF)—but for some reason, you'll almost never encounter capacitances expressed in millifarads. Even the millifarad constitutes a large unit for capacitance. You'll rarely see any component that has a capacitance of more than 1000 μF (which would equal 1 mF) in a real-world electrical system.

Still Struggling

The function of X_c versus f appears as a curve when we graph it in rectangular coordinates. The curve contains a *singularity* at $f = 0$; it "blows up negatively" as the frequency approaches zero. The function of X_c versus C also appears as a curve that attains a singularity at $C = 0$; it "blows up negatively" as the capacitance approaches zero. Summarizing:

- If we hold C constant, then X_c varies inversely with the negative of f.
- If we hold f constant, then X_c varies inversely with the negative of C.

Figure 10-9 illustrates these relations as graphs on a rectangular coordinate plane.

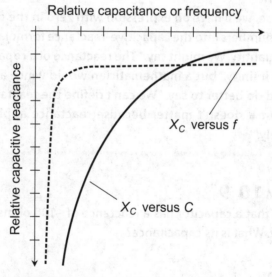

FIGURE 10-9 • Capacitive reactance X_c varies inversely with the negative of the capacitance C, and also inversely with the negative of the AC frequency f.

PROBLEM 10-7

A capacitor has a value of 0.00100 μF at a frequency of 1.00 MHz. What is the capacitive reactance?

✓ **SOLUTION**

We can apply the foregoing formula directly because we know the input data in microfarads (millionths) and in megahertz (millions):

$$X_c = -1/(6.2832 \times 1.00 \times 0.00100)$$
$$= -1/(0.0062832)$$
$$= -159 \text{ ohms}$$

PROBLEM 10-8

What will happen to the capacitive reactance of the above capacitor if the frequency decreases to zero, so that the power source provides DC rather than AC?

✓ **SOLUTION**

In this case, we'll have an expression with zero in the denominator if we plug the numbers into the capacitive-reactance formula, yielding a meaningless quantity. We might say, "The reactance of a capacitor at DC equals negative infinity," but a mathematician would wince at a statement like that. We'd do better to say, "We can't define the reactance of a capacitor at DC, but it doesn't matter because reactance applies to AC circuits exclusively."

PROBLEM 10-9

Suppose that a capacitor has a reactance of –100 ohms at a frequency of 10.0 MHz. What is its capacitance?

✓ **SOLUTION**

In this problem, we need to put the numbers in the formula and then use algebra to solve for the unknown C. Let's start with the equation

$$-100 = -(6.2832 \times 10.0 \times C)^{-1}$$

Dividing through by –100, we obtain

$$1 = (628.32 \times 10.0 \times C)^{-1}$$

When we multiply through by *C*, we get

$$C = (628.32 \times 10.0)^{-1}$$

$$= 6283.2^{-1}$$

$$= 0.00015915$$

which rounds to $C = 0.000159$ µF. Because we input the frequency value in megahertz, this capacitance comes out in microfarads. We might also say that $C = 159$ pF, remembering that 1 pF = 0.000001 µF.

TIP *The arithmetic for dealing with capacitive reactance can give you trouble if you're not careful. You have to work with reciprocals, so the numbers can get awkward. Also, you have to watch those negative signs. You can easily forget to include a minus sign when you should (I've done that more than once), and you might insert a negative sign when you shouldn't. The signs are critical when you want to draw graphs describing systems containing reactance. A minus sign tells you that you're working with capacitive reactance rather than inductive reactance.*

The Resistance versus Capacitive Reactance (RX_C) Quarter-Plane

In a circuit containing resistance and capacitive reactance, the characteristics work in two dimensions, in a way that "mirrors" the situation with the RX_L quarter-plane. We can place resistance and capacitive-reactance half-lines end-to-end at right angles to construct an RX_C *quarter-plane* as shown in Fig. 10-10. We plot resistance values horizontally, with increasing values toward the right. We plot capacitive reactance values vertically, with increasingly negative values as we move downward.

We can denote complex impedances Z containing both resistance and capacitance in the form

$$Z = R + jX_C$$

keeping in mind that values of X_C never go into positive territory. If we have a pure resistance, say $R = 3$ ohms, then the complex-number impedance equals $Z = 3 + j0$, which corresponds to the point $(3, j0)$ on the RX_C quarter-plane. If we have a pure capacitive reactance, say $X_C = -4$ ohms, then the complex impedance equals $Z = 0 + j(-4)$, which we can write more simply as $Z = 0 - j4$

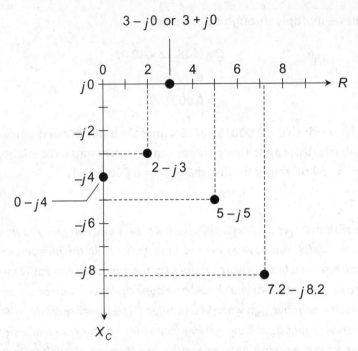

FIGURE 10-10 • The resistance versus capacitive reactance (RX_C) quarter-plane, showing some specific impedances.

and plot at the point $(0, -j4)$ on the RX_C quarter-plane. The points representing $Z = 3 + j0$ (which we can also express as $Z = 3 - j0$ because the two values are identical) and $Z = 0 - j4$, along with three others, appear on the RX_C quarter-plane in Fig. 10-10.

In practical circuits, all capacitors exhibit some *leakage conductance*. If the frequency goes to zero, that is, if the source produces DC, a tiny current will flow because no real-world dielectric material constitutes a perfect electrical insulator (not even a vacuum!). Some capacitors have almost no leakage conductance, but none are completely free of it. Conversely, all electrical conductors have a little capacitive reactance, simply because they occupy physical space, so we'll never see a mathematically pure conductor of AC, either. Therefore, the impedances $Z = 3 - j0$ and $Z = 0 - j4$ both represent theoretical idealizations.

Remember that the values for X_C indicate reactance values, not capacitance values. Reactance varies with the frequency in an RX_C circuit. If we raise or lower the AC frequency that we apply to a particular capacitor, the value of X_C changes. Increasing the AC frequency causes X_C to get *smaller negatively* (closer to zero). Reducing the AC frequency causes X_C to get *larger negatively* (farther from zero, or lower down on the RX_C quarter-plane). If the frequency goes all the way down to zero, then the capacitive reactance drops off the bottom of the plane, out of sight, and loses meaning. In that case we have two plates or sets of plates holding opposite electrical charges, but no "action"—unless or until we discharge the component.

Capacitors and DC Revisited

If the plates of a practical capacitor have large surface areas, are spaced close together, and are separated by a good solid dielectric, we will experience a sudden, dramatic bit of "action" when we discharge the component. A massive capacitor can hold enough charge to electrocute an unsuspecting person who comes into contact with its terminals. The well-known scientist and U.S. statesman *Benjamin Franklin* wrote about an experience of this sort with a "home-brewed" capacitor called a *Leyden jar*, which he constructed by placing metal foil sheets inside and outside a glass bottle and then connecting a high-voltage battery to them for a short while. After removing the battery, Franklin came into contact with both foil sheets at the same time and described the consequent shock as a "blow" that knocked him to the floor. Luckily for himself and the world, he survived.

Graphs of Complex Impedance

If we consider capacitive and inductive reactance as "different sides of the same coin," we can plot them together along a single, complete *imaginary-number line*. The two half-lines intersect at their back-end points, both of which represent zero reactance. If we orient this line vertically and then add a horizontal half-line to portray resistance, putting the back-end point of the resistance half-line exactly where the back-end points of the reactance half-lines intersect, we obtain a complete, working graphical scheme for portraying complex impedances.

The *RX* Half-Plane

Recall the quarter-plane for resistance R and inductive reactance X_L from the preceding sections. This quarter-plane corresponds to the upper-right quadrant of the complex-number coordinate plane, which you learned about in your pre-calculus or analytic geometry courses. Similarly, the quarter-plane for resistance R and capacitive reactance X_C coincides with the lower-right quadrant of the complex number plane. We represent resistance values as nonnegative real numbers. Reactances, whether they are inductive (positive) or capacitive (negative), correspond to imaginary numbers.

"What," you might ask, "about negative values for resistance? Why don't we include them?" In a literal sense, a passive electrical component (one that does not produce any energy of its own) can't have true negative resistance. We'll never encounter any material that exhibits less than no ohmic loss, or that conducts better than perfectly! In some circuits, a source of DC, such as a battery, can act as a "negative resistance" of sorts. In other cases, a component can draw decreasing current, rather than increasing current, as we increase the potential difference across it; some engineers refer to this property as "negative resistance." But in the *RX* (resistance-reactance) *half-plane* (Fig. 10-11), the resistance never attains any value less than zero.

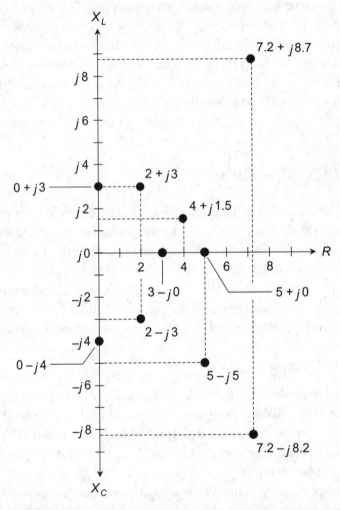

FIGURE 10-11 • The complete resistance versus reactance (*RX*) half-plane, showing some specific impedances.

Reactance in Review

Capacitive reactance extends inductive reactance, X_L, into the realm of negatives, in a way that cannot generally occur with resistance. Capacitors act like "negative inductors." Interesting things happen when we combine capacitors and inductors in a single circuit and then apply AC at various frequencies.

Reactance can vary from extremely large negative values, through zero, to extremely large positive values. Engineers and physicists always quantify reactances as imaginary numbers. In the mathematical model of complex impedance, capacitances and inductances manifest themselves "perpendicularly" to resistance. Alternating-current reactance occupies a different, and independent, dimension from DC resistance. The general symbol for reactance is X; we add subscripts to denote inductive reactance (X_L) and capacitive reactance (X_C).

Vector Representation of Impedance

We can portray any specific complex impedance Z as a unique complex number of the form

$$Z = R + jX$$

where R can attain any non-negative real-number value and jX can attain any imaginary-number value. We can therefore always plot a known complex impedance as a point in the RX half-plane, or as a vector with its back-end point at the coordinate origin $(0 + j0)$. We call such vectors *complex impedance vectors*, or simply *impedance vectors*.

Imagine how an impedance vector changes as we vary either R or X, or both. For example, any of the following three events will cause the impedance vector to grow longer:

- The reactance X remains constant while the resistance R increases.
- The resistance R remains constant while the inductive reactance X_L increases.
- The resistance R remains constant while the capacitive reactance X_C increases negatively.

Think of the point representing $R + jX$ moving around in the RX half-plane of Fig. 10-11, and imagine where the corresponding points on the resistance and reactance axes lie. You can locate specific points by drawing straight lines from the point $R + jX$ to the R and X axes, so the lines intersect the axes at right angles. Figure 10-11 shows several sample points and their complex-impedance values.

Now think of the points for R and X moving toward the right and left, or up and down, on their axes. Imagine what happens to the point $R + jX$, and the corresponding vector that "runs" or "points" from the coordinate origin to $R + jX$, in various scenarios.

Absolute-Value Impedance

Sometimes you'll read or hear that a device or component has an "impedance" of a certain number of "ohms." For example, in audio amplifiers, you might find "8-ohm" speakers and "600-ohm" input terminals. How can manufacturers quote a single number for a quantity that needs two numbers for its complete expression? We can answer this question in two parts.

First, single-number "impedance" specifications generally refer to devices that do not exhibit any reactance under their recommended operating conditions. Therefore, an "8-ohm" speaker really has a complex impedance of $8 + j0$, and a "600-ohm" input circuit has been designed to operate with a complex impedance at, or near, $600 + j0$.

Second, engineers sometimes talk about the length or magnitude of the impedance vector, calling this length a certain number of "ohms." If we talk about "impedance" this way, then theoretically we commit the sin of ambiguity, because infinitely many different vectors can have the same length in the RX half-plane.

The expression "$Z = 8$ ohms" can theoretically refer to any vector in the RX half-plane that measures eight units long (Fig. 10-12). So what happens when we apply audio-frequency AC signals to a real-world speaker whose "impedance" is supposedly "8 ohms?" That's a problem for the engineers who design speakers. If you have much experience with hi-fi audio systems, you know that speaker design and manufacture constitutes a fine art.

PROBLEM 10-10

Name seven different complex impedances having an absolute value of 10 ohms.

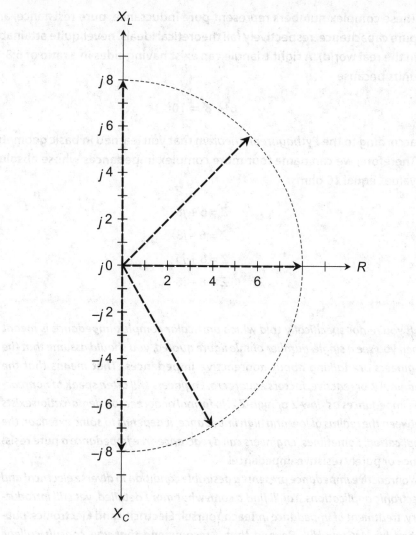

FIGURE 10-12 • All of these vectors represent the same absolute-value impedance, even though their complex components differ.

✔ SOLUTION

We can easily name three complex impedances such that the absolute-value impedance equals 10 ohms:

$$Z = 0 + j10$$

$$Z = 10 + j0$$

$$Z = 0 - j10$$

These complex numbers represent pure inductance, pure resistance, and pure capacitance, respectively (all theoretical ideals, never quite attainable in the real world). A right triangle can exist having sides in a ratio of 6:8:10 units because

$$6^2 + 8^2 = 10^2$$

according to the *Pythagorean theorem* that you learned in basic geometry. Therefore, we can name four more complex impedances whose absolute values equal 10 ohms:

$$Z = 6 + j8$$
$$Z = 6 - j8$$
$$Z = 8 + j6$$
$$Z = 8 - j6$$

TIP *If you're not specifically told which particular complex impedance is meant when you see a single-number ohmic figure quoted, you should assume that the engineers are talking about* nonreactive *impedances. That means that the imaginary, or reactive, factors equal zero. Engineers will often speak of nonreactive impedances as "low-Z or high-Z." No formal or agreed-on demarcation exists between the realms of low and high impedance; it depends to some extent on the application. Sometimes, engineers call a reactance-free impedance a* pure resistance *or purely resistive impedance.*

Nonreactive impedance presents a desirable condition in diverse electrical and electronic applications. You'll find a somewhat more detailed, yet still introductory, treatment of impedance in Teach Yourself Electricity and Electronics, *published by McGraw-Hill. Beyond that, I recommend that you consult college textbooks in electrical, electronics, and telecommunications engineering. You might also consider becoming an amateur radio operator and playing around with antenna systems! You'll get a lot of first-hand (and fun) experience with complex impedance that way.*

QUIZ

Refer to the text in this chapter if necessary. A good score is eight correct. Answers are at the back of the book.

1. **Regardless of the AC frequency, the complex-impedance vectors for a pure inductance of 20 mH and a pure capacitance of 100 μF**
   A. point in the same direction.
   B. point in opposite directions.
   C. point at right angles to each other.
   D. have equal length.

2. **What is the complex-number impedance of a 40-mH pure inductance?**
   A. $0 + j(4.0 \times 10^{-5})$
   B. $4.0 \times 10^{-5} + j0$
   C. $4.0 \times 10^{-5} + j(4.0 \times 10^{-5})$
   D. We need more information to determine it.

3. **Suppose that we connect four capacitors of identical capacitance together as shown in Fig. 10-13. The net capacitance C equals**
   A. half the capacitance of any one of the components individually.
   B. the capacitance of any one of the components individually.
   C. twice the capacitance of any one of the components individually.
   D. four times the capacitance of any one of the components individually.

4. **Imagine that we apply an AC signal having a frequency of exactly 5 MHz to the combination of capacitors shown in Fig. 10-13. Then we double the frequency to exactly 10 MHz. What happens to the capacitive reactance X_C of the entire circuit?**
   A. It becomes 1/4 of its previous value.
   B. It becomes 1/2 of its previous value.
   C. It stays the same.
   D. It doubles.

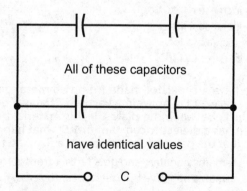

FIGURE 10-13 • Illustration for Quiz Questions 3 and 4.

5. Imagine that someone tells us (realistically or otherwise) that a certain electrical component exhibits a complex-number impedance of 50 + $j0$ at a frequency of 75 MHz. This person has just claimed that

   A. the component acts as a pure capacitive reactance at 75 MHz.
   B. the component acts as a pure inductive reactance at 75 MHz.
   C. the component acts as a pure resistance at 75 MHz.
   D. the person must be misinformed, because such a situation can never occur.

6. If we pass AC having an extremely low frequency through an inductor made from perfectly conducting wire, the reactance is extremely

   A. large in the negative sense.
   B. large in the positive sense.
   C. small in the positive sense.
   D. small in the negative sense.

7. If we pass AC having an extremely low frequency through a capacitor made from perfectly conducting plates separated by dry air, the reactance is extremely

   A. large in the negative sense.
   B. large in the positive sense.
   C. small in the positive sense.
   D. small in the negative sense.

8. As we decrease the frequency of an AC signal passing through a component having a fixed inductance, what happens?

   A. The component offers diminishing opposition to the AC.
   B. The component offers increasing opposition to the AC.
   C. The opposition to the AC does not change.
   D. It depends on the strength of the current in amperes.

9. Imagine that we have a coil wound on a plastic cylinder, so that the core comprises almost entirely air. Then we place a rod made of high-permeability ferromagnetic alloy inside the coil. How does this action affect the coil's inductive reactance X_L at a fixed AC frequency?

   A. It causes X_L to increase a lot.
   B. It causes X_L to increase a little.
   C. It causes X_L to decrease a lot.
   D. It causes X_L to decrease a little.

10. Imagine that we have a capacitor made from two metal disks, each having a radius of 75 cm, spaced 1 cm apart in a vacuum. Then we place a solid disk of material, 1 cm thick, between the plates. If this material behaves as an ideal dielectric, and if it has a dielectric constant of 1.5, what happens to the capacitive reactance X_C that the pair of disks exhibits at a fixed AC frequency?

    A. The value of X_C remains negative and gets farther from zero.
    B. The value of X_C remains negative but gets closer to zero.
    C. The value of X_C remains positive and increases.
    D. The value of XC remains positive but decreases.

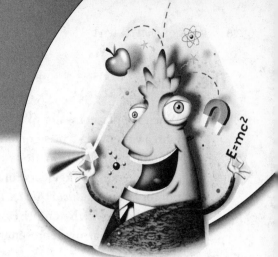

11

Semiconductors

The term *semiconductor* arises from the ability of certain materials to conduct electric currents "part time." A wide variety of elements and compounds, particularly solids, can function as semiconductors. Two primary types of semiconductors exist: *N type*, in which electrons constitute most of the charge carriers, and *P type*, in which electron absences called *holes* constitute most of the charge carriers.

CHAPTER OBJECTIVES

In this chapter, you will

- Analyze the behavior of a simple diode.
- Discover what diodes can do.
- See what happens inside a transistor.
- Learn how transistors can amplify AC signals.
- Find out why integrated circuits have revolutionized electronics.

The Diode

When we place a wafer of N-type semiconductor material in physical contact with a wafer of P-type semiconductor material, we obtain a *P-N junction* at the boundary. All P-N junctions exhibit special properties that electronics and communications engineers find useful. Figure 11-1 shows the schematic symbol for a *semiconductor diode*. The short, straight line in the symbol represents the *cathode*, which comprises a sample (sometimes called a *wafer*) of N-type material. The arrow represents the *anode*, which comprises a sample of P-type material.

The Ideal Diode

In an ideal semiconductor diode, electrons can flow freely from atom to atom in the direction opposite the arrow, but cannot easily flow in the direction that the arrow points. The P-N junction thereby acts as a "one-way current gate."

If we connect a battery and a resistor in series with a diode so that the anode goes toward the positive battery terminal, current flows (Fig. 11-2A). If we turn the battery around so that the anode goes toward the negative battery terminal, current does not flow (Fig. 11-2B). In the circuits shown by these two schematic diagrams, the resistor limits the maximum current that can pass through the diode. Without this so-called *current-limiting resistor* in the circuit, the diode P-N junction might literally burn up.

The above-described rules for semiconductor diodes hold true under most circumstances, and they represent ideal (perfect) scenarios. However, exceptions occur, one of them having to do with a phenomenon called *forward break-over*, and the other having to do with *avalanche effect*.

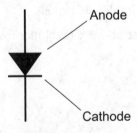

FIGURE 11-1 • The schematic symbol for a diode, showing the anode and cathode.

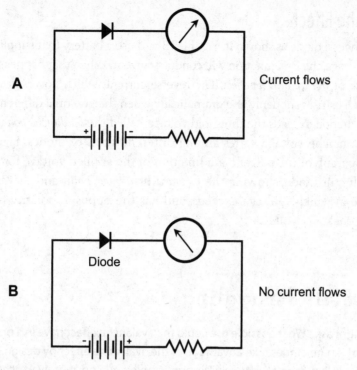

FIGURE 11-2 • At A, a positive anode allows current to flow. At B, a negative anode normally prevents current from flowing.

Forward Breakover

When we connect a diode in the manner shown by Fig. 11-2A, a certain minimum potential difference must exist across the P-N junction before it can conduct current. We call this threshold the *forward breakover voltage* (or simply the forward breakover). Depending on the particular semiconducting elements that compose the P-type and N-type semiconductor wafers, the forward breakover voltage of a diode can range from approximately 0.3 to 1 V.

In theory, the forward breakover voltages of multiple diodes add together as if the diodes were batteries. When we connect two or more diodes in series and orient all of the P-N junctions in the same direction, the forward breakover voltage of the combination theoretically equals the sum of the forward breakover voltages of each diode. In practice, slight variations occur depending on the nature of the external circuitry. If we connect two or more diodes (all of identical manufacturer's type) in parallel with their P-N junctions oriented the same way, the forward breakover voltage of the combination theoretically equals that of the diode that has the smallest forward breakover voltage.

Avalanche Effect

If we connect a diode as shown in Fig. 11-2B and use a battery that supplies a high enough voltage, the P-N junction will conduct current. When that happens, we witness a case of the *avalanche effect*. The reverse current, which hovers near zero at lower voltages, rises suddenly and dramatically when the potential difference across the P-N junction exceeds the threshold voltage, called the *avalanche voltage*.

The avalanche voltage varies among different kinds of diodes. Figure 11-3 portrays a graph of the current as a function of the applied voltage for a typical semiconductor diode, showing the *forward breakover point* and the *avalanche point*. The avalanche voltage exceeds, and has the opposite polarity from, the forward breakover voltage.

Still Struggling

You might ask, "What possible use could the avalanche effect have in a practical system?" Some diodes take advantage of the avalanche effect by design. These devices, called *Zener diodes*, exhibit predictable and constant avalanche voltages. Zener diodes enjoy various applications in electronics engineering, most notably for the purpose of *voltage regulation*.

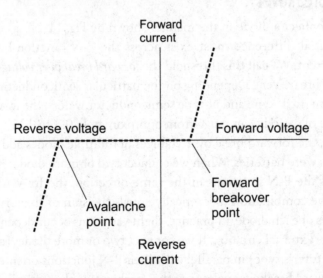

FIGURE 11-3 · A generic characteristic curve for a semiconductor diode.

Bias

When the N-type material carries a negative charge with respect to the P-type material in a P-N junction, electrons flow easily from N to P, and the diode conducts almost as well as a piece of copper wire would under the same conditions (except when the voltage remains less than the forward breakover). We call this condition *forward bias*. When we switch the polarity of the voltage source so that the N-type material carries a positive charge with respect to the P-type material, we have a state of *reverse bias*, and the diode conducts almost as poorly as an open switch would under the same conditions (except when avalanche effect occurs).

Under reverse-bias conditions, the electrons in the N-type material gravitate toward the positive charge pole, away from the P-N junction. In the P-type material, holes gravitate toward the negative charge, also away from the junction. In the vicinity of the P-N junction on either side of the actual boundary, we observe a relative shortage of electrons (in the N-type material) and holes (in the P-type material). The lack of charge carriers impedes conduction, producing a *depletion region*. As the reverse-bias voltage increases, the depletion region grows wider, as long as the potential difference across the junction doesn't reach or exceed the avalanche voltage.

Junction Capacitance

Under conditions of reverse bias, a P-N junction can act as a capacitor, because the depletion region in effect constitutes a thin dielectric layer. A special type of diode called a *varactor* is manufactured with this property specifically in mind. We can vary the *junction capacitance* of a varactor at will by adjusting the reverse-bias voltage, which affects the width of the depletion region. As the reverse voltage increases, the depletion region widens and the capacitance at the junction goes down. Conversely, as the reverse voltage decreases, the depletion region narrows and the capacitance goes up. A well-engineered varactor can exhibit capacitance that fluctuates rapidly, following voltage variations even at quite high frequencies. Of course, this behavior only takes place as long as the reverse-bias voltage never exceeds the avalanche voltage.

Rectification

A *rectifier diode* passes current in only one direction under ideal operating conditions. This property makes the device useful for converting AC electricity to

DC electricity. Rectifier diodes constitute the heart of a so-called *DC power supply*, which allows us to operate DC devices and systems, such as television sets and personal computers, from household AC utility power. If DC power supplies didn't exist, we'd have to rely on batteries to operate most of our familiar household appliances.

As we have seen, a semiconductor diode can act as a "one-way current gate," subject to the limitations imposed by forward breakover and avalanche breakdown. As long as neither of these effects take place, the diode conducts during a little less than half of the AC cycle. Depending on which way we connect the diode in the circuit, the P-N junction allows either the positive part or the negative part of the AC cycle to pass through, so that we obtain *pulsating DC* at the output.

Detection

A diode can recover the audio signal from certain types of *radio-frequency* (RF) AC signals. We call this effect *demodulation* or *detection*. If we want a diode to act as a detector in an RF system, the diode must exhibit extremely low junction capacitance, so that it works entirely as a rectifier, and does not behave to any extent as a capacitor.

Some RF diodes comprise subminiature versions of an archaic device called a *cat whisker*, in which a fine wire rested against a crystal of the mineral *lead sulfide*, commonly called *galena*. Modern components of this type, known as *point-contact diodes*, exhibit minimal junction capacitance. Therefore, as the AC frequency increases (within reason), the diodes keep acting like rectifiers and don't start to behave like capacitors. This property makes point-contact diodes useful in wireless communications devices and other systems that operate at high AC frequencies.

Gunn Diodes

Semiconductor-device manufacturers fabricate *Gunn diodes* from a compound known as *gallium arsenide*, symbolized GaAs. When we apply a certain voltage to a Gunn diode, the diode produces an ultra-high-frequency AC signal (it *oscillates*) because of the so-called *Gunn effect*, named after J. Gunn of International Business Machines (IBM) who first observed the phenomenon in the 1960s. Oscillation occurs as a result of a property that some engineers call *negative resistance*.

Still Struggling

Technically, the term "negative resistance" constitutes a misnomer because, as we have learned, nothing can conduct better than perfectly! In the context of the Gunn diode, the expression "negative resistance" refers to the fact that, under specific controlled conditions, the current through the P-N junction decreases as the applied voltage increases, contrary to the behavior of simple DC electrical devices.

IMPATT Diodes

The acronym *IMPATT* (pronounced "IM-pat") comes from the words *impact avalanche transit time*. We won't concern ourselves with the exact nature of this effect, except to note that it resembles negative resistance.

An IMPATT diode can act as an ultra-high-frequency *oscillator* (a circuit that generates AC signals at low power levels) just as a Gunn diode can, but the IMPATT diode consists of silicon rather than gallium arsenide. An IMPATT diode can also perform as a low-power AC *amplifier*.

Tunnel Diodes

Another type of diode that can oscillate at ultra-high AC frequencies is the *tunnel diode*, also known as the *Esaki diode*. It produces a very small amount of AC signal power.

Tunnel diodes work well as amplifiers in microwave receivers. This property holds true especially for GaAs devices, which act to increase the amplitudes of weak signals without introducing any unwanted *electronic noise*—extremely irregular AC disturbances that cover a large range of frequencies. (As an example of noise, consider the "hiss" that you hear in a radio receiver with the gain turned up but no station coming in.) Engineers strive to minimize electronic noise in most practical systems.

LEDs and IREDs

Depending on the exact mixture of semiconductors used in manufacture, *visible light* of any color, as well as *infrared* (IR), can arise when we drive current through a diode in the forward direction. A *light-emitting diode* (LED) generates visible light. An *infrared emitting diode* (IRED) produces energy at *near-infrared* (NIR) wavelengths, slightly longer than the wavelength of visible red light.

The intensity of energy emission from an LED or IRED depends to some extent on the forward current. As the current rises, the brightness increases up to a certain point. If the current continues to rise, no further increase in brilliance takes place. We then say that the LED or IRED has reached a state of *saturation*.

Injection Lasers

An *injection laser*, also called a *laser diode*, constitutes a special form of LED or IRED with a relatively large and flat P-N junction. The injection laser emits *coherent electromagnetic* (EM) *waves* if we drive enough current through it. Coherent waves all line up with each other and all have the same frequency, compared with the *incoherent waves* typical of most light-producing and IR-producing components.

Figure 11-4 illustrates a laser diode in simplified form. The component rests on a "foundation" called the *substrate*, which acts as the negative terminal and also conducts away excess heat. Mirrors sit at opposite ends of the N-type wafer. One of the mirrors (the one labeled in the drawing) reflects most, but not all, of the incident light or IR radiation. The opposite mirror (not shown) reflects all of the radiant energy that strikes it. The coherent rays emerge from the end of the wafer having the partially reflective mirror.

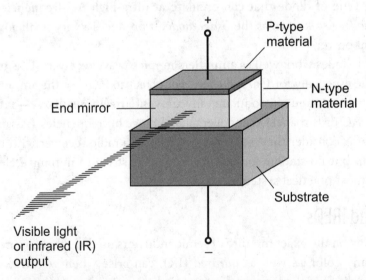

FIGURE 11-4 · Simplified cross-sectional drawing of an injection laser, also known as a laser diode.

Silicon Photodiodes

A silicon diode, specially constructed so that visible light can strike the P-N junction, forms a *photodiode*. We apply a voltage to the P-N junction in the reverse direction, so it ordinarily does not conduct. When visible light, IR, or *ultraviolet* (UV) rays strike the P-N junction, current flows. Within certain limits, the current varies in direct proportion to the intensity of the incident radiation. The *sensitivity* of the device depends on the wavelength of the incident radiation.

When visible, IR, or UV rays of varying intensity strike the P-N junction of a reverse-biased silicon photodiode, the output current follows along with the fluctuations in radiation strength. This property makes silicon photodiodes useful for receiving *modulated-light* signals in *fiberoptic* and *free-space laser* communication systems. The "following-along ability" of the device diminishes as the frequency of the fluctuations goes up.

Photovoltaic (PV) Cells

Some types of silicon diodes can generate DC all by themselves if sufficient IR, visible-light, or UV energy strikes their P-N junctions. We call this property the *photovoltaic effect*. All DC-generating *solar cells* function because of this phenomenon.

Photovoltaic (PV) cells have large P-N junction surface area, maximizing the amount of radiant energy that strikes the junction after passing through a thin layer of translucent P-type semiconductor material. A single silicon PV cell produces approximately 0.6 V DC in direct sunlight under *no-load conditions* (that is, when nothing is connected to it that can draw any current from it). The maximum amount of current that a PV cell can deliver depends on the surface area of the P-N junction. Figure 11-5 shows a simplified cut-away view of a PV cell.

We can connect multiple silicon PV cells in series-parallel combinations to provide solar power for electronic devices such as portable radios, notebook computers, and small television sets. A large assembly of such cells constitutes a *solar panel*. The DC voltages of the cells add when we connect them in series, just as the voltages of ordinary electrochemical cells do. A typical *solar battery* supplies approximately 6, 9, or 12 V DC. When we connect two or more identical sets of series-connected PV cells in parallel, the output voltage doesn't go up, but the solar battery can deliver more current. The current-delivering capacity increases in direct proportion to the number of parallel-connected cell sets.

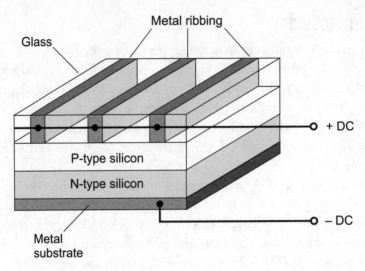

FIGURE 11-5 · Simplified cross-sectional drawing of a photovoltaic (PV) cell.

PROBLEM 11-1

In theory, how many silicon PV cells do we need to make a solar battery that supplies 13.8 V DC?

✓ SOLUTION

We can connect the PV cells in series so that the voltages add up arithmetically. Each cell produces approximately 0.6 V DC. Therefore, if we want to get 13.8 V DC, we must theoretically connect 13.8/0.6, or 23, silicon PV cells in series. (In practice, we'll likely have to use one or two extra cells, or one or two fewer cells, depending on external variables such as the load resistance, the ambient temperature, and the brightness of the light.)

The Bipolar Transistor

Bipolar transistors contain two P-N junctions connected together. Such a device can result from placing a thin layer of P-type semiconductor material between two thicker N-type layers, or from placing a thin layer of N-type material between two thicker P-type layers.

NPN and PNP

Figures 11-6A and 11-B illustrate a simplified rendition of an *NPN transistor* and the symbol that represents it in schematic diagrams. The P-type, or center,

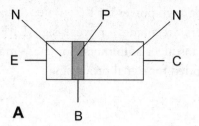 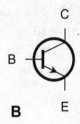

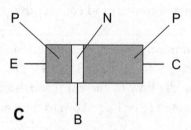 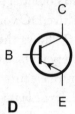

FIGURE 11-6 · Pictorial diagram of an NPN transistor (A), schematic symbol for an NPN transistor (B), pictorial diagram of a PNP transistor (C), and schematic symbol for a PNP transistor (D).

layer forms the *base*. The thinner of the two N-type semiconductor layers composes the *emitter*, and the thicker N-type layer forms the *collector*. In some diagrams these electrodes bear the labels B, E, and C, respectively, but the transistor symbol alone indicates which electrode is which (the arrow appears at the emitter). A *PNP transistor* (Figs. 11-6C and 11-6D) has two P-type layers, one on either side of a thinner N-type layer.

By examining the schematic symbol, we can easily tell whether we're looking at an NPN bipolar transistor or a PNP bipolar transistor. In the NPN symbol, the arrow at the *emitter-base* (E-B) *junction* points outward. In the PNP symbol, the arrow at the E-B junction points inward.

Generally, PNP and NPN transistors can perform identical tasks. The only difference between the two configurations lies in the polarities of the voltages that the electrodes receive, and the directions in which the currents flow among the electrodes. In many (but not all!) applications, we can directly replace an NPN device with a PNP device of the correct specifications (or vice versa), reverse the polarity of the power supply, and obtain a new circuit that works just about as well as the old one did.

Various kinds of bipolar transistors exist. Some are manufactured especially for use in RF amplifiers and oscillators; others are intended for use at audio frequencies (AF). Some can handle high power for RF wireless transmission or AF hi-fi amplification, and others are made for weak-signal RF reception, microphone preamplifiers, and transducer amplifiers. Some are manufactured for switching, and others are intended for signal processing.

NPN Biasing

When we design a circuit using an NPN transistor, we'll usually provide the collector with a positive voltage relative to the emitter. In the *common-emitter* configuration, perhaps the most often used arrangement, we place the emitter at or near zero potential (called *common ground*) and connect the collector to a source of positive DC voltage.

Figure 11-7 shows the basic common-emitter configuration. The power-supply or battery voltage typically ranges from about 3 to 50 V DC. We call the base the "control electrode" because the flow of current through the transistor depends on the *base bias* voltage, denoted E_B or V_B, relative to the *emitter-collector bias voltage*, denoted E_C or V_C.

Zero Bias

If we don't connect the base to any source of DC, or if we place the base at the same potential as the emitter, we call the condition *zero bias*. Under this condition, no appreciable current can flow between the emitter and the collector, unless we inject a signal voltage of sufficient amplitude (as we'll see in a moment).

With zero bias, the emitter-base (E-B) current I_B equals zero, and the E-B junction does not conduct. This condition, called *cutoff*, prevents current from

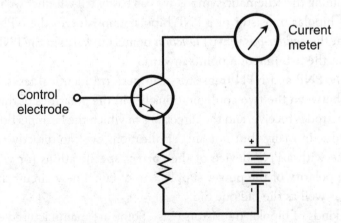

FIGURE 11-7 · Typical biasing of an NPN transistor.

flowing between the emitter and the collector unless we inject an AC signal or DC pulse at the base to change the situation. This signal or pulse must attain positive polarity for at least part of its cycle, and its positive peaks must rise to a voltage that exceeds the forward breakover voltage of the E-B junction.

Reverse Bias

Suppose that we connect a second battery to the base of the NPN transistor at the point marked "control electrode" in Fig. 11-7, so that the base attains a negative voltage with respect to the emitter. The addition of this new battery will cause the E-B junction to operate in a condition of *reverse bias*. Let's assume that this new battery does not have such a high voltage that avalanche breakdown takes place at the E-B junction.

We might inject an AC signal or DC pulse at the base electrode, but such a signal or pulse must cause conduction of the E-B junction for part of the cycle if we want our circuit to do anything. The positive peaks must overcome the sum of the reverse-bias battery voltage and the forward breakover voltage of the E-B junction. Otherwise the transistor will remain in a state of cutoff all the time, and we'll never observe any current at the collector.

Forward Bias

Suppose that we make the bias voltage at the base of an NPN transistor positive with respect to the emitter, starting at small levels and gradually increasing. We call this condition *forward bias*. If we keep the forward-bias voltage to levels smaller than the E-B junction's forward breakover voltage, no current flows in the collector. When the base bias voltage reaches or exceeds forward breakover, the E-B junction conducts, allowing current to flow between the emitter and the collector.

Normally, the base-collector (B-C) junction of a bipolar transistor operates in a state of reverse bias. Nevertheless, the emitter-collector (E-C) current, more often called *collector current* and denoted I_C, flows when the E-B junction conducts. A small rise in the positive-polarity signal at the base, attended by a small rise in the base current I_B, causes a large increase in I_C. This phenomenon allows a bipolar transistor to act as a signal or pulse amplifier.

Saturation

If we provide the E-B junction with forward bias and keep increasing it so that the base current I_B continues to rise, the collector current I_C will keep rising for

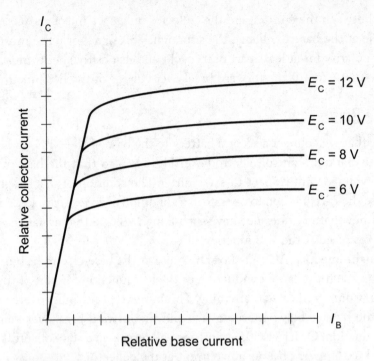

FIGURE 11-8 · A family of characteristic curves for a hypothetical NPN bipolar transistor.

a while. But eventually, I_C will no longer rise very much as we increase the forward bias. Ultimately, the I_C-versus-I_B function, or *characteristic curve*, of the transistor will level off.

Figure 11-8 shows a *family of characteristic curves* for a hypothetical bipolar transistor. The actual current values depend on the particular type of device. Power transistors yield higher currents; weak-signal transistors yield smaller currents. Where the curves level off, the transistor operates in *saturation*. When in that state, a transistor loses its ability to efficiently amplify AC signals, although it can still work as a high-speed electronic switch.

PNP Biasing

Engineers bias PNP transistors in a "mirror image" version of the biasing scheme for an NPN device, as shown in Fig. 11-9. You'll notice that Figs. 11-7 and 11-9 appear identical except for the polarity of the battery and the orientation of the arrow inside the transistor symbol. To overcome forward breakover at the E-B junction in a PNP bipolar transistor, an applied signal or pulse at the base must attain enough negative voltage at its peaks to overcome the forward breakover voltage of the E-B junction.

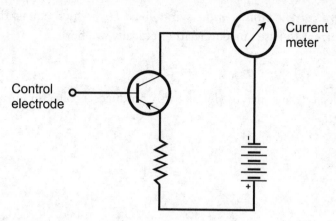

FIGURE 11-9 · Typical biasing of a PNP transistor.

Still Struggling

Either the PNP or the NPN device can serve as a "current valve." Small changes in the base current, I_B, induce large fluctuations in the collector current, I_C, when we operate the device in the steeply sloped region of the characteristic curve, where we see significant "rise over run." While the internal atomic activity differs in a PNP device as compared with an NPN device, the external circuitry "can't tell the difference" in many practical applications.

How a Bipolar Transistor Amplifies Signals

Because a small change in I_B causes a large change in I_C when we bias a bipolar transistor properly, a transistor can operate as a *current amplifier*. We can express the extent of such amplification in terms of what happens with either *static* (steady) or *dynamic* (varying) input signal current.

Static Current Amplification

The maximum obtainable current amplification factor of a particular bipolar transistor constitutes a specification called the *beta*. Depending on the method of transistor manufacture, the beta can range from a factor of a few times up to hundreds of times. We express the beta of a bipolar transistor in terms of the *static forward current transfer ratio*, symbolized H_{FE}, which equals the ratio of collector current to base current. Mathematically, we have

$$H_{FE} = I_C/I_B$$

For example, if a base current (I_B) of 1 mA produces a collector current (I_C) of 35 mA, then

$$H_{FE} = I_C/I_B$$
$$= 35/1$$
$$= 35$$

If I_B = 0.5 mA and I_C = 35 mA, then

$$H_{FE} = I_C/I_B$$
$$= 35/0.5$$
$$= 70$$

Dynamic Current Amplification

We can express the current amplification of a bipolar transistor in another way: as the ratio of the observed difference in I_C to a small difference in I_B that produces it. We call this specification the *dynamic current amplification* or *current gain*. In equations, we can abbreviate "the difference in" by writing an uppercase Greek letter Δ (delta). If we use the symbol B_d to represent the dynamic current amplification as defined here, we have

$$B_d = \Delta I_C/\Delta I_B$$

In any bipolar transistor, the highest values of B_d correspond to points where the characteristic curve has the steepest slope ("rise over run"). Geometrically, the dynamic current amplification at any given point on the curve equals the slope of a line tangent to the curve at that point.

When the *operating point* of a transistor lies on the steepest part of the characteristic curve, the device exhibits the largest possible current gain, as long as we don't let the input signal get too strong. This value of B_d coincides closely with the static current amplification H_{FE}. Because the characteristic curve constitutes an almost perfectly straight line in this region, the transistor can serve as a *linear amplifier* as long as we don't let the input signal get too strong. In a linear amplifier, the output signal waveform constitutes a faithful, although magnified, reproduction of the input signal waveform. The term "linear" arises from the fact that the transistor operates along the "straight-line" part of the characteristic curve.

If we shift the operating point into a part of the characteristic curve where the graph does not look like a straight line, the current gain decreases, and the amplifier becomes *nonlinear*. The same thing can happen if the input signal gets strong enough to drive the transistor into the nonlinear part of the characteristic curve during *any portion* of the signal cycle.

Gain versus Frequency

In any bipolar transistor, the current gain decreases as the signal frequency increases. We can't avoid this phenomenon (as much as some of us in the engineering business wish we could); it's inherent in the atomic nature of semiconductor materials. We can express the *gain-versus-frequency* behavior of a bipolar transistor in either of two ways.

The *gain bandwidth product*, abbreviated f_T, represents the frequency at which the current gain becomes equal to unity (1) with the emitter connected to ground. This means, in effect, that the transistor has no current gain; the output signal's current amplitude equals the input signal's current amplitude, even under ideal operating conditions.

The *alpha cutoff* represents the frequency at which the current gain becomes 0.707 times (that is, 70.7 percent of) its value at exactly 1 kHz (1000 Hz). Most bipolar transistors can function as current amplifiers at frequencies higher than the alpha cutoff, but no transistor can work as a current amplifier at frequencies higher than its gain bandwidth product.

 PROBLEM 11-2

Suppose that a bipolar transistor has a current gain, under ideal conditions, of 23.5 at an operating frequency of 1000 Hz. The alpha cutoff is specified as 900 kHz. What is the maximum possible current gain that the device can exhibit at 900 kHz?

SOLUTION

We multiply 23.5 by 0.707 to obtain 16.6. That's the maximum possible current gain that the transistor can produce at 900 kHz.

PROBLEM 11-3

Imagine that the peak-to-peak signal input current in the above-described transistor is 2.00 μA at a frequency of 1000 Hz. Further suppose that the operating conditions are ideal, and that the transistor never goes into the nonlinear part of the characteristic curve during any part of the input signal cycle. If we change the frequency to 900 kHz, what happens to the peak-to-peak signal output current?

✔ SOLUTION

First, let's remember that the current gain of the transistor is 23.5 at a frequency of 1000 Hz. Because the peak-to-peak signal input current equals 2.00 μA at this frequency, the peak-to-peak output signal current is

$$2.00 \ \mu A \times 23.5 = 47.0 \ \mu A$$

At 900 kHz, the peak-to-peak output signal current decreases to

$$0.707 \times 47.0 \ \mu A = 33.2 \ \mu A$$

The Field-Effect Transistor

The other major category of transistor, besides the bipolar device, is the *field-effect transistor* or *FET*. Two main types of FET exist: the *junction FET* (JFET) and the *metal-oxide FET* (MOSFET).

Principle of the JFET

In a JFET, the current varies because of the effects of an electric field within the device. Electrons or holes move along a current path called the *channel* from the *source* (S) electrode to the *drain* (D) electrode. This movement of charge carriers results in a drain current, I_D, that's normally the same as the source current, I_S.

The drain current depends on the voltage at the *gate* (G) electrode. As the gate voltage E_G changes, the effective width of the channel varies. Fluctuations in E_G cause fluctuations in the current through the channel. Small fluctuations in E_G can cause large variations in the flow of charge carriers through the JFET. This phenomenon allows the device to function as a *voltage amplifier*.

N-Channel and P-Channel

Figures 11-10A and 11-10B portray an *N-channel JFET* and with its schematic symbol. The N-type material provides the path for the current. Electrons

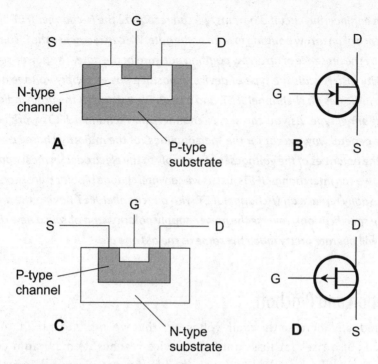

FIGURE 11-10 · Pictorial diagram of an N-channel JFET (A), schematic symbol for an N-channel JFET (B), pictorial diagram of a P-channel JFET (C), and schematic symbol for a P-channel JFET (D).

constitute the *majority charge carriers* (the ones, either electrons or holes, that exist in greater numbers within a particular sample of semiconductor material). We apply a positive DC voltage to the drain with respect to the source. The gate consists of P-type material. Another, larger section of P-type material, the substrate, forms a boundary on the side of the channel opposite the gate. A negative voltage on the gate produces an electric field that interferes with the flow of charge carriers through the channel. As we make the gate voltage E_G more negative, the electric field chokes off the current though the channel to an increasing extent, and the collector current I_D continues to decline.

A *P-channel JFET* (Figs. 11-10C and 11-10D) has a channel consisting of P-type semiconductor material. The majority charge carriers comprise holes rather than electrons. We place the drain at a negative DC potential with respect to the source. The gate and substrate consist of N-type semiconductor material. As we make the gate voltage E_G more positive, the electric field chokes off the current through the channel to an increasing extent, and the collector current I_D therefore gets smaller.

TIP   *In engineering circuit diagrams, we can recognize the N-channel JFET by the presence of an arrow pointing inward at the gate. We can recognize the P-channel JFET by the presence of an arrow pointing outward at the gate. The power-supply polarity also reveals the type of device. A positive power-supply voltage at the drain indicates an N-channel JFET, and a negative voltage at the drain indicates a P-channel type. As you can see, N-channel and P-channel JFETs work in the same general way, except for the "personalities" of the majority charge carriers and the polarities of the voltages that we apply to the electrodes. In most applications, we can interchange JFETs just as we can interchange bipolar transistors. We can usually replace an N-channel JFET with a P-channel JFET having the appropriate specifications, reverse the power-supply polarity, and obtain a new circuit that will operate pretty much the same as the old one did.*

Depletion and Pinchoff

If we gradually increase the drain voltage E_D that we apply to a JFET, the drain current I_D also rises (at least until the device reaches its maximum current-handling capability, at which point we had better not increase E_D any further for fear of burning the device out!). However, this simple relation holds true only as long as we keep the gate voltage E_G constant. But in most real-life electronic systems, the instantaneous gate voltage does *not* remain constant; it fluctuates from instant to instant in time as we apply an AC signal to the gate along with the DC gate bias E_G.

As E_G increases (negatively in an N-channel or positively in a P-channel), a *depletion region* develops within the channel. Charge carriers can't flow in the depletion region; when such a region exists, the charge carriers must pass through a narrowed channel. As we increase E_G, the depletion region widens, and the channel therefore narrows. If we make E_G high enough while leaving the drain voltage E_D constant, the depletion region completely obstructs the channel, and the channel no longer conducts current. We call this condition *pinchoff*.

How a JFET Amplifies Signals

Figure 11-11 shows a graph of the drain (and channel) current I_D as a function of the gate bias voltage E_G for a hypothetical N-channel JFET when no input signal exists at the gate electrode and the drain voltage E_D remains constant.

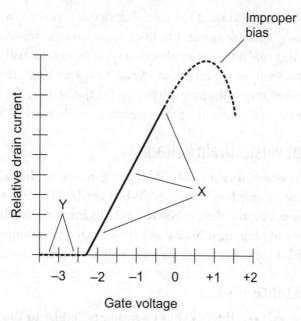

FIGURE 11-11 · Relative drain current as a function of gate voltage in a hypothetical N-channel JFET.

Drain Current versus Gate Voltage

When we set E_G at fairly large negative values, the JFET operates in a state of pinchoff. No current flows through the channel, so $I_D = 0$. As E_G gets less negative, the channel opens up, and drain current I_D begins to flow. As E_G gets still less negative, the channel gets wider and I_D increases. As E_G approaches the point where the source-gate (S-G) junction operates at exactly its forward breakover voltage, the channel conducts as well as it possibly can, and I_D reaches its maximum possible value. If E_G becomes even more positive, the S-G junction conducts just as a forward-biased diode would do. In this state, some of the channel current flows through the gate circuit instead of reaching the drain; it's as if the device has "sprung a leak"! This situation degrades the performance of the JFET, and therefore represents an undesirable operating condition.

The JFET works best as a weak-signal amplifier when we set the gate voltage, E_G, to values where the slope of the curve in Fig. 11-11 attains its greatest steepness, or "rise over run." (Remember that this drawing depicts the situation for an N-channel device. To get the graph for a P-channel JFET, reverse the polarity signs for the gate voltage along the horizontal axis.) Under these conditions, a small change in the gate voltage produces a large change in the drain current. The range marked X in the graph illustrates the optimum weak-signal *operating region*. We would want to bias the device in this zone to make it work, for example, as a sensitive input amplifier for a wireless receiver. For power amplification, this rule does not apply. If we want to make a fairly strong input signal more powerful, as we might do in the final amplifier circuit of a high-power radio communications transmitter, we get the best results when we bias the JFET at or beyond pinchoff, in the operating region marked Y.

Drain Current versus Drain Voltage

We can plot the drain current I_D in a JFET as a function of the drain voltage E_D for various values of gate bias voltage E_G. When we do this, we obtain a *family of characteristic curves* for the device. Figure 11-12 illustrates a family of characteristic curves for a hypothetical N-channel JFET. Engineers often use graphs like those in Figs. 11-11 and 11-12 to choose a JFET for a particular application.

Transconductance

In real-world operation, JFETs exhibit a property similar to the dynamic current amplification of a bipolar transistor. We call this parameter the *dynamic*

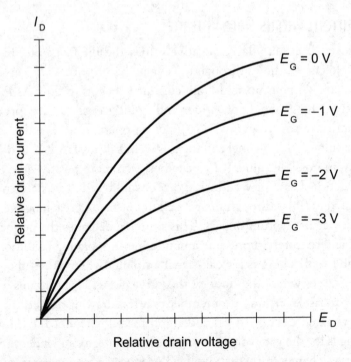

FIGURE 11-12 · Family of characteristic curves for a hypothetical N-channel JFET.

mutual conductance of the JFET. Some engineers call it the *transconductance*. Let's write G_m to symbolize it.

Refer again to the graph of Fig. 11-11. Suppose that we connect a JFET in a circuit so that the source exists at or near ground potential (a *common-source* configuration) and we set the gate voltage E_G at a certain value, causing a certain drain current I_D to flow. If the gate voltage changes by a small amount ΔE_G, then the drain current will change by a certain increment ΔI_D. We define the transconductance as the ratio of the drain current change to the gate voltage change. Mathematically, we have

$$G_m = \Delta I_D / \Delta E_G$$

Geometrically, the transconductance appears as the slope of a line tangent to the curve of Fig. 11-11 at some point.

As we can see by examining Fig. 11-11, the ratio $\Delta I_D / \Delta E_G$ varies depending on where we go along the curve. When we bias the JFET beyond pinchoff, as in the region marked Y, the slope of the curve equals zero. If we venture far enough into this range, no drain current flows, even if the gate voltage fluctuates by a small amount. If we want I_D to change when we apply a low-voltage AC signal to the gate, we must bias the JFET to allow some channel current to flow in the absence of an input signal. The region marked X shows the zone of greatest transconductance. That's where the curve has its maximum slope, and it's where we can obtain the best weak-signal amplification.

If we set the gate bias E_G in a JFET so that the device operates near the center of the region marked X in Fig. 11-11 under no-signal conditions, then small fluctuations in E_G produce large fluctuations in I_D, which in turn cause large voltage fluctuations across *resistive load* that we place in series with the line connecting the drain to the battery or power supply. Therefore, the JFET performs as a weak-signal voltage amplifier. In addition, we'll get linear amplification, as long as we don't let the signal at the gate get too strong.

PROBLEM 11-4

Examine Fig. 11-12. Note that the curves in the graph grow farther apart as the drain voltage E_D increases (as we move toward the right). Let's extrapolate off the right-hand end of this graph. From appearances, we can imagine that if the drain voltage E_D keeps increasing indefinitely, the curves will flatten out into horizontal lines, and they will no longer get any farther from each other. What can we infer about the ability of this particular JFET to amplify signals as its E_D increases indefinitely?

SOLUTION

When we operate a JFET at relatively low drain voltages E_D, a certain peak-to-peak AC gate signal voltage with a DC component (say, a signal that ranges from –2 to –1 V) produces a small change in drain current I_D. As E_D increases, the curves represented by the gate voltages $E_G = -2$ V and $E_G = -1$ V grow farther apart. The same peak-to-peak signal amplitude will produce larger fluctuations in the drain current I_D, so the amplification will increase. As we continue to make the drain voltage E_D larger, the curves represented by $E_G = -2$ V and $E_G = -1$ V level off, and the vertical distance between them no longer increases. Therefore, the amplification factor does not increase significantly once E_D exceeds this limiting value. Figure 11-13 shows how this happens. We will get the same result for all input signals (at the gate) with relatively small peak-to-peak voltages that fall within the ranges indicated by the curves. Of course, if we make E_D too large, we will destroy the JFET! Most JFETs are designed for operation with E_D values of no more than a few tens of volts.

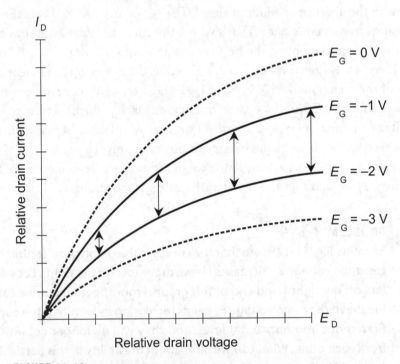

FIGURE 11-13 · Illustration for Problem 11-4.

The MOSFET

The acronym MOSFET (pronounced "MOSS-fet") stands for *metal-oxide-semiconductor field-effect transistor*. Figures 11-14A and 11-14B show a simplified cross-sectional drawing of an N-channel MOSFET along with the schematic symbol. Figures 11-14C and 11-14D illustrate a P-channel MOSFET cross-section and schematic symbol.

A Major Asset

When the MOSFET first appeared on the electronics market, its inventors and manufacturers called it an *insulated-gate FET* or *IGFET*. This term might

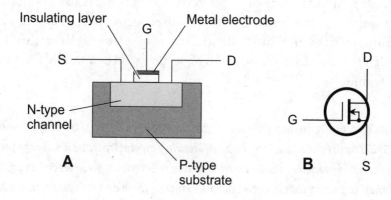

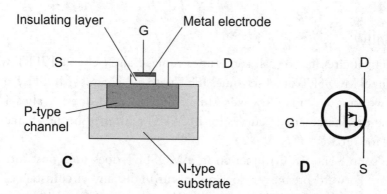

FIGURE 11-14 · Pictorial diagram of an N-channel MOSFET (A), schematic symbol for an N-channel MOSFET (B), pictorial diagram of a P-channel MOSFET (C), and schematic symbol for a P-channel MOSFET (D).

constitute a better description of the device than the currently accepted term does. A thin layer of dielectric material literally insulates the gate from the channel. As a result, all MOSFET devices exhibit extremely high input impedance. For this reason, a properly operating MOSFET draws essentially no current from the input signal source, making MOSFETs an excellent choice for the engineer who wants to design a weak-signal, low-power voltage amplifier.

A Major Problem

Electrostatic charge buildup presents a "mortal danger" to a MOSFET. When the discharge occurs through the dielectric layer between the gate and the channel (as it inevitably must), the dielectric breaks down permanently and the entire device will no longer function. When building, testing, or repairing circuits containing MOS devices, we must use special equipment to ensure that our hands don't acquire electrostatic charges. A humid environment does not offer significant protection against this hazard (although it can spawn a false sense of security in the mind of an inexperienced technician).

TIP   *As a radio-frequency (RF) technician in Miami, Florida, I destroyed several MOSFETs until I developed the habit of wearing an electrostatic discharge bracelet, securely wrapped around one wrist and connected to a good electrical ground. I learned to put on that bracelet before I started work on any system using MOS devices, and to keep it on until after I was finished.*

Flexibility

In practical circuits, we can sometimes replace an N-channel JFET with an N-channel MOSFET, or a P-channel JFET with a P-channel MOSFET, and still get a working circuit. However, this simplistic rule doesn't always apply because the characteristic curves for MOSFETs differ from the characteristic curves for JFETs.

The source-gate (S-G) junction in a MOSFET does not constitute a P-N junction. Forward breakover does not occur under any circumstances. If the S-G junction in a MOSFET conducts, it's because the S-G voltage has risen to such a high level that arcing has taken place, "punching" a conductive path through the dielectric material and thereby destroying the component.

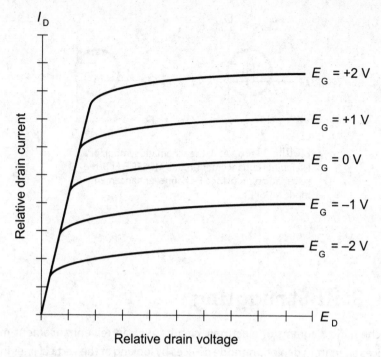

FIGURE 11-15 · Family of characteristic curves for a hypothetical N-channel MOSFET.

Figure 11-15 illustrates a family of characteristic curves for a hypothetical N-channel MOSFET. Note the subtle differences between these curves and the ones shown in Figs. 11-12 and 11-13.

Depletion Mode versus Enhancement Mode

In a JFET, the channel conducts with zero bias, that is, when the potential difference between the gate and the source equals zero. As the depletion region grows, charge carriers pass through a narrowed channel. We call this condition the *depletion mode*. A MOSFET can work in the depletion mode too. The drawings and schematic symbols of Fig. 11-14 pertain to so-called *depletion-mode MOSFETs*.

Metal-oxide-semiconductor technology allows a second mode of operation, entirely different from the depletion mode. An *enhancement-mode MOSFET* has a pinched-off channel at zero bias. We must apply a gate bias voltage E_G to the device in order to "create" a channel. If $E_G = 0$, then $I_D = 0$ in the absence of an input signal. Figure 11-16 shows the schematic symbols for N-channel and P-channel enhancement-mode MOSFETs.

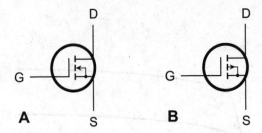

FIGURE 11-16 · At A, the schematic symbol for an N-channel enhancement-mode MOSFET. At B, the schematic symbol for a P-channel enhancement-mode MOSFET.

Still Struggling

In schematic diagrams of electronic circuits, we can tell enhancement-mode devices apart from depletion-mode devices by looking at the vertical lines inside the circles. Depletion-mode MOSFETs have solid vertical lines. Enhancement-mode devices have broken vertical lines.

Integrated Circuits

Most *integrated circuits* (ICs) look like plastic boxes with protruding metal pins. Common configurations include the *single inline package* (SIP), the *dual inline package* (DIP), and the *flatpack*. Another package looks like a transistor with too many leads. Engineers and technicians sometimes call this a *metal-can package* or *T.O. package*. We can symbolize an IC in schematic diagrams by drawing a triangle or rectangle and writing the component designator inside.

Compactness

Integrated-circuit devices and systems generally take up far less physical volume than equivalent circuits built up from discrete components. With ICs, we can construct more complicated circuits, and keep them down to more reasonable physical size, that we could ever hope to do with full-sized transistors, didoes, and resistors. Thus, for example, we can find battery-powered cell phones nowadays with dimensions so small that we run the risk of losing them in car seats or stacks of paper—with capabilities more advanced than the first full-scale industrial computers built in the mid-1900s, some of which took up whole rooms and consumed more power in a day than some households consume in a week!

High Speed

In an IC, *nanoscale* components and interconnections (having dimensions on the order of nanometers) allow for high switching speeds. Electric currents travel fast, but not infinitely fast. Charge carriers can get from one component to another more quickly in a small device than they can in a large one. Therefore, a small device can do more operations per unit time than a large one can, assuming the small device and the large one have equal complexity.

Low Power Requirement

Integrated circuits consume less power than equivalent discrete-component circuits. In battery-powered devices, this advantage takes on great importance. Because ICs draw so little current, they produce less heat than their discrete-component equivalents. As a result, ICs have superior energy efficiency and they experience relatively few of the chronic problems that plague equipment that gets hot with use, such as *frequency drift* and the internal generation of *electronic noise*.

Reliability

Systems using ICs fail less often, per component-hour of use, than systems that employ discrete components. In an IC, all of the component interconnections remain *hermetically sealed* inside a protective case, preventing corrosion or the intrusion of dust. The reduced failure rate translates into less downtime for the device as a whole.

Integrated-circuit technology lowers service costs, because repair procedures are simple when failures occur. Many systems use sockets for ICs, and replacing one involves nothing more than finding the faulty IC, unplugging it, and plugging in a new one. Technicians use special desoldering equipment to test and repair circuit boards that have ICs soldered directly to the foil.

Modular Construction

Modern IC appliances employ *modular construction*. Individual ICs perform defined functions within a circuit board (also called a *card*). The card, in turn, fits into a socket and has a specific purpose. Technicians use computers, programmed with customized software, to locate the faulty card in a system or appliance. The technician can remove and replace a bad card in a few seconds, minimizing the turnaround time in the service department. Other technicians can diagnose problems within individual cards to the component level and repair them at leisure, but the consumer doesn't have to wait for that!

QUIZ

Refer to the text in this chapter if necessary. A good score is eight correct. Answers are in the back of the book.

1. At the P-N junction of a reverse-biased semiconductor diode, conduction occurs
   A. regardless of the voltage.
   B. only if the voltage remains below a certain threshold.
   C. only if the voltage exceeds a certain threshold.
   D. under no conditions whatsoever.

2. Suppose that we connect four germanium diodes, each with a forward breakover voltage of 0.32 V, in series with their P-N junctions all oriented the same way. In theory, the entire combination exhibits a forward breakover voltage of
   A. 0.08 V.
   B. 0.32 V.
   C. 0.64 V.
   D. 1.28 V.

3. Fill the blank in the following sentence to make it true: "In a common-source JFET circuit, we define the _____ as the ratio of the change in drain current to the change in gate voltage."
   A. dynamic mutual conductance
   B. beta
   C. static forward current transfer ratio
   D. gain bandwidth ratio

4. Suppose that we forward-bias the E-B junction in a bipolar transistor to the point that an increase in the base current produces no change in collector current. Under these conditions, the transistor
   A. functions optimally as a weak-signal amplifier.
   B. operates in a state of avalanche breakdown.
   C. will shut down to protect itself from destruction.
   D. can't amplify signals.

5. As the depletion region in a semiconductor diode becomes wider, the capacitance at the P-N junction
   A. increases.
   B. decreases.
   C. remains constant.
   D. depends on the forward current.

6. Suppose that a bipolar transistor has a gain bandwidth product of 24 MHz. At 48 MHz, we should expect to get a maximum current gain of

   A. less than 1.
   B. 0.5.
   C. 2.0.
   D. zero.

7. Suppose that a bipolar transistor has a current gain of 10.0 at an operating frequency of 1.00 kHz. The manufacturer says that the alpha cutoff is 1.13 MHz. What is the maximum possible current gain that the device can produce at 1.13 MHz?

   A. 5.00
   B. 7.07
   C. 14.1
   D. 20.0

8. In a MOSFET, source-gate conduction indicates

   A. a state of saturation.
   B. forward breakover.
   C. avalanche breakdown.
   D. a defective or damaged component.

9. A JFET can operate as an effective weak-signal AC signal amplifier when

   A. small changes in the drain voltage produce large changes in the source voltage.
   B. small changes in the source current produce large changes in the drain current.
   C. small changes in the gate voltage produce large changes in the drain current.
   D. any of the above things happen.

10. Because of its small size, a typical IC

   A. can dissipate more power than an equivalent circuit built from discrete components.
   B. can operate at higher speeds than an equivalent circuit built from discrete components.
   C. costs more than an equivalent circuit built from discrete components.
   D. fails more often than an equivalent circuit built from discrete components.

Test: Part II

Don't look back at any of the text while taking this test. The correct answer choices appear at the back of the book. Consider having a friend check your score the first time you take this test, without telling you which questions you got right and which ones you missed. That way, you won't subconsciously memorize the answers in case you want to take the test again later.

1. **A sample of ferromagnetic material has permeability**
   A. equal to 0.
   B. greater than zero but less than 1.
   C. equal to 1.
   D. greater than 1.
   E. that's 0 or negative.

2. **Suppose that a component exhibits a complex impedance of $30 - j40$. What can we say about the reactance?**
   A. None exists.
   B. It's inductive.
   C. It's resistive.
   D. It's capacitive.
   E. It's equal to 50 ohms.

3. **The absolute-value impedance of the component described in Question 2 is**
   A. 30 ohms.
   B. 40 ohms.
   C. −40 ohms.
   D. $-j40$ ohms.
   E. 50 ohms.

4. **Imagine a pair of flat metal plates, both rectangular in shape and measuring 1.50 m by 2.00 m. We place them parallel to each other in a dry room, so that they're 10 mm apart and separated only by air. We measure the capacitance between the plates. Then we move the plates so that they're spaced 12 mm apart, but still parallel to each other. We measure the capacitance again and find that it has**
   A. decreased a lot.
   B. decreased a little.
   C. not changed.
   D. increased a little.
   E. increased a lot.

5. **Consider the same pair of flat metal plates as described in Question 4, after we've increased the spacing between them to 12 mm. We apply a 2-MHz AC signal to the plates and determine the *absolute-value impedance*. Then we apply a 20-MHz AC signal and determine the absolute-value impedance again. We find that the frequency change has caused the absolute-value impedance to**
   A. decrease a lot.
   B. decrease a little.
   C. remain the same.
   D. increase a little.
   E. increase a lot.

6. What happens to the inductive reactance of a 100-turn wire coil at an AC frequency of 75 kHz if we add five turns to the coil but change nothing else?

   A. It decreases a lot.
   B. It decreases a little.
   C. It does not change.
   D. It increases a little.
   E. It increases a lot.

7. Someone tells you that two sine waves have the same frequency and exist in
phase opposition. You conclude that the waves differ in phase by

   A. 0°.
   B. 90°.
   C. 180°.
   D. 270°.
   E. 360°.

8. Figure Test II-1 shows a vector diagram of two sine waves having the same frequency but different amplitudes. In this situation, what can we say about the
phase relationship between waves P and Q?

   A. They're 180° out of phase with Q lagging.
   B. They're 90° out of phase with Q lagging.
   C. They're in phase coincidence with Q lagging.
   D. They're 90° out of phase with P lagging.
   E. They're 180° out of phase with P lagging.

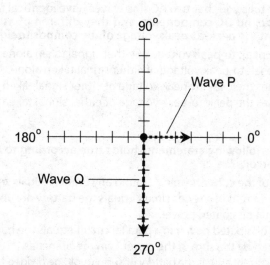

FIGURE TEST II-1 · Illustration for Part II Test
Question 8.

9. A simple DC circuit consists of a 7.500-V battery and an incandescent lamp that exhibits a resistance of 27.00 ohms with the battery connected. How much power does the lamp dissipate?

A. 97.20 W

B. 3.600 W

C. 2.083 W

D. 1.029 W

E. 0.2778 W

10. Suppose that a component exhibits 120 ohms of resistance and 50 ohms of inductive reactance at an AC frequency of 22 MHz. What's the complex impedance of the component at 22 MHz?

A. 130 ohms

B. $120 + j50$

C. $120 - j50$

D. $-120 + j50$

E. $-120 - j50$

11. We would likely observe significant inductive reactance in

A. a coil of wire.

B. an N-channel junction field-effect transistor (JFET).

C. an enhancement-mode, P-channel metal-oxide-semiconductor field-effect transistor (MOSFET).

D. a semiconductor diode operating in a state of saturation.

E. Any of the above

12. A technician tells you that two AC sine waves have identical frequencies, identical amplitude, no DC components, and they differ in phase by 180°. You can conclude that the peak-to-peak voltage of the composite signal equals

A. twice the peak-to-peak voltage of either signal taken alone.

B. half the peak-to-peak voltage of either signal taken alone.

C. 1.414 times the peak-to-peak voltage of either signal taken alone.

D. 2.828 times the peak-to-peak voltage of either signal taken alone.

E. zero.

13. Which of the following statements holds true according to Kirchhoff's current law for DC circuits?

A. The sum of the currents going around any branch equals zero.

B. The total current in a series circuit equals the battery or supply voltage divided by the total dissipated power.

C. The total dissipated power in a parallel circuit equals the battery or supply voltage divided by the sum of the currents in the branches.

D. The total current equals the battery or supply voltage divided by the net resistance.

E. The total current flowing into any branch point equals the total current flowing out of that point.

14. **Which of the following components or devices might we employ in the construction of a circuit intended to harness solar energy to refresh a rechargeable electrochemical battery?**

    A. A JFET
    B. A varactor diode
    C. A photovoltaic panel
    D. An electromagnet
    E. A rectifier diode

15. **In a triangular AC wave**

    A. the amplitude transitions occur instantaneously.
    B. the amplitude becomes more positive at a steady rate, and becomes more negative instantaneously.
    C. the amplitude becomes more negative at a steady rate, and becomes more positive instantaneously.
    D. the amplitude changes both positively and negatively at a steady rate.
    E. a graph of the waveform looks like the mathematical sine function.

16. **At most locations on the earth's surface, we observe a directional difference between true north (or geographic north) and "north" as a magnetic compass indicates. We call this discrepancy the geomagnetic**

    A. divergence angle.
    B. phase angle.
    C. inclination.
    D. declination.
    E. error angle.

17. **We can sometimes get a sample of matter in the gaseous state to act as a fair to good electrical conductor if we**

    A. ionize some of its atoms.
    B. change the isotope in some of its atoms.
    C. add protons to some of its atoms.
    D. remove protons from some of its atoms.
    E. reduce the relative humidity to zero.

18. **We can express magnetic flux density in terms of**

    A. ohms per volt.
    B. webers per meter.
    C. maxwells per centimeter squared.
    D. volts per ampere.
    E. watts per meter squared.

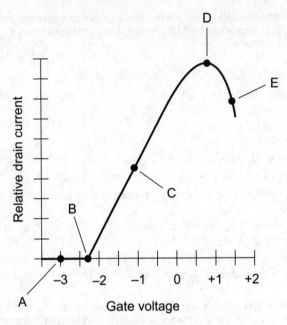

FIGURE TEST II-2 · Illustration for Part II Test Questions 19 and 20.

19. Which of the points in Fig. Test II-2 illustrates the best bias condition for a field-effect transistor if we want it to effectively amplify weak AC signals?

    A. Point A
    B. Point B
    C. Point C
    D. Point D
    E. Point E

20. Which of the points in Fig. Test II-2 illustrates a condition of pinchoff?

    A. Points A and B
    B. Points B and C
    C. Points C and D
    D. Points D and E
    E. Point E only

21. What's the positive peak amplitude of an AC sine wave that has a peak-to-peak amplitude of 10.8 V and a DC component of +2.0 V?

    A. +5.4 V
    B. +7.4 V
    C. +3.4 V
    D. +23.6 V
    E. +19.6 V

22. As we increase the frequency of AC through an inductor, we'll eventually reach a frequency at which the inductor behaves almost like

    A. a diode.
    B. a capacitor.
    C. an open circuit.
    D. a short circuit.
    E. a battery.

23. Suppose that a wire coil has an inductive reactance of 80 ohms at a frequency of 5.00 MHz. What's the inductive reactance of this same coil if we reduce the frequency to 2.50 MHz?

    A. 80 ohms
    B. 40 ohms
    C. 20 ohms
    D. 160 ohms
    E. 320 ohms

24. By definition, conventional DC flows

    A. in the same direction as the movement of the electrons.
    B. only through nonreactive components such as resistors, diodes, and transistors.
    C. only through components that exhibit capacitive reactance and no resistance.
    D. only through components that exhibit inductive reactance and no resistance.
    E. in the opposite direction from the movement of the negative charge carriers.

25. Imagine that a DC circuit carries a current of I amperes, has a battery that supplies E volts, and has a resistance of R ohms that dissipates P watts of power. Which, if any, of the following formulas A, B, C, or D contains an error?

    A. $P = EI$
    B. $I^2 = P/R$
    C. $E^2 = PR$
    D. $E = R/I$
    E. All of the formulas A through D are okay.

26. The magnetic lines of flux produced by a DC-carrying wire coil wound around a cylindrical ferromagnetic rod

    A. converge and diverge at the ends of the rod.
    B. radiate straight out perpendicular to the rod.
    C. remain entirely confined to the rod.
    D. appear as spirals in planes perpendicular to the rod.
    E. appear as circles in planes perpendicular to the rod.

27. Which of the following units can express the reciprocal of DC resistance?

    A. Anti-ohms
    B. Webers
    C. Gilberts
    D. Siemens
    E. Farads

28. Imagine three light bulbs connected in series with a 12.00-V battery. The first bulb consumes 8.000 W of power, the second bulb consumes 12.00 W, and the third bulb consumes 16.00 W. How much potential difference appears across the second bulb?

A. 2.667 V
B. 4.000 V
C. 5.333 V
D. 6.000 V
E. 7.500 V

29. Which of the following constitutes a major functional difference between an N-channel JFET and a P-channel JFET?

A. In general, P-channel devices can handle higher DC voltages than N-channel devices.
B. In general, P-channel devices can work at higher AC frequencies than N-channel devices.
C. All P-channel devices exhibit inductive reactance, while all N-channel devices exhibit capacitive reactance.
D. All P-channel devices exhibit capacitive reactance, while all N-channel devices exhibit inductive reactance.
E. The majority charge carriers in the channel of an N-channel device are electrons, while the majority charge carriers in the channel of a P-channel device are holes.

30. When we add resistance as a real-number quantity to reactance as an imaginary-number quantity, we get

A. absolute-value impedance.
B. negative impedance.
C. complex-number impedance.
D. magnetomotive impedance.
E. characteristic impedance.

31. Suppose that each increment on the horizontal axis in Fig. Test II-3 represents exactly 1 ms. What's the approximate period of the illustrated wave?

A. 12 ms
B. 1.5 ms
C. 3.0 ms
D. 6.0 ms
E. 0.75 ms

32. Suppose that each increment on the horizontal axis in Fig. Test II-3 represents exactly 1 ms. What's the approximate frequency of the illustrated wave?

A. 333 Hz
B. 1.33 kHz
C. 833 Hz
D. 670 Hz
E. 170 Hz

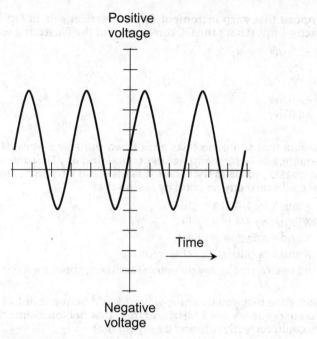

Positive
voltage

Time

Negative
voltage

FIGURE TEST II-3 · Illustration for Part II Test
Questions 31 through 36.

33. Suppose that each increment on the vertical axis in Fig. Test II-3 represents exactly
1 mV. What's the approximate positive peak amplitude of the illustrated wave?

   A. +1.0 mV

   B. +2.0 mV

   C. +3.0 mV

   D. +4.0 mV

   E. +6.0 mV

34. Suppose that each increment on the vertical axis in Fig. Test II-3 represents exactly
1 mV. What's the approximate negative peak amplitude of the illustrated wave?

   A. −1.0 mV

   B. −2.0 mV

   C. −3.0 mV

   D. −4.0 mV

   E. −6.0 mV

35. Suppose that each increment on the vertical axis in Fig. Test II-3 represents exactly
1 mV. What's the approximate peak-to-peak amplitude of the illustrated wave?

   A. 1.0 mV

   B. 2.0 mV

   C. 3.0 mV

   D. 4.0 mV

   E. 6.0 mV

36. Suppose that each increment on the vertical axis in Fig. Test II-3 represents exactly 1 mV. What's the DC component of the illustrated wave?

    A. −2.0 mV
    B. Zero
    C. +1.0 mV
    D. +2.0 mV
    E. +4.0 mV

37. Imagine that someone talks about two pure sine waves X and Y, both with a frequency of 24 MHz and neither containing any DC component. Suppose that she goes on to claim that wave X leads wave Y by 180°. From this information, you could correctly respond by saying that

    A. X lags Y by 1/4 of a cycle.
    B. X lags Y by 3/4 of a cycle.
    C. X and Y coincide in phase.
    D. X and Y oppose each other in phase.
    E. no one can make any definitive statement about their relative phase.

38. If someone tells you that sine wave X has a frequency of 24 MHz and sine wave Y has a frequency of 17 MHz, and then that person claims that X leads Y by 90°, you could correctly respond by saying that

    A. Y leads X by 1/4 of a cycle.
    B. Y leads X by 3/4 of a cycle.
    C. X and Y coincide in phase.
    D. X and Y oppose each other in phase.
    E. no one can make any definitive statement about their relative phase.

39. Imagine that we connect three inductors in parallel. Suppose that no mutual inductance exists among them, and they each individually exhibit a reactance of $0 + j90$ at a frequency of 200 kHz. What's the complex impedance of the parallel combination at a frequency of 400 kHz?

    A. $0 + j135$
    B. $0 + j15$
    C. $0 + j540$
    D. $0 + j60$
    E. We need more information to calculate it.

40. Suppose that a single-turn, air-core loop of wire carries 2.50 A of DC from a battery. Suppose that we rewind the same length of wire so that it forms an air-core coil with four turns instead of only one. If the current remains the same, what happens to the magnetomotive force?

    A. It doesn't change.
    B. It decreases to half of its former value.
    C. It decreases to 1/4 of its former value.
    D. It doubles.
    E. It becomes four times as great.

41. **Which of the following properties A, B, C, or D, if any, constitutes an important asset of integrated circuits compared with circuits built from discrete components such as transistors, resistors, and diodes?**

    A. Large mass, allowing for dissipation of high power
    B. Exceptional sensitivity to electrical signals and noise
    C. Minimal power consumption, maximizing battery life
    D. Ease of replacing individual components inside the device
    E. None of the above

42. **Imagine that a certain wave disturbance has a frequency of 0.00678 MHz. We might better denote this frequency as**

    A. 6780 kHz.
    B. 678 kHz.
    C. 67.8 kHz.
    D. 6.78 kHz.
    E. 678 Hz.

43. **A resistor has a value of 68 ohms and carries 34 mA of direct current. How much potential difference appears across this component?**

    A. 2.3 V
    B. 2.0 kV
    C. 0.5 mV
    D. 79 mV
    E. 2.0 V

44. **In a junction field-effect transistor (JFET), the channel current depends on all of the following factors** *except*

    A. the gate voltage.
    B. the width of the depletion region.
    C. the drain voltage.
    D. the source current.
    E. the thickness of the dielectric layer.

45. **What happens to the collector current in a bipolar transistor as we gradually increase the forward bias at the base, leaving the collector voltage constant and never applying any AC input signal to the device?**

    A. It remains constant.
    B. It decreases gradually until it reaches zero.
    C. It decreases gradually, reaching zero and then reversing.
    D. It increases indefinitely.
    E. It increases up to a certain value and then levels off.

46. **Suppose that we connect a 68-pF capacitor and a 47-pF capacitor in series. What is the net capacitance of the combination?**

    A. 115 pF
    B. 28 pF

C. 57 pF
D. 42 pF
E. We must know the applied AC frequency to calculate it.

47. Which, if any, of the lines or curves in Fig. Test II-4 reasonably portrays relative capacitive reactance as a function of relative capacitance or frequency?

A. Line A
B. Line B
C. Curve C
D. Curve D
E. None of them

48. Suppose that we connect four resistors in series. Three of them have values of 47 ohms, but we don't know the value of the fourth one. We connect a 105-V battery across the series combination, measure the current between the second and third resistors, and obtain a meter reading of 500 mA. From this information, we can calculate that the unknown resistor has a value of

A. 22 ohms.
B. 47 ohms.
C. 69 ohms.
D. 85 ohms.
E. 94 ohms.

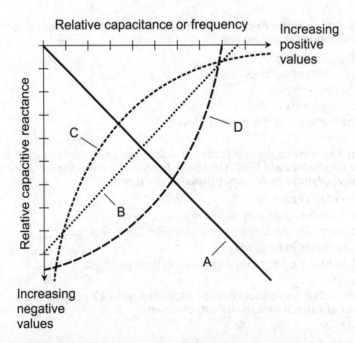

FIGURE TEST II-4 · Illustration for Part II Test Question 47.

49. **In an ideal forward-biased diode where the applied DC voltage remains lower than the forward breakover voltage,**

    A. no current flows.
    B. a small but significant current flows.
    C. the maximum possible current flows.
    D. an intermittent current flows.
    E. a fluctuating current flows.

50. **In a pure AC sine wave with no DC component, the rms voltage equals approximately**

    A. 0.354 times the peak-to-peak voltage.
    B. half the peak-to-peak voltage.
    C. 0.707 times the peak-to-peak voltage.
    D. 1.414 times the peak-to-peak voltage.
    E. twice the peak-to-peak voltage.

Part III

Waves, Particles, Space, and Time

chapter 12

Wave Phenomena

The universe teems with ripples, like the surface of a pond with an added dimension. Waves can occur in any medium we care to imagine (and doubtless in some realms that no human has yet imagined). Consider the following examples:

- The air during a musical concert
- The surface of a pond after a pebble falls in
- The surface of a lake on a windy day
- The surface of the ocean on the north shore of Oahu
- The tops of the stalks in a wheat field
- The surface of a soap bubble when you blow on it
- High clouds near the jet stream
- The earth's surface during a major quake
- The earth's interior after a major quake
- The human brain at any time
- A guitar string after you pluck it
- A utility power line carrying alternating current
- A radio or television transmitting antenna
- An optical fiber carrying a laser beam
- The electromagnetic field inside a microwave oven

CHAPTER OBJECTIVES

In this chapter, you will

- See how waves arise and travel.
- Learn how basic wave properties relate mathematically.
- Observe resonance and standing waves.
- Watch waves interact with physical barriers and other waves.
- Discover the dichotomy between waves and particles.

Intangible Waves

When you sit on a beach as the ocean swells roll in, you can sense their force and rhythm. Prehistoric humans no doubt spent hours staring at the waves on lakes and oceans, wondering where the rollers came from, why they were sometimes big and sometimes small, why they were sometimes smooth and other times choppy, why they sometimes moved with the wind and other times moved against the wind. Imagine children, 100,000 years ago, dropping a pebble into a pond or watching a fish jump, and noticing that the ripples emanating from the disturbance looked like ocean swells except smaller. Then ponder the discoveries scientists eventually made with the help of instruments and mathematics. The children and the scientists shared one important quality: fascination with natural phenomena.

Electromagnetic Waves

Think about the electromagnetic (EM) fields generated by wireless broadcasting. In the journal of cosmic history, these waves have existed in our corner of the universe for only a short while. Television programs have been aired for less than 2 millionths of 1 percent of the life of our galaxy.

Wireless broadcast waves don't arise directly out of nature. They come from specialized hardware systems, invented by a particular species of living thing, on the third planet in orbit around a medium-small star. Do other sentient beings who live on other planets in orbit around other stars employ similar machines to generate similar waves? If so, we haven't heard any of their signals and recognized them as such—yet!

Gravitational Waves

Some scientists think that the fabric of space swarms with *gravitational waves*, just as the sea surface roils with swells and chop. On an even larger scale, some cosmologists suggest that the birth, evolution, and demise of the known universe constitute a single cycle in an oscillating system, a wave with a period measured in billions (thousand-millions) or even trillions (million-millions) of earth years.

How many people walk around thinking they are submicroscopic specks on a particle in an expanding and collapsing bubble in the laboratory of the heavens? Not many; but even to them, gravitational waves, assuming they exist, remain invisible to the eye and defy the keenest mind's insight.

Perfect Waves

Surfers dream of riding perfect waves; engineers strive to synthesize them. A surfer might describe an ideal wave as "tubular," "glassy," and part of an "overhead set" in a "swell" on the north shore of Oahu. In the mind of a communications engineer, a perfect wave forms a sinusoid, which you learned about in Chapter 8.

Even lay people who have never heard of the sine function, can immediately recognize or easily remember the sine-wave shape. An acoustic sinusoid makes an unforgettable noise, concentrating all of the sound energy into a disturbance at a single frequency and manifesting as a "pure tone." A visual sinusoid has an unforgettable appearance. It concentrates visible light at a single wavelength, producing a brilliant hue. A perfect set of ocean waves can give a surfer an unforgettable thrill; it, too, concentrates a lot of energy at a single wavelength.

In Chapter 8, we described a sine wave in terms of a revolving object. As viewed from high above, the object moves in a perfect circle. As seen from edge-on, the object appears to move toward the left, speed up, slow down, reverse, move toward the right, speed up, slow down, reverse, move toward the left again, speed up, slow down, and reverse—repeatedly and indefinitely. Suppose that the object revolves at a constant rate of exactly one revolution per second. In that case, it travels through 180° of arc every half second, 90° of arc every 1/4 second, 45° of arc every 1/8 second, and 1° of arc every 1/360 second. A scientist or engineer would say that the object has an angular speed of 360 degrees per second (360°/s).

Graphing a Sine Wave

Imagine a graph of the above-described revolving object's position with respect to time as seen from some vantage point that lies in the plane of its revolution. Let's plot time horizontally on this graph, so that we see the past as we look toward the left, and we see the future as we look toward the right. One complete revolution of the object shows up as a sine wave. We can assign angular degree values along this wave, corresponding to the degrees around the circle as shown in Fig. 12-1.

Constant rotational motion, such as that of the object on the string, takes place all over the universe. The child whirling the object can't make it abruptly slow down and speed up, or instantly stop and change direction, or "chug-chug" around in discrete steps like a ratchet wheel. But once that mass starts moving, it doesn't take much energy to maintain that motion. Uniform

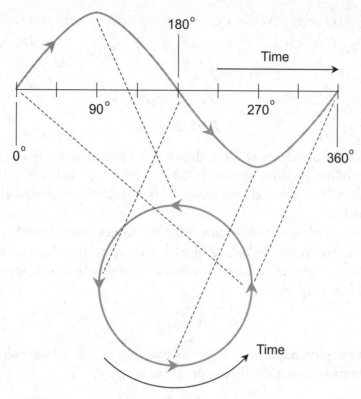

FIGURE 12-1 • Graphical representation of a sine wave as circular motion.

circular motion constitutes a theoretical ideal. No better scheme exists for whirling an object. A sinusoid also constitutes a theoretical ideal. There's no better way to make a wave.

Fundamental Properties

All waves possess three different but interdependent properties (among others). The *wavelength* represents the distance between identical points on two adjacent waves. We express wavelength in meters. The *frequency* represents the number of wave cycles that occur, or that pass a given point, per unit time. We express frequency in cycles per second, or hertz. The *propagation speed* represents the rate at which the disturbance travels through the medium. We express propagation speed in meters per second.

Period, Frequency, Wavelength, and Propagation speed

Sometimes, we'll find it easier to talk about a wave's *period* rather than its frequency. The period T (in seconds) of a sine wave equals the reciprocal of the frequency f (in hertz). We can express this relationship with two formulas, as follows:

$$f = 1/T = T^{-1}$$

and

$$T = 1/f = f^{-1}$$

If a wave has a frequency of 1 Hz, then it has a period of 1 s. If a wave has a frequency of one cycle per minute (1/60 Hz), it has a period of 60 s. If a wave has a frequency of one cycle per hour (1/3600 Hz), it has a period of 3600 s, or 60 min.

Scientists symbolize wavelength using the Greek letter lambda (λ). The period of a wave relates to the wavelength λ (in meters) and the propagation speed c (in meters per second) as follows: "Wavelength equals speed times period." Mathematically:

$$\lambda = cT$$

We can now state three more formulas that express the relationship among period, frequency, wavelength, and propagation speed:

$$\lambda = c/f$$
$$c = f\lambda$$
$$c = \lambda/T$$

PROBLEM 12-1

Imagine that, in the upper portion of Fig. 12-1, we plot the time in seconds (rather than the angle in degrees) on the horizontal axis. That means we replace 0° with 0.00 s on the horizontal scale, 90° with 0.25 s, 180° with 0.50 s, 270° with 0.75 s, and 360° with 1.00 s. Suppose that the "whirling speed" that the child imposes on the ball doubles, so that the ball completes 2 revolutions per second (2 rev/s) instead of 1 rev/s. What happens to the graph in Fig. 12-1, keeping in mind that we plot time on the horizontal scale instead of degrees?

✔ SOLUTION

Consider the formula $\lambda = cT$ given above. In this case, the period T gets cut in half. If T goes down to half its former value, so does λ, because c is a constant. If each horizontal division represents a constant amount of time, then the wave graph in our "modified" version of the upper portion of Fig. 12-1 gets "squashed horizontally" by a factor of 2. Instead of one complete wave cycle appearing along the length of the axis, two complete wave cycles appear.

Frequency Units

Audible acoustic waves repeat at intervals that comprise tiny fractions of a second. The lowest sound frequency that a human being with "good ears" can hear is approximately 20 cycles per second, or 20 hertz (20 Hz). The highest frequency an acoustic wave can have, and still be heard by a person with "good ears," is roughly three orders of magnitude (1000 times) higher, or 20,000 Hz (20 kHz).

Radio waves travel in different media than sound waves. They rarely attain frequencies lower than a few thousand hertz, and their highest frequencies range into the trillions of hertz. *Infrared* (IR) and *visible light* waves occur at frequencies much higher than the radio waves. *Ultraviolet* (UV) *rays, x rays,* and *gamma* (γ) *rays* range into *quadrillions* (thousand-million-millions) and *quintillions* (million-million-millions) of hertz.

To denote high frequencies, scientists and engineers employ frequency units of kilohertz (kHz), megahertz (MHz), gigahertz (GHz), and terahertz (THz). Each unit is 1000 times higher than the previous one in this succession. As you learned in Chapter 8, these units of frequency relate as follows:

$$1 \text{ kHz} = 1000 \text{ Hz}$$

$$1 \text{ MHz} = 1000 \text{ kHz} = 10^6 \text{ Hz}$$

$$1 \text{ GHz} = 1000 \text{ MHz} = 10^9 \text{ Hz}$$

$$1 \text{ THz} = 1000 \text{ GHz} = 10^{12} \text{ Hz}$$

More about Speed

The fastest wave speed ever measured is 299,792 kilometers (186,282 miles) per second in a vacuum. We can round this figure off to 300,000 km/s or 3.00×10^8 m/s. It's the proverbial "speed of light," the absolute maximum speed with which anything can travel, at least according to "conventional" theory. (Some esoteric experiments suggest that certain effects propagate faster, but over long distances, 3.00×10^8 m/s represents the "ultimate cosmic speed limit" as far as we know.) Lesser disturbances than light, in media humbler than intergalactic space, poke along at far lower speeds.

In air at sea level, sound waves travel at about 335 m/s. Aeronautical engineers and aircraft pilots refer to this speed as *Mach 1*. When you speak to someone across a room, your voice travels at Mach 1. Sound in air propagates at Mach 1 regardless of the frequency, and regardless of the intensity (within reason). The exact figure varies a little, depending on the altitude, the temperature, and the relative humidity, but 335 m/s represents a good number to remember.

Still Struggling

Electromagnetic waves don't always race along at the cosmic speed limit. In glass or under water, light waves propagate at speeds significantly less than 3.00×10^8 m/s. Radio waves slow down slightly when they pass through the earth's ionosphere. These variations in speed affect the wavelength, even if the frequency remains constant.

 PROBLEM 12-2

One nanometer (1 nm) equals 10^{-9} m. Suppose a light beam has a wavelength of 500 nm in free space, and then enters a new medium through which light propagates at only 2.00×10^8 m/s. What happens to the wavelength if the frequency remains constant?

 SOLUTION

Note the formula above that defines wavelength in terms of speed and frequency:

$$\lambda = c/f$$

The speed decreases to 200/300, or 2/3, of its initial value, so the wavelength shortens to 2/3 of its initial span. The wavelength in the new medium therefore equals 500 nm × 2/3, or 333 nm.

Amplitude

In addition to the frequency or period, the wavelength, and the propagation speed, waves have another property: *amplitude*, also called *intensity*. In practical scenarios, amplitude corresponds to the strength or the height of a wave, the relative "vertical" distance between *peaks* (wave maxima) and *troughs* (wave minima). When we hold all other factors constant, the amplitude of a wave increases as the energy increases, and vice-versa.

The energy in a light wave varies in direct proportion to the amplitude, in direct proportion to the frequency, and in inverse proportion to the wavelength. The same rule holds true for γ rays, x rays, UV, IR, and radio waves. For waves on the surface of a liquid, however, these relations do not precisely apply. Amplitude is sometimes, but not always, an exact indicator of the energy contained in a wave disturbance.

A wave disturbance can vary in intensity without affecting the frequency, the wavelength, the period, or the propagation speed. For example, we might have two lasers, one powerful and the other not so powerful, both of which emit visible red light at a specific wavelength. The two lasers produce energy at the same frequency, and the rays travel at the same speed through any given medium. However, the wave amplitude from the stronger laser exceeds the wave amplitude from the weaker laser.

Seiche and Harmonics

If you have a bathtub, you know about *seiche* (pronounced "saysh"). You can make any enclosed or semienclosed body of water slosh back and forth at a rate that depends on the size and shape of the container. In your tub, you can easily establish a "slosh period" of 1 or 2 seconds. Give the water a little push, and then another, and then another. Keep this process going at a certain regular repetitive rate, and soon you'll have water all over your bathroom floor. The same thing can happen in a swimming pool during an earthquake, but with a much longer period. When waves approach each other from opposite directions and then pass through each other, the peaks and troughs attain exaggerated heights and depths, as shown in Fig. 12-2.

Anyone who plays a musical instrument such as a clarinet, flute, trumpet, or trombone knows about harmonics. You can blast out a note with certain keys pressed or with the slide at a given position; if you tighten your lips, you can sound a note *one octave* higher. We call the higher-frequency note the *second harmonic* of the first note. The chamber of the instrument contains twice as many wave peaks and troughs at the higher note, as compared with the

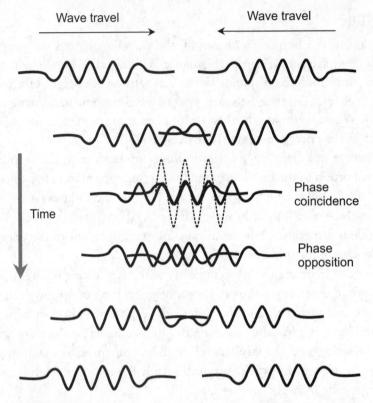

FIGURE 12-2 · When waves collide, their mutual effects can add or subtract.

lower note. If you try hard enough, you might get the instrument to produce a note at three times the original, or *fundamental*, frequency. We call that the *third harmonic*.

Still Struggling

When the frequency of a particular wave Y equals *n* times the frequency of some original or fundamental wave X (where *n* equals a whole number larger than 1), we say that Y is at the *nth harmonic* of X. Mathematically, no limit exists as to how far this frequency-multiplication process can continue (Fig. 12-3). When the frequency of one wave equals a harmonic of the frequency of another wave, we say that the two waves are *harmonically related*.

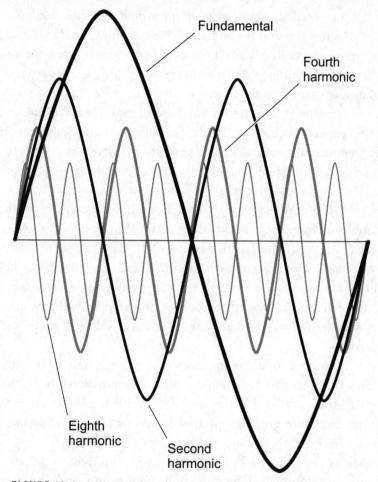

FIGURE 12-3 • Resonant effects occur at wavelengths representing whole-number fractions of the wavelength of the fundamental.

Resonance

You can demonstrate harmonics if you have a piece of rope about 10 m long. Tie one end of the rope securely to an immovable object such as a fence post or a hook in a wall. Hold the other end of the rope, and back off until you've pulled it tight. Then start pumping the rope up and down: slowly at first, then gradually faster. At a certain pumping speed, the rope will "get into the rhythm." It will seem to help you with your pumping! You've produced a condition of *resonance*.

Keep the rope oscillating at this speed for awhile, and then suddenly double the rate at which you pump the end of the rope. If you keep things going at the new, higher speed, you'll get a full wave cycle to appear along the rope. The wave will reverse itself in phase each time you pump, and its curvature will attain a familiar shape: the sinusoid.

Keep on pumping at this rate for a minute or two. Then, if you can, double the pumping speed once more. This experiment requires some conditioning and coordination, but eventually you'll get two complete wave cycles to appear along the length of the rope. You've reached the second harmonic of the previous oscillation, and resonance is taking place again.

If you're strong and fast enough, and if you have enough endurance, you might double the frequency another time, getting four complete wave cycles to appear along the rope (the fourth harmonic). If you're a professional athlete, maybe you can double it even one more time, getting eight waves (the eighth harmonic). Theoretically, no limit exists to the number of wave cycles that can appear between the shaker and the anchor. In the real world, of course, the diameter and elasticity of the rope impose a limit, as does the physical ability of the shaker!

When you pump a rope, the impulses that you generate have *longitudinal motion*; they travel lengthwise along the rope. The individual molecules in the rope undergo *transverse motion*; they move from side to side (or up and down). The waves result from impulses alternately adding and canceling in phase as they bounce back and forth along the length of the rope.

Stop shaking the rope and let it come to rest. Pull it tight, and then give it a quick, hard, single pump. A lone pulse will shoot from your hand toward the far end of the rope, and then reflect from the anchor and travel back toward you. As the pulse propagates, its amplitude decays. Hold your hand steady as the pulse returns. The pulse energy will be partially absorbed by your arm, and partially reflected from your hand, heading down toward the far end again. After several reflections, the wave will die down to nothing. Some of its energy will have dissipated in the rope. Some energy will have gone into the object at the rope's far end. Your body will have reabsorbed some of the energy that originally came from it. Even the air in the room will have taken up a little bit of the original energy.

Standing Waves

Start shaking the rope rhythmically once again. Set up waves along its length just as you did before. Send sine waves down the rope. At certain shaking frequencies, the pulses reflect back and forth between your hand and the anchor so that their effects add together: each point on the rope experiences a force upward, then downward, then up again, then down again. The reflected impulses reinforce, exaggerating the vertical motion of the rope. *Standing waves* appear.

Standing waves derive their name from the fact that they do not, in themselves, travel anywhere. They simply "stand there." But they can nevertheless acquire tremendous power. Some points along the rope move up and down a lot, some move up and down a little, and others stand absolutely still. *Loops* constitute the points where the rope moves up and down to the greatest extent. *Nodes* constitute the points where the rope doesn't move up or down at all. Two loops and two nodes exist in a single, complete standing-wave cycle. If you look at the situation closely, you'll see that the loops and nodes occur at evenly spaced intervals along the rope.

 PROBLEM 12-3

How distant is a standing-wave loop from the nearest node on either side, in terms of degrees of phase?

 SOLUTION

As you remember from Chapter 8, a complete wave cycle contains 360° of phase. From the above description, two loops and two nodes exist in a complete cycle, all equally spaced from each other. Therefore, every loop lies exactly 1/4 cycle, or 90° of phase, away from the nearest node on either side. Conversely, any given node lies 90° of phase away from the nearest loop on either side.

Irregular Waves

Not all waves present themselves as sine waves. Some waves have abrupt transitions; unlike the smooth sinusoid, they jump or jerk back and forth. If you've ever used an *oscilloscope*, you've likely seen all sorts of waveforms such as the *square wave*, the *ramp wave*, the *sawtooth wave*, and the *triangular wave*. You learned about these waveforms in Chapter 8. You can generate any of them with an electronic music synthesizer.

Each waveform exhibits its own unique mathematical perfection, but you'll never see square waves, ramp waves, sawtooth waves, or triangular waves on the surface of the ocean! *Irregular waves*, however, come in myriad shapes, like fingerprints or snowflakes. The world's lakes, rivers, and oceans teem with irregular waves. In the "real world of waves," simplicity rarely occurs.

Most musical instruments produce irregular waves, like the chop on the surface of a lake. They comprise complicated combinations of sine waves at various harmonically related frequencies. We can break any periodic waveform, no matter how irregular, down into sinusoidal components, although the mathematics that define this process can get arcane. Cycles superimpose themselves on longer cycles, which in turn superimpose themselves on still longer cycles, *ad infinitum*.

Even the square wave, ramp, sawtooth, and triangular waves, with their straight edges and sharp corners, constitute composites of smooth sinusoids that exist in precise proportions. Waves of this sort treat human ears and brains "more kindly" than pure acoustic sine waves.

TIP *If you have access to a high-end music synthesizer or signal generator, try setting the device to produce square, ramp, sawtooth, triangular, and irregular waves, and listen to the differences in the way they sound. Even when all have the same frequency, the timbre, or "tone" of the sound, varies considerably.*

Wave Interaction

Two or more waves can combine to produce interesting effects and, in some cases, remarkable patterns. We can exaggerate amplitudes, alter waveforms, and generate entirely new waves. Common wave-interaction phenomena include *interference*, *diffraction*, and *heterodyning*.

Interference

Imagine yourself as a surfer who spends the Northern Hemisphere winter months on the North Shore of Oahu, Hawaii. In the maritime sub-Arctic, storms parade across the North Pacific Ocean, spinning off from a parent vortex near the Kamchatka peninsula. The so-called *Kamchatka low* remains strong and stable, spawning storm systems that swoop southeastward and vent their fury on distant shores. Swells from these storms propagate across the entire Pacific. The swells arrive on the beaches of Oahu and break over coral and sand, often reaching heights in excess of 5 m (16 ft).

Trade winds blow from east to west, producing smaller swells across the big ones. Gusts of wind and local squalls add chop. On a good day, the storm swells dominate, so you can ride the big breakers without being bumped around by the chop. On a bad day, waves pile onto waves in a haphazard way. The main swell contains every bit as much power as it does on a good day, but the wave interaction—the interference—makes surfing difficult.

When a great distance separates two major marine storms, with each storm producing significant swells, the composite waveforms get particularly interesting. Such conditions attract more scientists than surfers. This type of situation can occur during the winter on the North Shore of Oahu, but we find it more often in the tropics during hurricane season (late summer and early autumn). Tropical storms produce some of the largest surf on this planet. When a hurricane prowls the sea, swells radiate in expanding circles from the storm's central vortex. If a vast distance separates two storms of similar size and intensity, complex swell patterns can span millions of square miles of ocean surface. Between the storms, swells alternately cancel and reinforce each other, producing wild seas.

Interference patterns created by multiple wave sources appear at all scales, from swells at sea to sound waves in a concert hall, from radio broadcast towers to holographic apparatus. The slightest change in the relative positions or wavelengths of two sources can make a profound difference in the way the composite pattern emerges, as shown in Figs. 12-4 and 12-5.

Rogue Waves

Imagine two massive tropical storms sweeping around the Atlantic Basin, steered by currents in the upper atmosphere. The interference pattern produced by their swells evolves from moment to moment. Multiple crests and troughs conspire along a front hundreds of kilometers long, generating a so-called *rogue wave* that can damage, or even capsize, ocean liners and freighters. Veteran sailors may tell you incredible stories about walls of water that break in the open sea, seeming to defy the laws of hydrodynamics and striking fear into stout hearts.

Wave interference on the high seas, while potentially frightening in its proportions, doesn't readily lend itself to scientific observation. Patterns sometimes reveal themselves to observers flying in aircraft at high altitudes. Sophisticated radar devices can portray subtleties of the ocean surface, but the high seas do not offer a favorable environment for controlled experiments. You can't expect to venture out in a boat, sail into storm-swell interference patterns, and return with meaningful data, although you might return with a mind full of stories to tell your grandchildren (if you survive).

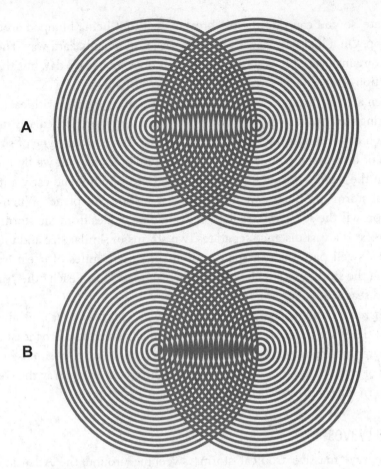

FIGURE 12-4 · A slight source displacement can dramatically change an interference pattern. Notice the difference between the pattern formed by the intersecting lines at A, compared with the pattern at B.

Even a child can conduct memorable small-scale experiments with wave convergence and interference. Soap bubbles, with their rainbow-colored surfaces, seem tailor made for this purpose. Visible-light waves add and cancel across the visible spectrum, reflecting from the inner and outer surfaces of the soap film, teasing the eye with red, green, violet, then red again. Adults can play with wave interference, too. Any building with a large rotunda provides a perfect venue.

TIP   *According to legend, long ago in the halls of Congress, a few elected officials were able to eavesdrop on certain supposedly private whisperings, because the vast dome overhead reflected and focused the sound waves from one politician's mouth to another's ears.*

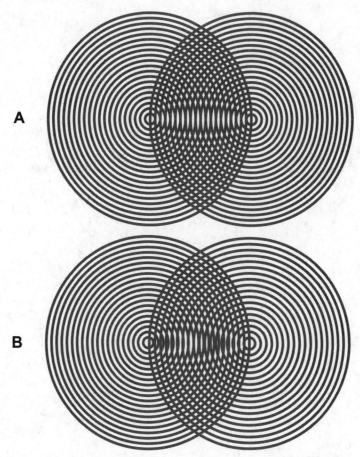

FIGURE 12-5 • Identical wavelength (at A) versus a 10 percent difference in wavelength (at B). The pattern changes significantly, even when the wavelength changes only slightly.

Diffraction

Waves can gang up, fight each other, and travel in illogical directions at unreasonable speeds. Waves can also propagate around obstacles with "sharp corners" because of the wave's ability to *diffract*. When a wave disturbance encounters a "sharp" obstruction, the obstruction behaves as a new source of energy at the same wavelength (Fig. 12-6). The phenomenon can occur repeatedly. Anyone who lives in a city for a long enough time learns that you can run from noise, but you can't escape it! You can blame this maddening effect, at least partially, on diffraction.

Long waves diffract more readily than shorter waves. As an edge or corner becomes sharper relative to the wavelength of sound, diffraction occurs more

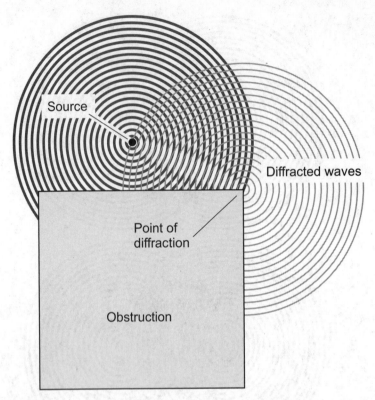

Source

Diffracted waves

Point of
diffraction

Obstruction

FIGURE 12-6 · Diffraction allows wave disturbances to "go around corners."

efficiently. As the frequency decreases, the wavelength increases, so all edges and corners become, in effect, sharper. This phenomenon is not unique to sound waves. It happens with water waves, as any surfer knows. It happens with radio waves, explaining why you can hear the broadcast stations on your car radio, especially on the AM broadcast band where the electromagnetic (EM) waves measure hundreds of meters long, even when buildings or hills loom between you and the transmitter. It happens with visible light waves too, although only under certain conditions.

When the wavelength of a disturbance greatly exceeds the dimensions of an obstruction, the waves diffract so efficiently that they pass the object as if it's not there at all. A flagpole has no effect on low-frequency sound waves. Ocean surf likewise ignores the pilings of a pier. Longwave radio transmissions can easily pass around concrete-and-steel buildings (but not easily get inside!).

Still Struggling

All waves can diffract around a corner—if that corner is "sharp" enough compared to the wavelength. One of the tests that scientists use to verify the wavelike nature of a disturbance involves finding out whether or not the effect can "go around a sharp corner."

Heterodyning

No matter what the mode, and regardless of the medium, waves can combine to produce other waves. When this happens with sound, we call it *beating*; when it happens with radio signals, we call it *heterodyning* or *mixing*. Two sound waves having slightly different frequency (or, as a musician would say, slightly different *pitch*) will beat to form a new wave at a much lower frequency, and another wave at a higher frequency. Figure 12-7 shows examples of wave beating in which the low-frequency disturbances present themselves clearly:

- In drawing A, the waves, shown by sets of vertical lines, differ in frequency by 10 percent (f and $1.1f$).

- In drawing B, the waves differ in frequency by 20 percent (f and $1.2f$).

- In drawing C, the waves differ in frequency by 30 percent (f and $1.3f$).

Beat and heterodyne waves always occur at frequencies equal to the sum and difference between the frequencies of the waves that produce them. If you play two notes together, one at 1000 Hz and another at 1100 Hz, the beat notes appear at 100 Hz and 2100 Hz. If you play notes at 1000 Hz and 1200 Hz, you get beat notes at 200 Hz and 2200 Hz. If you hold one note steady and continuously vary the frequency of the other, the beat notes rise and fall in pitch.

Engineers discovered *radio-frequency* (RF) *heterodyning* in the early part of the 20th century. Under certain conditions, two high-frequency AC signals combine to produce a new signal at the difference frequency, and another new signal at the sum frequency. Engineers can easily design and construct circuits to produce this effect. In fact, it's difficult to prevent, and can sometimes cause trouble in wireless equipment.

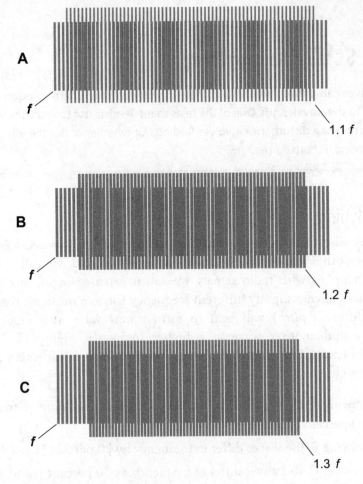

FIGURE 12-7 • Two waves can beat to form a new wave at the difference frequency. At A, the two waves differ in frequency by 10 percent; at B, by 20 percent; at C, by 30 percent.

How to Find Beat Frequencies

Given two waves having different frequencies f and g (in hertz) where $g > f$, they beat or heterodyne together to produce new waves at frequencies x and y (also in hertz) according to the formulas

$$x = g - f$$

and

$$y = g + f$$

These formulas also apply for frequencies in kilohertz, megahertz, gigahertz, and terahertz—provided, of course, that we stick with the same units throughout any given calculation.

TIP   *Using a couple of loud horns, you (and a couple of friends in you local band) can do an experiment to demonstrate sound-wave beating. When two people play trumpets at different pitches in the treble clef at the same time, you'll hear a low-frequency hum in addition to the original notes. The hum constitutes the lower-pitched beat note. You will not likely notice the higher-pitched note, even though it exists and will render on an oscilloscope. The low-frequency beat note will seem to come from an indeterminate direction, producing "acoustical confusion" in your mind.*

 **PROBLEM 12-4**

Suppose that you use an acoustic synthesizer to generate two sound waves, one at 500 Hz and another at 2.500 kHz. What are the beat frequencies?

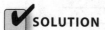 **SOLUTION**

Convert the frequencies to hertz. Then let $f = 500$ Hz and $g = 2500$ Hz. According to the above formulas, you get beat frequencies x and y, as follows:

$$x = g - f$$
$$= 2500 - 500$$
$$= 2000 \text{ Hz}$$

and

$$y = g + f$$
$$= 2500 + 500$$
$$= 3000 \text{ Hz}$$

If you want to get particular about significant digits, you can call these beat frequencies 2.00 kHz and 3.00 kHz, respectively.

Wave Mysteries

Sometimes it seems as if, the more we explore the mysteries of wave phenomena, the greater and more numerous those mysteries become. For every question that we think we've answered, we get new questions.

Lengthwise versus Sideways

When waves propagate through matter, the atoms or molecules oscillate to and fro, up and down, or back and forth. The nature of the particle movement differs from the nature of the wave as it travels. The atoms or molecules rarely move more than a few meters—sometimes less than a centimeter—but the wave disturbance can travel thousands of kilometers. Sometimes the particles vibrate in line with the direction of wave travel, giving us a *compression wave*, also called a *longitudinal wave*. In other instances, the particles move at right angles to the direction of propagation, producing a *transverse wave*. Figure 12-8 illustrates the distinction between these two wave propagation modes.

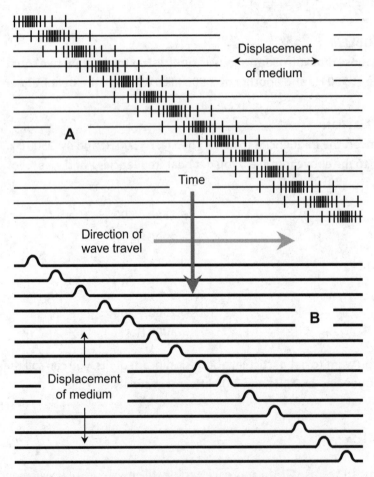

FIGURE 12-8 · In a longitudinal wave (A), particles vibrate parallel to the direction in which the disturbance travels. In a transverse wave (B), particles vibrate perpendicular to the direction in which the disturbance travels.

Sound waves in air occur only in the longitudinal mode, but sound waves in steel can propagate in the transverse mode as well. The waves on the surface of the ocean are mainly transverse (up and down) although a little longitudinal motion takes place among the water molecules. However, when large swells break on a beach, a great deal of longitudinal motion can occur (shoreward and seaward), as anyone who has experienced an undertow knows.

Still Struggling

What mysterious "stuff" actually wags or wiggles or compresses or stretches when a wave travels through a particular medium? That's a difficult question to answer. It depends on the medium, and on the nature of the wave disturbance. When sound waves travel through the air, through water, through steel, or through any other physical medium, individual atoms or molecules "do the waving."

Force Fields

When waves travel through a vacuum, they manifest as force fields in the case of EM waves, or as undulations in space and time known as *gravitational waves*. Scientists needed many years, if not centuries, to realize and accept the fact that waves can propagate through a perfect vacuum—through absolutely empty space—without any apparent medium to "carry" them.

Both EM and gravitational waves constitute transverse disturbances. The electric and magnetic force fields in radio waves, IR, visible light, UV, x rays, or γ rays alternate at right angles to each other and at right angles to the direction of propagation. This phenomenon occurs in three dimensions, so we can envision it in our "mind's eye." Gravitational waves, however, cause time and space themselves to undulate, defying most people's attempts to envision what actually happens.

Corpuscles of Light

The theory of EM-wave propagation constitutes a relatively recent addition to the storehouse of physics knowledge. *Isaac Newton*, the 17th-century English physicist and mathematician known for his theory of gravitation and his role (along with the German mathematician *Gottfried Wilhelm Leibniz*) in the invention of calculus, believed that visible light consists of submicroscopic particles. Today, scientists know that *corpuscles* of EM energy, called *photons*, have momentum, and they exert measurable pressure on objects they strike. We can break the energy in a beam of light down into packets of a certain minimum size, but no smaller.

We need not search very long to find complications with the so-called *corpuscular theory of light* (also called the *particle theory*). At a boundary between air and water, photons do inexplicable things. Have you ever thrust a fishing pole into a lake, or looked into the deep end of a swimming pool and seen 4 m of water look like 1 m? Then you know that light doesn't always travel in straight lines. Photons change direction abruptly when they pass at a sharp angle from water to air or vice versa (Fig. 12-9). What force gives them the "sideways push"? When light rays pass through a glass prism, the photons change course in the same way, but with an additional mysterious twist: the extent to which they deflect depends on their color.

Alternatives to the Corpuscular Theory

Some of Newton's colleagues thought that he had constructed an overly simplistic description of the nature of light, so they embarked on a "knowledge quest"

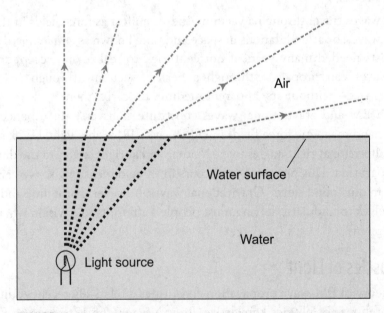

FIGURE 12-9 • If light rays comprise particles, what mysterious force pushes them sideways at the water surface?

for alternative theories. *Christian Huygens*, a Dutch physicist fond of optics, was one of the first to suggest that visible light constitutes a wave disturbance, like the ripples on a pond or the vibrations of a violin string. Today, even lay people speak of "light waves" as if the two words go together naturally, but few scientists in the 1600s sensed the connection. Huygens kept at his research, and showed that light waves interfere with each other in the same way as waves on water, and in the same way as waves from musical instruments.

The *wave theory of light* offers a good explanation of why light rays change direction at the surface of a lake or pool. When a light beam strikes the surface, the beam abruptly deflects at an angle, as shown in Fig. 12-10. The extent of the bending depends on the angle at which the wavefronts strike the surface. Wavefronts parallel to the boundary don't change direction at all. Wavefronts striking the water surface at a small angle change direction slightly. As the *angle of incidence* increases, so does the extent of the deflection. Wavefronts striking the water surface at a large enough angle from within the water don't pass through the surface boundary, but instead, they bounce back into the water as if from a flat mirror.

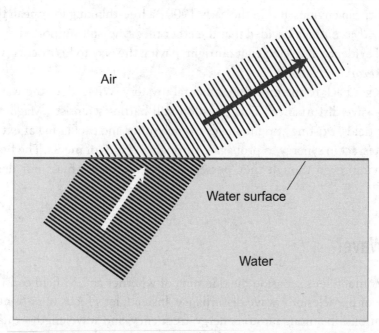

FIGURE 12-10 • Light waves change speed and wavelength when they strike a boundary between media having different indices of refraction.

What "Does the Waving"?

Huygens's colleagues had trouble accepting his wave theory, even when they saw the demonstrations with their own eyes. They asked bothersome questions such as the following:

- What "does the waving" in a light-wave disturbance?
- When light passes through the atmosphere, does the air vibrate?
- If light makes air vibrate, why can't we hear light?
- When light waves enter water, does the water undulate?
- If light makes water undulate, why doesn't it disturb the surface?
- If visible light comprises a wave disturbance, why does a glass jar with all the air pumped out appear transparent, rather than opaque? After all, nothing exists inside an evacuated jar to "carry the disturbance"!

In the 1800s, some scientists tried to answer questions such as these by postulating the existence of a *luminiferous ether*, a medium that supposedly permeated all of space, perhaps even the space between atoms of matter such as the rocks inside the earth. In their attempts to accommodate the existence of this strange "stuff," theoretical physicists did mental gymnastics that would render any politician envious. But in the early 1900s, a free-thinking European theorist named *Albert Einstein* decided that the ether theory wasn't supported by experimental evidence. He rejected it outright, paving the way to his discovery of the *special theory of relativity*.

Let's get back to the most fundamental mystery: What "does the waving" in an EM-wave disturbance? This question still baffles scientists. Magnetic and electric fields, existing at right angles to each other and oscillating at extremely high rates, act in synergy to propagate through all kinds of media. The fields "do the waving" even though they possess neither definable mass nor definable volume.

Particle or Wave?

For a moment, let's set aside the dilemma of whether an EM field consists of a barrage of particles or a wave disturbance. Instead, let's focus our attention on the relationship among photon energy, frequency, and wavelength—quantities that we can define and measure.

Energy, Frequency, and Wavelength

We can calculate the energy contained in a single photon of visible light or any other form EM radiation using the formula

$$e = hf$$

where e represents the energy (in joules) in the photon, f represents the frequency of the EM-wave disturbance (in hertz), and h represents a number known as *Planck's constant*, a tiny quantity approximately equal to 6.6261×10^{-34}.

If we know the wavelength λ (in meters) and the propagation speed c (in meters per second), then

$$e = hc/\lambda$$

When we rearrange this formula to express the wavelength of a photon in terms of the energy it contains, we obtain

$$\lambda = hc/e$$

For EM rays in free space, the product hc equals approximately 1.9865×10^{-25}, based on the approximation $c \approx 2.99792 \times 10^8$ m/s.

PROBLEM 12-5

How much energy does a photon contain in an EM disturbance having a wavelength $\lambda = 550$ nm in free space?

SOLUTION

First, let's convert 550 nm to meters. When we do that, we get $\lambda = 5.50 \times 10^{-7}$ m. Then we can use the formula for the energy e in terms of wavelength λ to get

$$e = hc/\lambda$$
$$= (1.9865 \times 10^{-25})/(5.50 \times 10^{-7})$$
$$= 3.61 \times 10^{-19} \text{ J}$$

PROBLEM 12-6

What's the wavelength of an EM disturbance where each photon has 1.000×10^{-25} J of energy in free space?

**SOLUTION**

We can use the formula for wavelength λ in terms of energy *e* to obtain

$$\lambda = hc/e$$
$$= (1.9865 \times 10^{-25})/(1.000 \times 10^{-25})$$
$$= 1.9865 \text{ m}$$

This EM disturbance presents itself as a signal in the *very-high-frequency* (VHF) portion of the radio-frequency (RF) spectrum.

Double-Slit Experiments

If a beam of light gets dim enough, its photons emanate from the source at intervals that we can measure in seconds, minutes, hours, days, or years. If a beam of light gets brilliant enough, its photons rain down at the rate of trillions per second. We can detect these particles and determine their energy content as if they were tiny bullets traveling at 2.99792×10^{8} m/s in free space. But the corpuscular theory of EM radiation does not explain *refraction* of the sort that occurs at the surface of a body of water. The corpuscular theory also fails to explain the beating and interference effects that we observe with visible light and high-speed subatomic particles. The so-called *double-slit experiment* offers a classic demonstration of the wavelike nature of visible light.

Around the year 1800, an English physicist named Thomas Young directed a beam of light, having a certain color and coming from a nearly perfect-point source, at a barrier with two narrow slits cut in it. Beyond the barrier, Young placed a flat sheet of photographic film. The light would, he supposed, pass through the two slits and land on the film, producing a pattern of some sort. "Common sense" seemed to dictate the following two notions:

- If light comprises discrete corpuscles, then the pattern on the film should show up as two bright, vertical lines.
- If light comprises a series of waves, then the pattern on the film should appear as alternating bright and dark bands.

When Young actually did the experiment, alternating bright and dark bands appeared (Fig. 12-11), practically proving that visible light constitutes a wave disturbance. These *interference bands* showed that light rays diffracted as they passed through the slits. The crests and troughs from the two diffracted rays

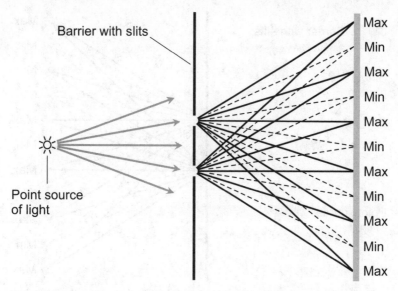

FIGURE 12-11 • When photons pass through a pair of slits in a barrier, a wavelike interference pattern results.

alternately added together and canceled out as they arrived at various points on the film. This phenomenon would naturally happen with a wave disturbance, but not with stream of corpuscles—or, at least, not with any sort of particle ever imagined up to that time.

Nevertheless, the outcomes of other experiments strongly suggested that light has a particle nature. Scientists couldn't dismiss or ignore those results. What about the pressure that visible light exerts? What about the discovery that visible-light energy breaks down into packets having a certain minimum "size" (energy content)? Might light somehow constitute both a wave disturbance and a particle stream?

Imagine that we hurl photons *one by one*, say at intervals of a second or longer, at a barrier with two slits and allow the photons to land on a sensitive surface. Scientists have conducted experiments of this sort, and interference bands have always appeared on the surface, no matter how weak the beam. Even if experimenters make the beam so dim that only one photon hits the surface every minute, the pattern of light and dark bands appears after a period of time long enough to expose the film, as shown in Fig. 12-12. This pattern changes depending on the distance between the two slits, but for a given slit separation and a given film distance from the slits, experimenters get the same pattern regardless of the beam intensity.

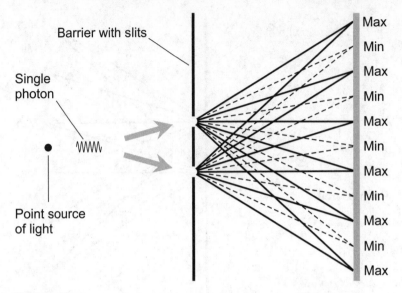

FIGURE 12-12 • How can "wave-particles" pass one at a time through a pair of slits, and nevertheless produce an interference pattern on a projection screen?

Still Struggling

What happens in experiments like the one just described? Do photons "know" where to land on the film, based on the wavelength of the light they represent? How can a single photon, passing through one slit, "ascertain" the separation between the slits, thus "knowing" where it "can" and "cannot" land on the film? Do single photons somehow split in two and pass through both slits at the same time? If so, how does the photon "know" that it's approaching a barrier with slits (instead of open space) before it gets to that barrier? Does some effect take place backward in time, so the photons from the light source "know" that they're about to pass through a barrier, and even "know" the physical details of that barrier? The very act of asking such questions smacks of metaphysics, so objective scientists get uneasy with such matters. But the phenomena occur nevertheless, daring us to explain why.

QUIZ

Refer to the text in this chapter if necessary. A good score is eight correct. Answers are at the back of the book.

1. Consider a beam of IR radiation whose wavelength is 1300 nm. How much energy does a single photon contain?

    A. 2.852×10^{-19} J
    B. 1.528×10^{-19} J
    C. 7.224×10^{-22} J
    D. We need more information to calculate it.

2. The particle theory of light works well when we want to explain

    A. the interference patterns produced by rays passing through multiple slits.
    B. the refraction of rays through a glass lens.
    C. the propagation speeds of rays through different transparent media.
    D. the physical pressure that a ray produces when it strikes a dust particle.

3. What is the frequency of an EM wave measuring 21 cm long as it propagates through interstellar space?

    A. 1.4 GHz
    B. 6.2 GHz
    C. 7.0 GHz
    D. We need more information to calculate it.

4. Which of the following types of waves can propagate in the transverse mode?

    A. Swells on the ocean
    B. Sound waves in steel
    C. Pulses along a rope
    D. All of the above

5. Suppose that two sound waves, one having a frequency of 775 Hz and the other having a frequency of 995 Hz, beat together. At which of the following frequencies does a beat note occur?

    A. 220 Hz
    B. 885 Hz
    C. 1.77 kHz
    D. More than one of the above

6. Imagine that sound waves propagate at 500 m/s in the atmosphere on the surface of a certain planet. An acoustic wave having a period of 0.155 s in that medium measures

    A. 77.5 m long.
    B. 12.9 m long.
    C. 310 m long.
    D. 3.23 km long.

7. Most musical instruments produce acoustic waves whose shapes are

    A. rectangular.

    B. irregular.

    C. sinusoidal.

    D. transverse.

8. In a vacuum, an EM wave 150 m long has a frequency of

    A. 200 kHz.

    B. 450 kHz.

    C. 2.00 MHz.

    D. 4.50 MHz.

9. Suppose that we dangle a rope measuring 3.00 km long from a hot-air balloon flying 4.00 km above the ground, so that the bottom end of the rope hangs free. We shake the rope at a constant rate of one complete cycle every 1.67 s. At what speed do the resulting waves travel downward along the rope?

    A. 5.01 km/s

    B. 1.79 km/s

    C. 557 m/s

    D. We need more information to calculate it.

10. In which of the following situations would you expect to observe the *least* amount of diffraction?

    A. Visible-light waves around a utility pole.

    B. Ocean waves around the piling of a pier.

    C. Low-pitched sound waves around the corner of a house.

    D. Radio waves at 830 kHz around a flagpole.

Forms of Radiation

Isaac Newton believed that visible light comprises tiny corpuscles that we recognize today as photons. But light, as we've learned, also has wavelike characteristics. The *particle/wave dichotomy* applies to all forms of radiant energy.

CHAPTER OBJECTIVES

In this chapter, you will

- Compare static and fluctuating electric and magnetic fields.
- Explore the electromagnetic spectrum.
- Find out what ELF "radiation" is (and what it isn't).
- Watch radio waves propagate through the atmosphere and around the earth.
- Learn about waves beyond the visible spectrum.
- Analyze radioactivity and discover exotic high-speed particles.

Electromagnetic Fields

Radiant energy manifests as a combination of electrical and magnetic effects. *Electric fields* always surround charged objects, including subatomic particles such as electrons and protons. Magnetic poles or moving charged particles produce *magnetic fields*. When an electric or magnetic field attains sufficient strength, its measurable effects extend for a considerable distance around the object responsible. If the fields vary in intensity with the passage of time, we have an *electromagnetic (EM) field*.

Static Fields

You've doubtless observed the attraction between opposite poles of magnets, and the repulsion between like poles. Similar effects occur with electrically charged objects. These forces seem to operate over limited distances under laboratory conditions because such fields rapidly weaken, as the distance between poles increases, to less than the smallest intensity that we can detect. In theory, however, all electric and magnetic fields extend into space indefinitely.

As a constant, direct electric current flows in a wire, that current produces a magnetic field around the wire. So-called *lines of magnetic flux* appear perpendicular to the direction of the current. If a constant voltage difference exists between two nearby objects, we get an electric field that produces *lines of electric flux* parallel to a straight line connecting the centers of the charged objects. When the intensity of a current or voltage changes with time, things get more complicated—and more interesting.

Fluctuating Fields

A fluctuating current in a wire, or a variable voltage between two nearby objects, gives rise to an EM field. The fluctuating electric (E) and magnetic (M) fields act in synergy so that they "leapfrog" through space. As a result, an EM field can travel farther than an E or M field alone before it "dies off." The E and M lines of flux in an EM field intersect at right angles everywhere in space. The EM field propagates in a direction perpendicular to the plane defined by the E and M lines of flux (Fig. 13-1).

In order for an EM field to exist around a length of current-carrying wire, the electrons in the wire must change their speed from moment to moment. In other words, they must accelerate along the length of the wire. We can easily

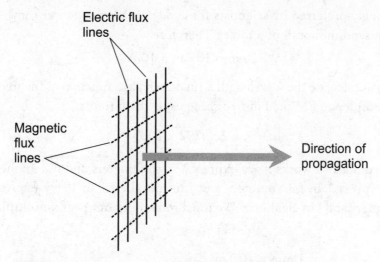

FIGURE 13-1 · An EM wave consists of fluctuating, mutually perpendicular electric and magnetic lines of flux. The field propagates at right angles to both sets of flux lines.

make this acceleration take place by forcing AC through the wire. An EM field can also result from certain other charge-carrier behaviors, such as the bending of an electron beam by a strong E field or M field.

Frequency and Wavelength

Electromagnetic waves travel through free space (a vacuum or dry air) at the speed of light: approximately 2.99792×10^8 m/s (1.86282×10^5 mi/s). We can often round this figure up and express it to three significant figures as 3.00×10^8 m/s. The wavelength of an EM field in free space decreases as the frequency increases. For example:

- An EM field at 1.00 kHz has a wavelength of 300 km.
- An EM field at 1.00 MHz has a wavelength of 0.300 km or 300 m.
- An EM field at 1.00 GHz has a wavelength of 0.300 m or 300 mm.
- An EM field at 1.00 THz has a wavelength of 0.300 mm or 300 μm.

The frequency of an EM field can get much higher than 1 THz; some of the most energetic known rays have wavelengths of 0.00001 *Ångström* (10^{-5} Å). The Ångström represents exactly 10^{-10} m. Some scientists use Ångströms to quantify extremely short EM wavelengths. You'd need a microscope of great magnifying power to see an object with a diameter of 1 Å. Another unit, increasingly preferred by scientists these days, is the *nanometer* (nm), which is a thousand-millionth of a meter. Therefore

$$1 \text{ nm} = 10^{-9} \text{ m} = 10 \text{ Å}$$

We can calculate the wavelength λ (in meters) as a function of the frequency f (in hertz) for an EM field in free space using the formula

$$\lambda = 2.99792 \times 10^8/f$$

This formula also works if we express λ in millimeters and f in kilohertz, or if we express λ in micrometers and f in megahertz, or if we express λ in nanometers and f in gigahertz. We must remember our prefix multipliers, as follows:

$$1 \text{ mm} = 10^{-3} \text{ m}$$
$$1 \text{ μm} = 10^{-3} \text{ mm} = 10^{-6} \text{ m}$$
$$1 \text{ nm} = 10^{-3} \text{ μm} = 10^{-6} \text{ mm} = 10^{-9} \text{ m}$$

We can calculate the frequency f (in hertz) as a function of the wavelength λ (in meters) for an EM field in free space by transposing f and λ in the above formula, obtaining

$$f = 2.99792 \times 10^8/\lambda$$

As in the preceding case, this formula will work for f in kilohertz and λ in millimeters, for f in megahertz and λ in micrometers, and for f in gigahertz and λ in nanometers.

TIP   *Remember that in formulas such as those just described, all units go with their reciprocal counterparts. Thousands go with thousandths; millions go with millionths; billions go with billionths; trillions go with trillionths.*

Many Forms

The discovery of EM fields led ultimately to the variety of wireless communications systems that we take for granted today. *Radio waves* are not the only form of EM radiation. As the frequency increases above that of conventional radio, we encounter new forms. First come the so-called *microwaves*. Then come infrared (IR) rays, often (inaccurately) called "heat rays." After that come visible light, ultraviolet (UV) rays, x rays, and gamma (γ) rays.

In the opposite and less-often-imagined sense, EM fields can exist at frequencies far below those of radio signals. In theory, an EM wave can go through one complete cycle every hour, day, year, thousand years, or million years. Some astronomers suspect that stars and galaxies generate EM fields with periods of years, centuries, or millennia. We might even speculate that the whole universe produces an EM field that goes through a complete cycle every few billion (thousand-million) years.

The EM Wavelength Scale

We can take advantage of a *logarithmic graph scale* to illustrate the range of EM wavelengths. We need this type of scale because of the extreme range over which EM wavelengths can exist. The left-hand portion of Fig. 13-2 shows a logarithmic scale covering 20 *orders of magnitude* (spanning 20 powers of 10), portraying

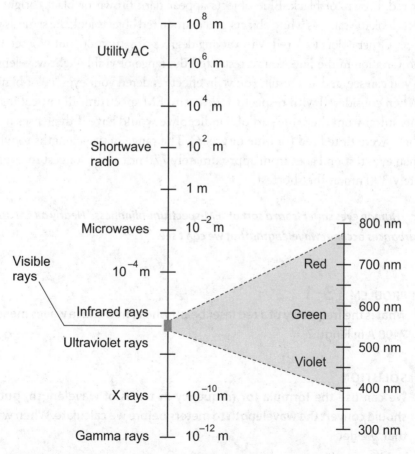

FIGURE 13-2 · At left, the EM spectrum at wavelengths ranging from 10^8 down to 10^{-12} m. At right, an exploded view of the visible-light portion of the EM spectrum.

wavelengths from 10^8 m down to 10^{-12} m. Utility AC appears near the top of this scale; it exhibits a free-space wavelength of many kilometers. The γ-rays appear near the bottom; they have microscopic wavelengths. As we can see, the *visible-light spectrum* takes up only a tiny sliver of the total *EM spectrum*. The right-hand scale in Fig. 13-2 renders the visible wavelengths in nanometers (nm).

The Visible "Window"

To get some idea of how the visible-light "window" actually compares in span to the greater EM spectrum, find a red or blue colored piece of glass or cellophane and look through it at a daylight scene. Such a color filter greatly restricts your view of your surroundings, because only a narrow range of visible wavelengths can pass through it.

When you view a scene through a red filter, for example, everything looks red, dark red, brown, or black. Blue objects appear dark brown or black; bright red objects look the same as white objects, and maroon red objects look the same as gray objects. Other colors look red with varying degrees of saturation, but you see little or no variation in the hue. You've restricted the range of visible-light wavelengths that you can see, and as a result, you've in effect rendered your eyes "color-blind."

When considered with respect to the entire EM spectrum, all optical instruments suffer from the same sort of handicap we would have if the lenses in our eyeballs were tinted red (or blue or green). The range of wavelengths to which human eyes respond goes from approximately 770 nm at the longest to approximately 390 nm at the shortest.

TIP   *Human eyes suffer from a sort of "EM-spectrum-blindness." Nearly all EM disturbances occur at wavelengths that we can't see.*

PROBLEM 13-1

What is the frequency of a red laser beam whose free-space waves measure 7400 Å in length?

SOLUTION

We can use the formula for frequency in terms of wavelength, but we should convert the wavelength to meters before we calculate. When we do that, we get

$$7400 \text{ Å} = 7400 \times 10^{-10} \text{ m}$$
$$= 7.400 \times 10^{-7} \text{ m}$$

Now we can find the frequency in hertz as

$$f = 2.99792 \times 10^8 / \lambda$$
$$= 2.99792 \times 10^8 / (7.400 \times 10^{-7})$$
$$= 4.051 \times 10^{14} \text{ Hz}$$

That's a frequency of 405.1 THz. Just for fun, let's compare this frequency with a typical frequency-modulated (FM) broadcast signal at 100 MHz. When we divide out the ratio, we find that the red light beam has a frequency in excess of 4,000,000 times that of the FM broadcast signal.

PROBLEM 13-2

What is the wavelength of the EM field produced in free space by the AC in a common utility line? Take the frequency as 60.0000 Hz (accurate to six significant figures).

 SOLUTION

We can use the formula for wavelength in terms of frequency directly to obtain

$$\lambda = 2.99792 \times 10^8 / f$$
$$= 2.99792 \times 10^8 / 60.0000$$
$$= 4.99653 \times 10^6 \text{ m}$$

which is a span of almost 5000 km, half the distance from the earth's equator to the north geographic pole as measured over the surface.

ELF Fields

Many electrical and electronic devices produce EM fields. Some of these fields have wavelengths much longer than standard broadcast and communications radio signals. A field of this sort alternates at an *extremely low frequency* (ELF); that's how the term *ELF fields* has arisen.

What ELF Is (and Isn't)

The *ELF spectrum* begins, technically, at the lowest possible frequencies (less than 1 Hz) and extends upward to approximately 3 kHz. This range corresponds to wavelengths longer than 100 km. The most common ELF field in the modern world has a frequency of 60 Hz. These ELF waves constantly emanate from all live utility wires in the United States. In some countries they have a frequency of 50 Hz.

The term *extremely low frequency radiation*, and the media attention it sometimes gets, leads some lay people to unreasonably fear this form of EM energy. An ELF field bears no resemblance whatsoever to x rays or γ rays, which can cause sickness and death if received in large doses. Neither does ELF energy resemble UV radiation (which has been linked to skin cancer) or intense IR radiation (which can inflict serious burns).

Still Struggling

An ELF field can't make anything radioactive. Some biologists suspect that long-term exposure to strong ELF fields might have an adverse effect on human health, but a conclusive verdict on the issue has not come in.

Computer Monitors and ELF

The common *cathode-ray tube* (CRT), of the sort used in older desktop computer displays and some television sets, represents an ELF source that has received a lot of publicity. Other parts of a computer produce little EM energy. Popular *liquid-crystal displays* (LCD panels) emit essentially none.

In the CRT, the characters and images appear on the screen when an electron beam strikes a phosphor coating on the inside of the glass. The electrons change direction as the beam sweeps rapidly from left to right, and from top to bottom, on the screen. The sweeping motion results from the effects of powerful, low-frequency magnetic fields produced by deflecting coils. Because of the orientations of the coils inside the CRT, more EM energy "radiates" from the sides of a CRT monitor than from the front. Therefore, assuming that a CRT monitor actually poses a health hazard, people sitting off to the side of the appliance run a greater risk of overexposure than someone looking at the screen from directly in front.

You'll sometimes see appliances marketed with claims to eliminate, or greatly reduce, ELF fields. Some such schemes work well; others don't work at all. *Electrostatic screens*, that you can place in front of a CRT monitor glass to keep it from attracting dust, will not stop ELF fields. *Glare filters* won't do anything either. Physical distance provides effective "shielding" from ELF fields, especially for people sitting next to (rather than in front of) old-fashioned CRT desktop computer monitors. The ELF field dies off rapidly with distance from the monitor cabinet. In an office environment, computer workstations using CRT monitors should sit at least 1.5 m (about 5 ft) apart. You should keep at least 0.5 m (about 18 in) away from the front of your own monitor.

Because LCD displays have become affordable in recent years, and because all new computers can operate with them, few people use CRTs anymore. Therefore, ELF fields from computer displays pose far less "danger" today than they did in, say, 1990. Nevertheless, the ELF issue has attracted the attention of fear-mongers and quacks, as well as the interest of legitimate scientists. Don't succumb to unsubstantiated media hype. If you are concerned about ELF "radiation" in your home or work environment, consult someone whose word you can trust, such as a professional engineer.

RF Waves

We call an EM disturbance a *radio-frequency* (RF) *wave* if its wavelength falls within the range of 100 km to 1 mm, corresponding to a frequency range of 3 kHz to 300 GHz.

Formal RF Band Designators

Communications engineers split the so-called *RF spectrum* or *radio spectrum* into eight *bands*, each representing one order of magnitude in terms of frequency and wavelength: the *very-low*, *low*, *medium*, *high*, *very-high*, *ultra-high*, *super-high*, and *extremely high frequencies*, abbreviated as VLF, LF, MF, HF, VHF, UHF, SHF, and EHF, respectively. Table 13-1 denotes the RF bands in terms of frequency and free-space wavelength.

Some of the RF bands have alternative names. Engineers sometimes call energy at VLF and LF *longwave radio* or *long waves*. Energy in the HF range sometimes goes by the nickname *shortwave radio* or *short waves*, even though the wavelengths far exceed those of most EM waves in today's wireless systems. Super-high-frequency and extremely high-frequency RF waves are also known as *microwaves*.

Radio-frequency waves propagate through the earth's atmosphere and through space in various ways, depending on the wavelength. The upper atmosphere, where the gases tend to ionize easily, dramatically affects waves at VLF, LF, MF, and HF. The lower atmosphere, where weather phenomena prevail, can bend, reflect, or scatter waves at VHF, UHF, SHF, and EHF.

TABLE 13-1   The bands in the radio-frequency (RF) spectrum. Each band spans one order of magnitude (power of 10) in terms of frequency and wavelength.

Designator	Frequency	Wavelength
Very low frequency (VLF)	3–30 kHz	100–10 km
Low frequency (LF)	30–300 kHz	10–1 km
Medium frequency (MF)	300 kHz–3 MHz	1 km–100 m
High frequency (HF)	3–30 MHz	100–10 m
Very high frequency (VHF)	30–300 MHz	10–1 m
Ultra high frequency (UHF)	300 MHz–3 GHz	1 m–100 mm
Super high frequency (SHF)	3–30 GHz	100–10 mm
Extremely high frequency (EHF)	30–300 GHz	10–1 mm

Earth's Ionosphere

The atmosphere of our planet becomes less dense with increasing altitude. Therefore, we observe more solar energy at high altitudes than we do at the surface. High-speed subatomic particles, UV rays, and x rays ionize the rarefied gases in the upper atmosphere. Ionized regions occur at specific altitudes and compose the so-called *ionosphere*. The ionosphere causes *absorption* (loss) and *refraction* (bending) of radio waves, making long-distance communication possible at some radio frequencies.

Ionization in the upper atmosphere occurs in four "fuzzy" layers. We call the lowest of these the *D layer*. It hovers at an altitude of about 50 km (30 mi), and ordinarily exists only on the daylight side of the earth. The D layer contributes nothing to long-distance wireless communication. In fact, at MF during the daytime, the D layer absorbs EM energy, degrading or preventing long-distance wave propagation.

The *E layer*, which forms at about 80 km (50 mi) above the surface, also exists mainly during the day, although nighttime ionization sometimes takes place. The E layer can facilitate medium-range radio communication at certain radio frequencies, especially between about 20 and 200 MHz.

Wireless communications engineers call the uppermost ionized regions the *F1 layer* and the *F2 layer*. The F1 layer, normally present only on the daylight side of the earth, forms at about 200 km (125 mi) altitude. The F2 layer, which exists more or less around the clock, floats roughly 300 km (180 mi) above the surface. These layers commonly return EM signals to the earth at frequencies above 10 MHz or so.

On the dark side of the earth, when the F1 layer disappears, the F2 layer becomes the *F layer*. This layer can return EM signals to the earth in the standard AM broadcast band, as well as at shortwave frequencies up to 10 or 20 MHz. The F1, F2, and F layers can take credit for the evolution of shortwave broadcasting and amateur radio during the 20th century. Under ideal conditions, the F1, F2, and F layers allow wireless communications and broadcasting to take place between points on the surface separated by thousands of kilometers, with no intervening infrastructure whatsoever.

Figure 13-3 illustrates the relative altitudes of the ionospheric D, E, F1, and F2 layers above the earth's surface. All of these layers have some effect on the way radio waves propagate at very low, low, medium, and high frequencies. (Sometimes, but less often, ionospheric effects occur into the VHF portion of the radio

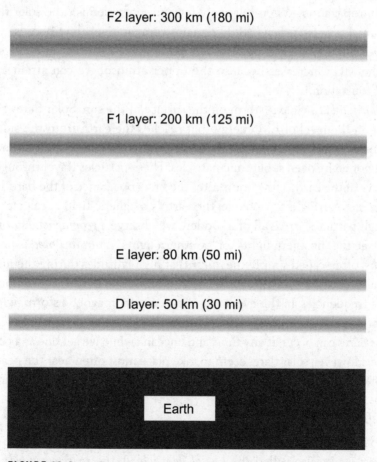

FIGURE 13-3 · Layers of the earth's ionosphere. These ionized regions affect the behavior of EM waves at some radio frequencies.

spectrum.) The ionosphere not only facilitates long-distance wireless communication between points on the earth's surface; it also prevents radio waves at frequencies below approximately 5 MHz from reaching the surface from outer space.

Solar Activity

The number of sunspots that we see on the solar disk fluctuates from year to year. The variation follows a periodic and dramatic pattern—a waveform itself! Astronomers and communications engineers refer to this fluctuation of sunspot numbers as the *sunspot cycle*. Sunspot maxima occur at intervals of roughly 11 years. Sunspot numbers generally rise more rapidly than they decline, and the absolute maximum and minimum sunspot counts vary from cycle to cycle.

The sunspot cycle affects propagation conditions at radio frequencies up to about 70 MHz for F1- and F2-layer propagation, and 150 to 200 MHz for E-layer propagation. When relatively few sunspots blemish the solar disk, the *maximum usable frequency (MUF)* remains comparatively low, because the ionization of the upper atmosphere is not dense. At or near the time of a sunspot peak, the MUF increases because the upper atmosphere can attain a higher level of ionization.

A *solar flare* is a violent storm on the surface of the sun. Solar flares cause an increase in the level of EM energy that reaches the earth from the sun over a wide range of frequencies. Solar flares also cause the sun to emit an increased quantity of high-speed subatomic particles. These particles travel through space and arrive at the earth a few hours after the first appearance of the flare. Because the particles carry electric charge, the earth's magnetic field accelerates them. If enough particles arrive all of a sudden, we observe a *geomagnetic storm*. Then people see the "northern lights" or "southern lights" (*Aurora Borealis* or *Aurora Australis*, often called simply the *aurora*) at high latitudes during the night. We might also notice a sudden deterioration or cessation of radio communication at some frequencies. In the event of an extreme geomagnetic storm, wire communications circuits and even the electric utility grid might misbehave.

Solar flares can occur at any time, and once in awhile we get one as a complete surprise. However, solar flares seem to take place most often near the peak of the sunspot cycle, when average sunspot numbers hover near their maximum.

Ground-Wave Propagation

In radio communication, the *ground wave* consists of three distinct components: the *direct wave* (also called the *line-of-sight wave*), the *reflected wave*, and the *surface wave*.

The direct wave travels in a straight line. It plays a significant role only when the transmitting and receiving antennas lie at two points on a so-called *line of sight*—a straight line that never dips below the earth's surface. At most radio frequencies, EM fields pass easily through visually opaque objects such as trees and frame houses. However, concrete-and-steel structures cause *attenuation* (signal loss) in the direct wave at VHF and above.

A radio signal can reflect from the earth's surface, or from certain structures such as concrete-and-steel buildings. The reflected wave combines with the direct wave (if any) at the receiving antenna. Sometimes the direct and reflected waves arrive at the receiver in phase opposition, in which case the received signal appears weak or nonexistent even if the transmitting and receiving antennas lie along a line of sight. This effect occurs mostly at frequencies well above 30 MHz (wavelengths considerably less than 10 m).

The surface wave travels in contact with the earth's surface, which actually forms part of the circuit. This effect takes place only with *vertically polarized* EM fields (those in which the electric flux lines align perpendicular to the earth's surface) and at frequencies below about 15 MHz (wavelengths longer than about 20 m). At frequencies much above 15 MHz, surface waves don't propagate efficiently because the earth's surface acts like a high-value resistor. At frequencies from about 9 kHz up to 300 kHz, the surface wave can travel for hundreds or even thousands of kilometers because the earth's surface acts as a good conductor.

TIP   *Some people call the surface wave the "ground wave," but that expression constitutes a technical misnomer.*

Sporadic-E Propagation

At certain radio frequencies, the ionospheric E layer can return EM waves to the earth. This phenomenon occurs intermittently, and conditions can change rapidly. Because of the unstable nature of this mode, communications engineers and radio operators refer to it as *sporadic-E propagation*. It usually takes place at frequencies between approximately 20 and 150 MHz. Occasionally, we will see it happen at frequencies up to around 200 MHz. Communication typically occurs over distances up to several hundred kilometers, but once in a while a circuit path length will exceed 2000 km (1200 mi).

The standard FM broadcast band sometimes "falls under the spell" of sporadic-E propagation. When that happens, you'll hear unfamiliar stations on your FM radio! You should not be surprised if you receive a station from a city in another state, or even in another country, that you would never hear under ordinary conditions. Sporadic-E propagation can easily be mistaken for long-distance propagation phenomena that take place entirely in the lower atmosphere, independently of the ionosphere.

Auroral Propagation

In the presence of unusual solar activity, the Aurora Borealis and Aurora Australis can reflect radio waves at some frequencies. We call this phenomenon *auroral propagation*. The aurora occur at altitudes of 25 km (40 mi) to 400 km (250 mi) above the earth's surface. Theoretically, auroral propagation can occur between any two points on the surface from which the same part of the aurora lie on a line of sight. Auroral propagation seldom occurs when either the transmitter or the receiver operates from a location less than 35° north or south of the geographic equator. Auroral propagation can take place at frequencies considerably higher than 30 MHz, and it's often accompanied by deterioration in "shortwave radio" ionospheric propagation by means of the E and F layers.

Meteor-Scatter Propagation

When a meteor from space enters the upper part of the atmosphere, the meteor produces an ionized trail as a result of the friction between itself and the air molecules. Such an ionized region reflects EM energy at certain wavelengths. This phenomenon, known as *meteor-scatter propagation*, can result in over-the-horizon radio communication or reception. The ionized meteor trails reflect the EM waves in much the same way as the aurora does.

A meteor produces a trail that persists for a few tenths of a second up to several seconds, depending on the size of the meteor, its speed, and the angle at which it enters the atmosphere. This short amount of time allows for the exchange of relatively little information, but during a *meteor shower*, ionization can take place on an almost continuous basis.

Amateur radio operators have observed meteor-scatter propagation at frequencies well into the VHF range. Communication distances can range from just beyond the horizon up to about 2000 km (1200 mi), depending on the altitudes of the ionized trails, the orientations of the trails (which depend on the direction from which the meteors arrive), the location of the transmitting station, and the location of the receiving station.

Tropospheric Bending

The lowest 13 to 20 km (8 to 12 mi) of the atmosphere, where most weather phenomena occur, constitute the *troposphere*. At EM wavelengths shorter than about 15 m (frequencies above 20 MHz), refraction and reflection can take place within and between tropospheric air masses of different density. The air also produces some scattering of EM energy at wavelengths shorter than about 3 m (frequencies above 100 MHz). All of these effects go by the collective name of *tropospheric propagation*. This mode can result in effective, continuous wireless communication over distances of hundreds of kilometers under ideal conditions.

A common type of tropospheric propagation takes place when radio waves refract in the lower atmosphere. We see this effect quite often along and near *weather fronts*, where warm air lies above cool, more dense air. The cooler air has a higher index of refraction than the warm air, causing EM fields to veer downward at a considerable distance from the transmitter. We call this phenomenon *tropospheric bending*. It can cause anomalies in reception of FM broadcast signals similar to the effects of sporadic-E propagation.

Troposcatter

At frequencies above about 100 MHz, the troposphere scatters radio waves to some extent. The scattering allows over-the-horizon communication at VHF, UHF, and microwave frequencies. We call this mode *tropospheric scatter* or *troposcatter*. Dust and clouds in the air increase the scattering effect, but some troposcatter occurs regardless of the weather. Troposcatter takes place mostly at low altitudes where the air has the greatest density, but some scattering can occur at altitudes up to about 16 km (10 mi). Troposcatter can provide reliable communication over distances of several hundred kilometers between stations employing high-power transmitters and sensitive receivers.

Figure 13-4 illustrates how scattering and bending occur in the lower atmosphere. Imagine the transmitting station at the lower left, and the receiving station at the lower right. A *temperature inversion* exists in this example; a warm air mass overlies a cooler one. The inversion exaggerates the bending. If the boundary between the cool air near the surface and the warm air above it is well-defined enough, *EM-wave reflection* can occur at the boundary in addition to the bending. If the inversion covers a large geographic area, signals can "bounce" repeatedly between the inversion boundary and the surface, providing exceptional long-range communication, especially over salt water.

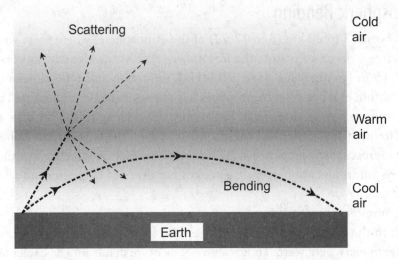

FIGURE 13-4 · The troposphere can bend and scatter radio waves at some frequencies.

Ducting

The term *duct effect* (or simply *ducting*) refers to a form of tropospheric propagation that takes place at approximately the same frequencies as tropospheric bending and scattering. This form of propagation nearly always happens close to the surface, sometimes at altitudes of less than 300 m.

A *duct* forms when a layer of cool air becomes "sandwiched" between two layers of warmer air, a situation that sometimes arises along and near weather fronts in the temperate latitudes. Ducts can also form above water surfaces during the daylight hours, and over land surfaces at night. Radio waves can get "trapped" inside the region of cooler air, in much the same way as light waves remain "trapped" inside an *optical fiber*.

TIP *Ducting can facilitate over-the-horizon communication at VHF and UHF between wireless stations separated by hundreds of kilometers. However, a duct can disappear within seconds, even after persisting for hours. A duct can also change altitude slightly but suddenly, necessitating an adjustment in the antenna height at the transmitting station, the receiving station, or both.*

PROBLEM 13-3

Imagine that you use a handheld wireless transceiver, also called a "walkie-talkie," to converse with someone across town. You stand on a hill so that you can see your counterpart's house. The two of you are located well within the manufacturer's rated communications range for the radios.

Nevertheless, your friend can hardly hear you, and you can hardly hear her. You take a few steps forward, backward, to the right, or to the left, and the signal gets strong. What might cause this effect?

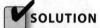

SOLUTION

Initially, the direct wave and the reflected wave from your friend's radio antenna arrive in phase opposition at your antenna. Therefore, the two waves almost completely cancel each other out. Moving a few meters in any direction remedies this problem, causing the direct and reflected waves to arrive at your antenna in a more favorable phase relationship, so that you and your friend can conduct a conversation with ease.

Beyond the Radio Spectrum

The shortest RF waves, according to conventional definitions, measure approximately 1 mm long, corresponding to a frequency of 300 GHz. As the wavelength falls below 1 mm and gets steadily shorter, we encounter the IR, visible, UV, x-ray, and γ-ray spectra, in that order.

Infrared

The longest IR waves measure roughly 1 mm in length, while the longest visible-light waves (which we see as red) span a little less than 0.001 mm. Therefore, the *IR spectrum* covers three orders of magnitude in terms of wavelength and frequency. The prefix *infra-*, which means "below," arises from the fact that the IR frequencies lie immediately below the visible red frequencies.

Our bodies sense IR radiation as "heat rays." The waves do not literally constitute heat, but IR rays can warm up objects that they strike, and by which they are at least partially absorbed. The sun constitutes a brilliant source of IR radiation. Other sources include incandescent light bulbs, glowing embers, and electrical heating elements. If you have an electric stove and you switch on one of the burners to "low," you can feel the IR radiation from it even though the element appears black to the eye.

We can detect IR rays using special films that work in most old-fashioned 35-mm film cameras. Some high-end photographic cameras have focus numbers for IR as well as for visible light printed on their lens controls. Glass transmits IR at the shorter wavelengths (so-called *near IR*) but blocks IR at the longer wavelengths (*far IR*). When you take an IR photograph in visible-light darkness, warm objects show up clearly. That's how *night-vision* apparatus works.

The fact that glass transmits near IR but blocks far IR allows glass greenhouses to maintain interior temperatures much higher than that of the external environment. The same phenomenon heats up your car or truck interior if you leave its windows closed on a sunny day. We can take advantage of this so-called *greenhouse effect* to help us heat energy-efficient homes and office buildings. We can equip large windows having direct southern exposures (in the northern hemisphere) or northern exposures (in the southern hemisphere) with blinds that we can open up on sunny winter days and close on cloudy days and at night.

Infrared radiation at low and moderate levels poses no danger and no known health risk. It can help relieve the discomfort of joint injuries and muscle strains. At high intensity, however, IR radiation can cause serious burns. In structural or forest fires, the powerful IR rays can scorch the clothing off a person and then literally cook the body. The most extreme earthly IR radiation comes from volcanoes, nuclear bombs, and catastrophic meteorite impacts. The IR burst from a 20-megaton hydrogen bomb (equivalent to 2×10^7 tons of conventional explosive) would instantly kill every exposed living organism within a radius of several kilometers.

In some portions of the IR spectrum, the atmosphere of our planet appears opaque. In the near IR between about 770 nm (the visible red) and 2000 nm, the lower atmosphere exhibits fair to good transparency in dry, clear weather. Water vapor causes IR attenuation between wavelengths of approximately 4500 and 8000 nm. Carbon dioxide (CO_2) gas interferes with the transmission of IR at wavelengths ranging from about 14,000 to 16,000 nm. Rain, snow, fog, and dust also interfere with the propagation of IR.

Still Struggling

The presence of significant CO_2 in the atmosphere may keep the earth's surface warmer than it would be if there were less CO_2. This large-scale version of the greenhouse effect has given rise to widespread controversy and confusion. Many people, including some of the world's top scientists, suspect or believe that increasing CO_2 levels in the atmosphere result largely from recent human activity, and will lead to extensive *climate change* ("global warming") within the next few decades. Some people refute this hypothesis, claiming that insufficient evidence exists. The debate has acquired political overtones, with ideology overriding reason in some quarters. If you plan to become a climatologist, maybe you can help humanity resolve this "hot issue"!

Visible Light

The visible portion of the EM spectrum lies within the wavelength range of 770 to 390 nm. Emissions at the longest wavelengths appear red. As the wavelength decreases, we see orange, yellow, green, blue, indigo, and violet, in that order. Visible light transmits fairly well through the atmosphere at all wavelengths. Scattering increases toward the blue, indigo, and violet end of the spectrum. This scattering explains why the sky appears blue during the daytime. Long-wavelength light scatters the least, so the sun often appears red or orange when it is on the horizon. Red is the preferred color for terrestrial line-of-sight laser communications systems for this same reason. Rain, snow, fog, smoke, and dust interfere with the transmission of visible light through the air. In the next chapter, you'll learn how visible light behaves under various conditions.

Ultraviolet

As the wavelength of an EM disturbance falls below 390 nm, the energy contained in each individual photon increases. The *UV spectrum* of wavelengths starts at about 390 nm (on the long end) and extends down to about 1 nm (on the short end). At a wavelength of approximately 290 nm, the atmosphere becomes highly absorptive. At shorter wavelengths the air remains essentially opaque, protecting the environment against damaging UV radiation from the sun. *Ozone* (molecules consisting of three oxygen atoms) in the upper atmosphere contributes to this effect.

Ordinary glass almost completely blocks energy at UV wavelengths, so cameras with glass lenses won't work if you want to take UV photographs. Instead, you must employ a so-called *pinhole camera*, severely limiting the amount of energy that passes into the detector. While a camera lens has a diameter of several millimeters or centimeters, the aperture in a pinhole camera measures less than 1 mm across. As the diameter of the hole decreases, the camera can resolve more and more details, but its sensitivity decreases—a classical trade-off in laboratory science.

Another type of device, called a *spectrophotometer*, can sense UV and allow us to measure UV intensity at various wavelengths. In this apparatus, a transparent sheet of glass or plastic with thousands of parallel, closely spaced, opaque lines, called a *diffraction grating*, disperses EM energy into its constituent wavelengths from IR through the visible and into the UV range. If we move the sensing device back and forth, we can single out energy at any desired

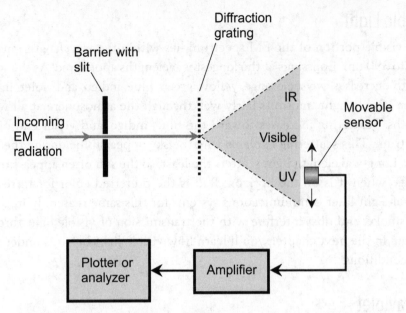

FIGURE 13-5 • Functional diagram of a spectrophotometer, which scientists use to sense and measure IR, visible, and UV radiation.

wavelength for analysis. Figure 13-5 illustrates the principle of operation. At the extremely short wavelength end of the UV spectrum (*hard UV*), the detection device may consist of a *radiation counter* similar to the apparatus employed for the detection of x rays and γ rays. For photographic purposes, ordinary camera film will work at the longer UV wavelengths (*soft UV*). If we want to make photographs in the hard UV range, we must use special film that resembles x-ray film.

Ultraviolet rays possess an interesting property that you can observe using a "black light." Most hobby shops sell lamps of this sort. They have cylindrical form, and superficially you might mistake them for small fluorescent tubes. (The incandescent "black light" bulbs sold in department stores don't produce much UV.) When subjected to UV, certain substances glow brightly in the visible range. We call this phenomenon *fluorescence*. Art stores sell acrylic paints that glow in various colors when UV strikes them, producing dramatic effects in visible-light-dimmed rooms. The phosphor coatings on cathode-ray tubes (CRTs) fluoresce under UV. So do certain living organisms such as scorpions. If you live in the desert, go outside some night with a "black light" and switch it on. You'll find out straightaway whether or not you have scorpions for company. Whatever you do with a "black light" lamp, however, *never* look directly at it. The UV radiation can damage your eyes even though you can't see it.

Most of the radiation from the sun occurs in the IR and visible parts of the EM spectrum. If the sun were a much hotter star, producing more energy in the UV range, life on any earthlike planet in its system would have evolved in a different way, if at all. Excessive exposure to UV, even in the relatively small amounts that reach the earth's surface on bright days, can cause skin cancer and eye cataracts over time. Evidence suggests that excessive exposure to UV suppresses the activity of the immune system, rendering people and animals more susceptible to infectious diseases. (A little exposure can be good, however, because it provides essential vitamin D to the human body.) Some scientists have observed that the so-called *ozone hole* in the upper atmosphere, prevalent in the southern hemisphere, has grown over the past century. If the "hole" gets larger or more transparent to UV, it might have a significant impact on the evolution of future life forms.

X Rays

The x-ray spectrum consists of EM energy at wavelengths from approximately 1 nm down to 0.01 nm. (Various texts disagree on the exact location of the "dividing line" between the hard UV and x-ray regions.)

The discovery of x rays occurred accidentally in 1895 when a physicist named *Wilhelm Roentgen* conducted experiments involving electrical currents in gases at low pressure. If the current attained sufficient intensity, the high-speed electrons produced mysterious radiation when they struck the anode (positively charged electrode) in the tube. Scientists tagged these rays with an "x" because of their mysterious behavior, which no one had ever witnessed before. The rays routinely penetrated barriers opaque to visible light and UV. A phosphor-coated object happened to lie in the vicinity of the tube containing the gas, and Roentgen noticed that the phosphor glowed. Subsequent experiments showed that the rays possessed so much inherent energy that they passed through the skin and muscles in the human hand, casting shadows of the bones on a phosphor-coated surface or a photographic film.

A modern *x-ray tube* operates by accelerating electrons to high speed and forcing them to strike a heavy metal anode (usually made of tungsten). Figure 13-6 illustrates a simplified functional diagram of an x-ray tube.

As the wavelength of x rays decreases, we find it more difficult to direct and focus them because of their increasing penetrating power. A piece of paper with a tiny hole works well for UV photography, but in the x-ray spectrum, the

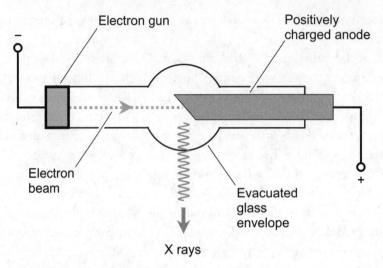

FIGURE 13-6 · Functional diagram of an x-ray tube.

radiation passes right through the paper. Even aluminum foil appears relatively transparent to x rays. But if x rays strike a reflecting surface at a nearly grazing angle, and if the reflecting surface consists of suitable material, some focusing occurs. As the wavelength decreases, the angle, relative to the surface, at which the waves strike must decrease if we want reflection to take place. At the short-est x-ray wavelengths, the angle between the ray and the surface must remain smaller than about 1° of arc. Figure 13-7A shows this grazing reflection effect. Figure 13-7B illustrates how an x-ray observing device achieves its focusing. The focusing mirror tapers in the shape of an elongated *paraboloid*. As parallel x rays enter the aperture, they strike the inner reflective surface at a grazing angle. The x rays come to a focal point where a radiation counter or detector can pick them up.

X rays cause ionization of living tissue. The cumulative effect can result in damage to cells over a period of years. This fact explains why x-ray technicians in doctors' and dentists' offices used to work behind barriers lined with lead. Otherwise, those personnel would have received dangerous cumulative doses of x radiation. It only takes a few millimeters of lead to block virtually all x rays. Less dense metals and other solids can also block x rays, but these barriers must have greater thickness. The important factor is the *total mass* through which the radiation must pass. Sheer physical displacement can also reduce the intensity of x radiation, which diminishes according to the square of the distance from the source.

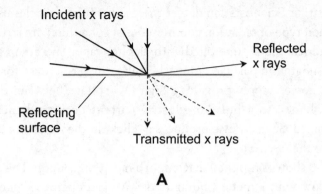

A

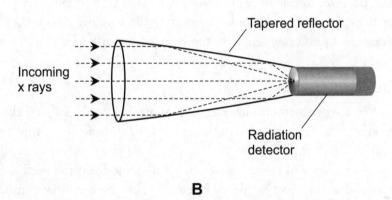

B

FIGURE 13-7 · At A, x rays reflect from a surface only when they strike the surface at a grazing angle. At B, a functional diagram of an x-ray focusing and observing device.

Gamma (γ) Rays

As the wavelength of EM radiation drops even below that of x rays, the penetrating power increases until focusing becomes practically impossible. The cutoff point where the x-ray region ends and the γ-ray region begins is approximately 0.01 nm (10^{-11} m). Gamma rays can, in theory, get shorter than this without limit. The γ classification represents the most energetic of all known EM fields. Short-wavelength gamma rays can penetrate several centimeters of solid lead, or more than a meter of concrete. They inflict even more damage to living tissue than x rays. Gamma rays come from radioactive materials, both natural (such as *radon*) and human made (such as *plutonium*).

Radiation counters offer the primary means of detecting and observing sources of γ rays. Gamma rays dislodge particles from the nuclei of some of the

atoms they strike. Scientists can detect these subatomic particles using a counter. A common type of radiation counter consists of a thin wire strung within a
sealed, cylindrical metal tube filled with alcohol vapor and argon gas. When a
high-speed subatomic particle enters the tube, the gas ionizes for a moment.
An external power supply provides a high DC voltage between the wire and
the outer cylinder, so a pulse of electric current occurs whenever the gas
becomes ionized. Such a pulse produces a "click" in the output of an amplifier
connected to the device.

Figure 13-8 shows simplified diagram of a radiation counter. The cylinder has
a glass window with a metal "sliding door." We can open the "door" to admit
particles having relatively low energy, and close the door to allow only the most
energetic particles to gain entry. High-speed subatomic particles, tiny yet massive for their size, have no trouble penetrating the window glass if they move
fast enough. All γ rays can penetrate the closed door with ease.

Cosmic Particles

If you sit in a room without any radioactive materials present, and then you
switch on a radiation counter with the window of its tube open, you'll notice
an occasional click from the device. Particles enter the tube even in everyday
environments! Some of these "rogue particles" come from the earth because
radioactive elements exist in the surface rocks almost everywhere (usually in

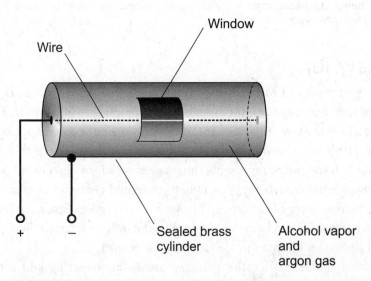

FIGURE 13-8 · Simplified diagram of a radiation counter.

small quantities). Some of the radiation comes indirectly from space. *Cosmic particles* strike atoms in the earth's upper atmosphere, and these atoms eject other high-speed subatomic particles that arrive at the counter tube.

In the early 1900s, physicists noticed radiation apparently coming from space. They found that the strange background radiation increased in intensity when observations were made at high altitude; the radiation level decreased when observations were taken from underground or underwater. This space radiation has been called *secondary cosmic radiation* or *secondary cosmic particles*. The actual particles from space, known as *primary cosmic particles*, almost never penetrate far into the atmosphere before they collide with, and break up, the nuclei of atoms. If we want to observe primary cosmic particles, we must send our instruments in a rocket-propelled spacecraft to the necessary altitude.

While the radiation in the EM spectrum comprises photons traveling at the speed of light, primary cosmic particles travel a little bit slower than light, but still mighty fast. At such high speeds, the protons, neutrons, and other particles gain mass because of *relativistic effects*, giving the particles so much momentum that they're not affected very much by the earth's magnetosphere. For this reason, primary cosmic particles approach the earth in nearly perfect straight-line paths, unlike solar particles that travel at much lower speeds and undergo considerable deflection.

By observing the trails of primary cosmic particles in a device called a *cloud chamber* aboard a low-orbiting spaceship, astronomers can ascertain the direction from which the particles have come. Tracing back into the heavens, they can pinpoint the source, sometimes even identifying a particular celestial object as the likely origin of the particles. Over time, cosmic-particle maps of the heavens can be generated and compared with celestial maps for radio waves, IR, visible light, UV, x rays, and γ rays.

PROBLEM 13-4

How much energy exists in a single photon in a stream of γ rays at a wavelength of 0.00100 nm?

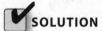

SOLUTION

Recall the formula for photon energy e in terms of wavelength (λ), the speed of EM propagation in free space (c), and Planck's constant (h):

$$e = hc/\lambda$$

For EM rays in free space, the product *hc* equals approximately 1.9865×10^{-25}. (You can go back and check the text in the previous chapter if you've forgotten the derivation.) The wavelength of 0.00100 nm equals 1.00×10^{-12} m. Therefore, the energy *e*, in joules, contained in each photon of the γ ray is

$$e = (1.9865 \times 10^{-25}) / (1.00 \times 10^{-12})$$
$$= 1.9865 \times 10^{-13} \text{ J}$$

We should round this result off to 1.99×10^{-13} J.

Radioactivity

The nuclei of most familiar chemical elements retain their identities, and remain unchanged, indefinitely. We call them *stable nuclei*. However, the nuclei of certain isotopes of specific elements change with time; we call them *unstable nuclei*. As unstable nuclei disintegrate, they emit high-energy photons (particularly γ rays) and various subatomic particles. The term *radioactivity* refers to any EM or particle radiation produced by disintegrating unstable nuclei.

Forms

Radioactivity, also called *ionizing radiation* because it can strip electrons from atoms, occurs in various forms. Gamma rays (which we've already discussed) constitute the most deadly form of ionizing radiation. Other forms include *alpha* (α) *particles* and *beta* (β) *particles*. Less common forms of ionizing radiation also exist, such as high-speed protons and antiprotons, high-speed neutrons and antineutrons, and the nuclei of atoms heavier than helium.

Alpha particles comprise helium 4 (He-4) nuclei propagating at high speeds. An He-4 nucleus consists of two protons and two neutrons. An α particle carries positive electric charge because no electrons surround it. Therefore, all α particles are ions. They have significant mass, so if they attain high enough speeds, they can acquire considerable kinetic energy. An α particle traveling at a sizable fraction of the speed of light (known as *relativistic speed*) attains increased mass compared with the mass of an He-4 nucleus at rest, giving the particle more kinetic energy than the simple formula would dictate. You'll learn about *relativistic mass increase*, and other effects, in Chapter 15. Modest physical barriers can block most α particles; they do not possess great penetrating power.

β particles are high-speed electrons or positrons (anti-electrons). We can denote any β particle consisting of an electron, also called a *negatron* because it has a negative electrical charge, by writing β−. We can denote any β particle consisting of a positron, which carries a positive charge, by writing β+. All β particles have nonzero mass at rest, or when they travel at nonrelativistic speeds. However, as happens with all types of particles and objects, the kinetic energy in a β particle increases relativistically if it travels at near-light speed.

Natural Sources

In nature, radioactivity comes from certain isotopes of elements with atomic numbers up to, and including, 92 (uranium). We call these atoms *radioactive isotopes*. An isotope of carbon, known as *carbon 14* (C-14), has eight neutrons. Atoms of C-14 are unstable; over time they decay into carbon 12 (C-12) atoms, which have six neutrons. Other examples of unstable atoms include *hydrogen 3* (H-3), also known as *tritium*, which has a nucleus consisting of one proton and two neutrons; *beryllium 7* (Be-7), with a nucleus containing four protons and three neutrons; and *beryllium 10* (Be-10), with a nucleus containing four protons and six neutrons.

In some instances, the most common isotope of a naturally occurring element exhibits radioactivity. Common examples include radon, radium, and uranium. We can think of the barrage of cosmic particles from deep space as a form of radioactivity, but these particles can themselves create radioactive isotopes when they strike stable atoms in the earth's upper atmosphere.

Human Made Sources

Radioactivity arises from a variety of human activities. The most well-known human source of radioactivity in the early years of atomic research was the *fission bomb*, also called the *atom bomb*. Its modern descendant is the vastly more powerful *hydrogen fusion bomb*. Such a weapon, when detonated, produces an immediate, intense burst of ionizing radiation. The high-speed subatomic particles produced in the initial blast, especially if the explosion occurs at or near the ground, cause large amounts of material to become radioactive. The resulting radioactive dust, called *fallout*, precipitates back to the earth's surface over periods of days, weeks, and months.

Nuclear *fission reactors* contain radioactive elements. We can harness the thermal energy from the decay of these elements to generate electric power. Some byproducts of fission exhibit radioactivity, but we can't use them to generate more power. These materials compose *radioactive waste*, also called *atomic waste*. Disposal of this material presents humankind with an awful environmental conundrum, because some of it takes *centuries* to decay to safe levels of radioactivity. If scientists and engineers can ever develop a *fusion reactor* and put it into production, humanity will enjoy a vast improvement over the fission reactor. Controlled hydrogen fusion generates no radioactive waste.

Physicists can produce radioactive isotopes by bombarding the atoms of certain elements with high-speed subatomic particles or energetic γ rays. We can accelerate charged particles to relativistic speeds using a *particle accelerator*, also known informally as an *atom smasher*. A *linear particle accelerator* has a long, evacuated tube that employs a high-voltage power source to accelerate charged particles such as protons, α particles, and electrons to speeds so great that they can disrupt atomic nuclei that they happen to strike. A *cyclotron* consists of a large, ring-shaped chamber that uses alternating magnetic fields to accelerate charged subatomic particles to relativistic speeds.

Decay and Half-Life

Radioactive substances gradually lose "potency" as time passes. Unstable nuclei degenerate one by one. Sometimes, an unstable nucleus decays into a stable nucleus in a single event. In other cases, an unstable nucleus changes into another unstable nucleus, which later degenerates into a stable one.

Imagine an extremely large number of radioactive nuclei all collected together in a single mass. Suppose that we measure the length of time required for each nucleus to degenerate, and then we average all the results. We call the average decay time the *mean life*, symbolized by the lowercase Greek letter tau (τ).

Some radioactive materials give off more than one form of emission. For any given ionizing radiation form (α particles, β particles, γ rays, or other), we obtain a separate *decay curve*, or function of intensity versus time. A radioactive decay curve always starts out at a certain value and tapers down toward zero. Some decay curves decrease rapidly and others decrease slowly. However, the characteristic shape is always the same; we can always define a decay curve in terms of a time span called the *half-life*, symbolized $t_{1/2}$.

Suppose that we measure the intensity of a certain radiation type at time t_0. After a period of time $t_{1/2}$ has passed, the intensity decreases to half the level it exhibited at time t_0. After the half-life passes again (total elapsed time $2t_{1/2}$), the intensity goes down to 1/4 of its original value. After yet another half-life passes (total elapsed time $3t_{1/2}$), the intensity goes down to 1/8 of its original value. In general, after n half-lives pass from the initial time t_0 (total elapsed time $nt_{1/2}$), the intensity goes down to $1/(2^n)$, or 0.5^n, times its original value. If the original intensity is x_0 units and the final intensity is x_f units, then

$$x_f = 0.5^n x_0$$

Figure 13-9 illustrates the general contour of a radioactive decay curve. The half-life, $t_{1/2}$, can vary tremendously, depending on the particular radioactive substance involved. Sometimes it's a tiny fraction of 1 s; in other cases it's millions of years. For each type of radiation emitted by a material, we obtain a separate value of $t_{1/2}$, and therefore a separate decay curve.

We can also define radioactive decay in terms of a number called the *decay constant*, symbolized by the lowercase Greek letter lambda (λ), just like wavelength. The decay constant equals the natural logarithm of 2 (approximately 0.69315) divided by the half-life in seconds. Therefore

$$\lambda = 0.69315/t_{1/2}$$

When you want to calculate the decay constant, make sure that you express the half-life $t_{1/2}$ in seconds. This precaution will ensure that you get a result for the decay constant expressed in the proper units (s^{-1}).

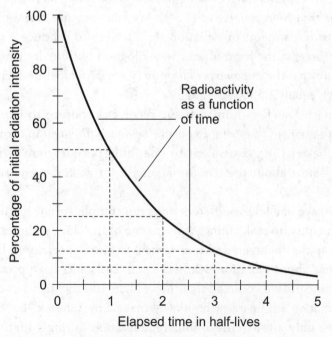

FIGURE 13-9 · General contour of a radioactivity decay curve.

The decay constant equals the reciprocal of the mean life in seconds. Mathematically, we have

$$\lambda = 1/\tau$$

and

$$\tau = 1/\lambda$$

From these equations, we can see that the mean life τ relates to the half-life $t_{1/2}$ according to the formulas

$$\tau = t_{1/2}/0.69315 = 1.4427\ t_{1/2}$$

and

$$t_{1/2} = 0.69315\ \tau$$

Units and Effects

Scientists use various units to quantify overall radiation exposure. The unit of radiation in the International System of Units is the *becquerel* (Bq), representing one *nuclear transition per second* ($1\ s^{-1}$). We can measure radiation exposure according to the amount of radiation that it takes to produce a coulomb of electrical charge, in the form of ions, in a kilogram of pure dry air. The SI unit for this quantity is the *coulomb per kilogram* (C/kg). An older unit, known as the *roentgen* (R), equals 2.58×10^{-4} C/kg.

When matter such as human tissue receives exposure to ionizing radiation, we can quantify the total exposure or *dose equivalent* in terms of a unit called the *sievert* (Sv), equivalent to one joule per kilogram (J/kg). Sometimes you'll hear about the *rem* (an acronym for *radiation equivalent man*); 1 rem = 0.01 Sv.

The existence of all these different units can complicate any discussion about radiation quantity. To make things worse, some of the older, technically obsolete units such as the roentgen and rem remain in use, especially in lay people's conversations about radiation, while the standard units have proven slow to gain acceptance. Have you read that "More than 100 roentgens of exposure to ionizing radiation within a few hours will make a person sick," or "People typically receive only a few rems of ionizing radiation during a lifetime"? Statements like that appeared in civil-defense documents during the 1960s after the

Cuban missile crisis, when fears of worldwide nuclear war led to the installation of air-raid sirens and fallout shelters throughout the United States.

When human beings receive excessive amounts of ionizing radiation in a short time, physical symptoms such as nausea, skin burns, fatigue, and dehydration commonly occur. In extreme cases, internal ulceration and bleeding can cause death. When people get too much radiation gradually over a period of years, cancer rates increase. Genetic mutations also become more likely than normal, giving rise to increased incidence of birth defects.

Practical Uses

Radioactivity has numerous constructive applications in science, industry, and medicine. The most well-known is the nuclear fission reactor, which gained popularity during the mid-1900s for generating electricity on a large scale. Fission reactors fell into disfavor toward the end of the 20th century because of the dangerous waste products they produce, but *energy crises* in recent years have caused scientists, engineers, and politicians to give the technology a second look.

In medicine, radioactive isotopes help doctors diagnose illness, locate tumors inside the body, measure rates of metabolism, and examine the structures of internal organs. Controlled doses of radiation can sometimes retard or destroy cancerous growths. Industrial engineers can use ionizing radiation to measure the dimensions of thin sheets of metal or plastic and to destroy bacteria and viruses that might contaminate food and other matter consumed or handled by people. Other applications include the irradiation of freight and mail to protect the public against the danger of biological attack.

Geologists and biologists use *radioactive dating* to estimate the ages of fossil samples and archeological artifacts. The element most commonly used for this purpose is carbon. When a sample forms, or during the life of the specimen under observation, a certain proportion of C-14 atoms prevails among the total carbon atoms. These C-14 atoms gradually decay into C-12 atoms. By measuring the radiation intensity and determining the proportion of C-14 in archeological samples, anthropologists can get an idea of how long ago the world's great civilizations arose, thrived, and declined.

PROBLEM 13-5

Suppose that the half-life of a certain radioactive substance is 100 days. We measure the radiation intensity as x_0 units. What will the intensity x_{365} be after 365 days?

 SOLUTION

To determine the intensity after 365 days, we can use the equation

$$x_{365} = 0.5^n x_0$$

where n equals the number of half-lives elapsed. In this case, we have

$$n = 365/100$$
$$= 3.65$$

Therefore

$$x_{365} = 0.5^{3.65} x_0$$

To determine the value of $0.5^{3.65}$, we need a calculator that has an x^y function key. We obtain

$$x_{365} = 0.0797 x_0$$

 PROBLEM 13-6

What's the decay constant of the substance described in Problem 13-5?

 SOLUTION

We can determine the formula for the decay constant, λ, in terms of half-life, $t_{1/2}$. In this case, $t_{1/2} = 100$ days. We must convert this figure to seconds. We can calculate the number of seconds in a day as

$$1 \text{ day} = 24 \times 60 \times 60$$
$$= 8.64 \times 10^4 \text{ s}$$

Therefore

$$t_{1/2} = 8.64 \times 10^4 \times 100$$
$$= 8.64 \times 10^6 \text{ s}$$

It follows that

$$\lambda = 0.69315/(8.64 \times 10^6)$$
$$= 8.02 \times 10^{-8} \text{ s}^{-1}$$

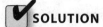 **PROBLEM 13-7**

What's the mean life of the substance described in Problem 13-5? Express the answer in seconds and in days.

 **SOLUTION**

The mean life, τ, equals the reciprocal of the decay constant. To obtain τ in seconds, we must divide the numbers in the above equation with the numerator and denominator interchanged, getting

$$\tau = (8.64 \times 10^6)/0.69315$$

To convert this figure to days, we must divide it by 8.64×10^4. When we do that, we get approximately 144 days.

Neutrinos

Perhaps the strangest known form of radiation takes the form of high-speed particles somewhat resembling high-speed electrons, but lacking electrical charge. In 1931, the physicist *Wolfgang Pauli* inferred the existence of such a particle on the basis of atomic theory. In 1932, *Enrico Fermi* christened the particle the *neutrino*. Various experiments done during the middle and late 20th century confirmed the existence of neutrinos. Scientists originally believed that neutrinos lack mass, but eventually, experiments revealed that they possess a tiny amount of mass. Neutrinos stream through the cosmos at almost the speed of light, spanning intergalactic distances and penetrating most material objects with ease—even whole planets! The neutrino has an antimatter counterpart, known as the *antineutrino*.

Sources and Characteristics

All stars, including our own sun, emit neutrinos. Exploding stars called *supernovae* constitute intense neutrino sources. Nuclear reactors, decaying radioactive minerals in the earth's interior, and the interaction of cosmic rays with atoms in the earth's upper atmosphere can also produce neutrinos. The particles exist in incredible numbers; literally trillions of them pass through your body every second with no effect. Nevertheless, neutrinos remain so elusive that scientists must go to considerable trouble to detect even a single one.

Detection

The earliest neutrino detectors consisted of large tanks containing liquefied chlorine. Later detectors used purified water. One of the most well-known neutrino detectors is the *Super-Kamiokande* in Hida, Japan, which holds 50,000 tons of water. The detector must have large mass and volume, because neutrinos rarely interact with the atoms of anything through which they pass. Scientists place neutrino detectors deep underground so that the sheer mass of the earth's crust blocks all other cosmic particles. Using this arrangement, neutrino observers can have confidence that any detected particle coming from space is indeed a neutrino.

In 2007, the U.S. National Science Foundation proposed the construction of a new underground laboratory in the old Homestake gold mine in Lead, South Dakota. If and when completed, the *Sanford Underground Laboratory*, also known as the *Deep Underground Science and Engineering Laboratory* (DUSEL), will house a neutrino detector, along with hardware for conducting other scientific and engineering experiments, at 2 km below the surface—twice the depth of the Super-Kamiokande. The facility will give astronomers and cosmologists a chance to observe neutrinos of cosmic origin with less "background noise" than ever before.

When a neutrino interacts with a water molecule, we observe a burst of visible light called *Cerenkov radiation*. Cerenkov radiation occurs when certain subatomic particles pass through a substance at a speed greater than the speed of visible light in that substance. It's the optical equivalent of a *sonic boom*, but extremely faint. The interior surface of the detector tank in the Super-Kamiokande is lined with more than 11,000 individual photodetectors whose output is amplified by electronic circuits and analyzed with computers. If a neutrino interacts with an atom in the tank, a cone-shaped burst of Cerenkov radiation strikes the photodetector array, and the computers can calculate the direction in space from which the neutrino arrived.

QUIZ

Refer to the text in this chapter if necessary. A good score is eight correct. Answers are at the back of the book.

1. When radio waves refract downward in the earth's lower atmosphere making over-the-horizon communication possible, we refer to the effect as
   A. a surface wave.
   B. ducting.
   C. tropospheric bending.
   D. troposcatter.

2. Neutrinos can pass easily through
   A. a sheet of paper.
   B. a brick wall.
   C. a mountain.
   D. All of the above

3. Which layer of the ionosphere can interfere with wireless communication, especially at medium frequencies during the daytime?
   A. The D layer
   B. The E layer
   C. The F1 layer
   D. The F2 layer

4. How much energy does a single photon contain in a UV beam having a wavelength of 200 nm?
   A. We need more information to calculate it.
   B. 1.00×10^{-25} J
   C. 3.97×10^{-32} J
   D. 9.93×10^{-19} J

5. If a photon carries 2.50×10^{-15} J of energy, then it has an EM wavelength of
   A. 0.0795 nm.
   B. 126 μm.
   C. 49.7 cm.
   D. We need more information to calculate it.

6. Which of the following sequences portrays EM fields in order of increasing free-space wavelength?
   A. Visible light, γ rays, UV rays, IR rays
   B. Radio waves, ELF radiation, IR, visible light
   C. γ Rays, x rays, UV rays, visible light
   D. UV rays, x rays, γ rays, ELF radiation

7. What EM frequency corresponds to a wavelength of 40 m in free space?
   A. 13 kHz
   B. 7.5 MHz
   C. 12 MHz
   D. 13 MHz

8. If a sample of radioactive material loses 75% of its radiation intensity after 10 years, then it has a half-life of
   A. 2.5 years.
   B. 5.0 years.
   C. 7.5 years.
   D. 20 years.

9. Which of the following types of radiation has photons containing the *least* energy?
   A. X rays
   B. ELF radiation
   C. Visible light
   D. IR radiation

10. Which of the following types of radiation has photons containing the *most* energy?
    A. X rays
    B. ELF radiation
    C. Visible light
    D. IR radiation

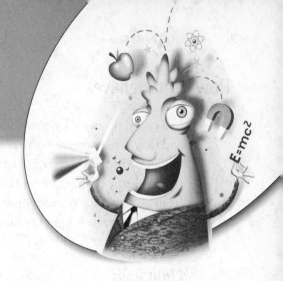

chapter 14

Optics

Until a few hundred years ago, the human eye alone served as the apparatus for optical observation of natural phenomena. As experimenters developed telescopes, microscopes, and other devices, humanity's visual horizons expanded (and shrank at the same time!).

CHAPTER OBJECTIVES

In this chapter, you will

- Watch light rays reflect and refract.
- Analyze refraction at boundaries between different media.
- Discover how lenses and mirrors can focus and spread light rays.
- Learn how telescopes and microscopes work.
- Evaluate the performance of telescopes and microscopes.

How Light Behaves

In a vacuum, visible light rays always take the shortest path between two points, and they always travel at the same speed relative to an observer. However, if the medium through which light passes differs from a vacuum—and especially if the medium changes along the way—these axioms fail. If a ray of light goes from air into glass or from glass into air, for example, the ray takes a sharp turn at the boundary. A light ray also changes direction when it reflects from a mirror or any other "shiny" surface.

Light Rays

We can call a thin shaft of light, such as that which passes from the sun through a pinhole in a piece of cardboard, a *ray* or *beam*. In a more technical sense, a ray or beam represents the path that an individual photon follows through space, air, glass, water, or other medium.

Visible light possesses wavelike and particle-like properties. This duality has interested physicists for well over a century. In some situations, the so-called *particle model* or *corpuscular model* explains light behavior quite well, and the wave model falls short. In other scenarios, things work out the other way around.

No one has ever actually seen a ray of light, although its path shows up in a smoky or foggy environment. Our eyes can only see the effects produced when a ray of light strikes something. Nevertheless, scientists can make generalizations and predictions about how light rays can and will behave.

Reflection

Prehistoric people knew about *reflection*. It would not take a stone age human very long to figure out that the "phantom in the pond" was an *image* of himself or herself, and not another person looking up from underwater! Any smooth, shiny surface reflects some or all of the light that strikes it. Any ray striking a flat, shiny surface reflects away at the same angle as that from which it arrives. You've heard the expression, "The *angle of incidence* equals the *angle of reflection*." Figure 14-1 illustrates this principle, which we call the *law of reflection*.

In optics, we express both the angle of incidence and the angle of reflection relative to a *normal line* (also called an *orthogonal* or *perpendicular*). In Fig. 14-1, we denote these angles as q. Their measure can range from 0°, where the light ray strikes at a right angle with respect to the surface, to almost 90°, a grazing angle relative to the surface.

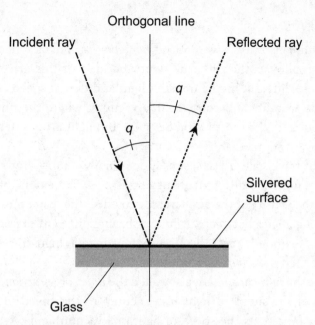

FIGURE 14-1 · When a ray of light reflects from a shiny, flat surface, the angle of incidence equals the angle of reflection. Here, we call both angles q.

Still Struggling

What, you ask, happens when a reflective surface has an irregular contour? In that case, the law of reflection applies for each ray of light striking the surface at a specific point. In such a case, we consider the reflection with respect to a flat plane passing through the point and lying *tangent* to the surface at that point. When many parallel rays of light strike a curved or irregular reflective surface at many different points, each ray obeys the law of reflection, but the reflected rays don't all emerge parallel. In some cases they converge, in other cases they diverge, and in still other cases they scatter.

Refraction

Prehistoric humans must have noticed *refraction* as well as reflection. A clear pond looks shallower than its true depth because of this effect. Refraction occurs because different media transmit light at different speeds. The speed of light remains absolute and constant in a vacuum, where the photons always travel at about 299,792 km/s (186,282 mi/s). But light propagates more slowly through many other media.

In air, light travels almost as fast as it does in a vacuum—but not quite! We can usually ignore the difference, but not always. The extent of the slowing increases as the density of the air increases, an effect that can produce refraction at near-grazing angles between air masses having different densities. In water, glass, quartz, diamond, and other transparent media, light travels quite a lot more slowly than it does in a vacuum.

We define the *refractive index*, also called the *index of refraction*, of a medium as the ratio of the speed of light in a vacuum to the speed of light in that medium. If c represents the speed of light in a vacuum and c_m represents the speed of light in a medium called M, then we can calculate the index of refraction for M (let's call it r_m) using the formula

$$r_m = c/c_m$$

We must always use the same units when expressing c and c_m. According to this definition, the index of refraction for any transparent material must always equal 1 or more.

As the index of refraction for a transparent substance increases, light rays refract more when they cross the boundary between that substance and air or a vacuum at any fixed angle of incidence larger than 0°. When the angle of incidence equals 0°, so that the ray arrives perpendicular to the surface, refraction never takes place. Different types of transparent material have different refractive indices. Quartz refracts more than glass, and diamond refracts more than quartz. The high refractive index of diamond produces the characteristic multicolored "shine" of the "precious stones."

 PROBLEM 14-1

Suppose that a transparent substance has a refractive index of 1.50 for yellow light. How fast does yellow light travel through this medium?

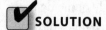

SOLUTION

We can use the above formula and "plug in" the refractive index and the speed of light in a vacuum. Let's express the speeds in kilometers per second, and round off c to 3.00×10^5 km/s. We can calculate the speed of yellow light in the clear substance, c_m, by setting up the equation

$$1.50 = 3.00 \times 10^5 / c_m$$

When we multiply through by c_m, we obtain

$$1.50\, c_m = 3.00 \times 10^5$$

Dividing through by 1.50 gives us

$$c_m = 3.00 \times 10^5 / 1.50$$
$$= 2.00 \times 10^5 \text{ km/s}$$

Light Rays at a Boundary

Figure 14-2 shows what happens to a ray of light traveling generally upward when the refractive index of the first (lower) medium exceeds that of the second (upper)

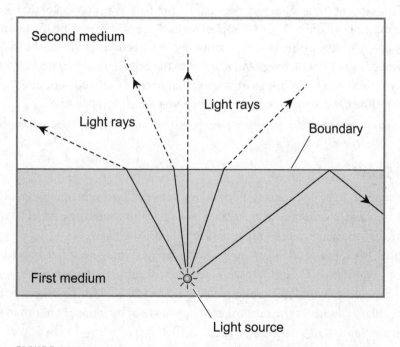

FIGURE 14-2 · Rays of light bend as they cross a boundary between media having different properties.

medium. A ray striking the boundary at a right angle (that is, angle of incidence of 0°) crosses the boundary without changing direction. But a ray changes direction when it hits the boundary at some other angle; the greater the angle of incidence, the sharper the turn. When the angle of incidence reaches a certain *critical angle*, then the light ray doesn't refract at the boundary. Instead, it reflects completely back into the first medium. We call this phenomenon *total internal reflection*.

A ray of light traveling in the opposite sense from that shown in Fig. 14-2, originating in the second (upper) medium and striking the boundary at a grazing angle, deflects downward. This effect causes distortion of landscape images when viewed from underwater. You'll witness it if you go skin diving or SCUBA diving outdoors and look up! You'll see the sky, trees, hills, buildings, people, and everything else above the horizon within a circle of light that distorts the scene like a wide-angle lens.

Still Struggling

What happens if the refracting boundary is not flat? In a situation of that sort, the principle shown by Fig. 14-2 applies for each ray of light striking the boundary at any specific point. We must consider the refraction with respect to a flat plane passing through the point, tangent to the boundary at that point. When many parallel rays of light strike a curved or irregular refractive boundary at many different points, each ray obeys the same principle individually.

Snell's Law

When a ray of light encounters a boundary between two substances having different indices (or indexes) of refraction, we can determine the extent to which the ray deflects according to an equation called *Snell's law*.

Figure 14-3 illustrates how Snell's law operates. Imagine a flat boundary B between two media M_r and M_s, whose indices of refraction are r and s, respectively. Imagine a ray of light crossing the boundary as shown. The ray refracts at the boundary whenever the ray strikes at any angle of incidence other than 0°, as long as the indices of refraction, r and s, differ.

Suppose that $r < s$, so that the light passes from a medium having a relatively lower refractive index to a medium having a relatively higher refractive index.

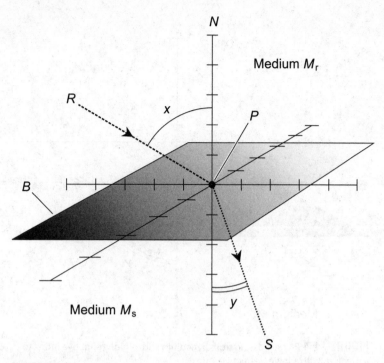

FIGURE 14-3 · A ray passing from a medium with a low refractive index to a medium with a higher refractive index.

Let N represent a line passing through a specific point P on the boundary plane B, such that N is normal to B at P. Let R represent a ray of light traveling through M_r that strikes B at P. Let x denote the angle that R subtends relative to N at P. Let S denote the ray of light that emerges from P into M_s. Let y denote the angle that S subtends relative to N at P. Under these conditions, we'll find that line N, ray R, and ray S all lie in a single plane perpendicular to the boundary plane B, and angle y is smaller than angle x ($y < x$). The two angles x and y have equal measure if, but only if, the ray R strikes the boundary at an incidence angle of $0°$. Everything relates mathematically according to the two equivalent equations

$$\sin y / \sin x = r/s$$

and

$$s \sin y = r \sin x$$

Now examine Fig. 14-4. Again, let B denote a flat boundary between two media M_r and M_s, whose absolute indices of refraction are r and s, respectively.

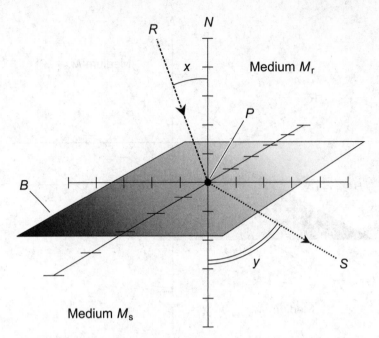

FIGURE 14-4 · A ray passing from a medium with a high refractive index to a medium with a lower refractive index.

This time, suppose that $r > s$; that is, the ray passes from a medium having a relatively higher refractive index to a medium having a relatively lower refractive index. Let N, B, P, R, S, x, and y denote the same quantities as they did in the example shown by Fig. 14-3. Then, just as before, we observe that $x = y$ if and only if ray R strikes B at an incidence angle of $0°$. If the angle of incidence exceeds $0°$, then line N, ray R, and ray S all lie in a single plane perpendicular to the boundary plane B, and angle y is larger than angle x ($y > x$). Snell's law holds here, just as it did before. The equations remain exactly the same. Once again, we have

$$\sin y / \sin x = r/s$$

and

$$s \sin y = r \sin x$$

Determining the Critical Angle

In the scenario of Fig. 14-4, the light passes from a medium having a relatively higher index of refraction r into a medium having a relatively lower index s. Therefore, $r > s$. As the incidence angle x increases, the measure of angle y

approaches 90°, and ray S gets closer to the boundary plane B. When x gets large enough (somewhere between 0° and 90°), angle y reaches 90°, and ray S lies exactly in plane B. If angle x increases still more, ray R undergoes total internal reflection at the boundary plane B, the boundary acts like a mirror, and none of the light penetrates into the lower medium M_s.

The critical angle equals the largest incidence angle that ray R can subtend, relative to the normal line N, without reflecting internally. The measure of the critical angle equals the *Arcsine* of the ratio of the indices of refraction. If we call the critical angle x_c, then

$$x_c = \sin^{-1}(s/r)$$

where $\sin^{-1}$ represents the Arcsine function. We can also write

$$x_c = \text{Arcsin}(s/r)$$

where Arcsin represents the Arcsine function.

PROBLEM 14-2

Suppose that we place a laser device beneath the surface of a freshwater pond whose surface is perfectly flat, smooth, and horizontal. Fresh water has a refractive index of approximately 1.33. Air has a refractive index of 1.00. If we shine the laser beam upward so that it strikes the surface at an angle of 30.0° relative to the normal, at what angle, also relative to the normal, will the beam emerge from the surface into the air?

SOLUTION

We must have a good scientific calculator to work out this problem (and most of those that follow). Let's envision the situation in Fig. 14-4 upside down. In that case, M_r represents the water and M_s represents the air. We have indices of refraction

$$r = 1.33$$

and

$$s = 1.00$$

Angle x measures 30.0°. We want to find the measure of angle y. We can invoke Snell's law, plug in the numbers, and solve for y. Let's start with the general equation

$$\sin y / \sin x = r/s$$

Inputting the known values for x, r, and s, we get

$$\sin y / (\sin 30.0°) = 1.33 / 1.00$$

Because $\sin 30.0° = 0.500$, we have

$$\sin y / 0.500 = 1.33$$

We can multiply through by 0.500 to get

$$\sin y = 1.33 \times 0.500$$
$$= 0.665$$

Finally, we can take the Arcsine of both sides to get

$$\sin^{-1} (\sin y) = \sin^{-1} 0.665$$
$$= 41.7°$$

From basic trigonometry, we remember that when we apply the Arcsine function to the sine of a quantity, we obtain the original quantity. Therefore, by substitution, we get our answer as

$$y = \sin^{-1} 0.665$$
$$= 41.7°$$

PROBLEM 14-3

What's the critical angle x_c for light rays shining upward from beneath a freshwater pond?

SOLUTION

We can use the formula for determining the critical angle, taking the values for r and s from Problem and Solution 14-2. We start with

$$x_c = \sin^{-1} (s/r)$$

Once again, we know that

$$r = 1.33$$

and

$$s = 1.00$$

When we plug in the known values for *s* and *r*, we get

$$x_c = \sin^{-1}(1.00/1.33)$$
$$= \sin^{-1} 0.752$$
$$= 48.8°$$

Remember that we must define all angles with respect to a line normal to the surface, not with respect to the surface itself.

Color Dispersion

The index of refraction for a substance depends on the wavelength of the light passing through it. Glass slows down light to the greatest extent at the shortest visible wavelengths (blue and violet), and least at the longest visible wavelengths (red and orange). Physicists call this wavelength-dependent variation of refractive index *color dispersion*. It's responsible for the well-known behavior of a typical glass prism. The more the glass slows the light, the more the light deflects as it passes through the prism. This effect explains why a prism casts a *color spectrum* (or "rainbow") when white light passes through, as shown in Fig. 14-5.

Color dispersion interests (and sometimes bothers) scientists involved in optics. We can use a prism to construct a *spectrometer*, a device for examining the intensity of visible light at specific wavelengths. (Fine gratings will also serve this purpose.) Color dispersion degrades the quality of white-light images viewed through simple lenses. The effect multiplies when we combine lenses to make telescopes and microscopes. When color dispersion takes place in a telescope, for example, the images of distant celestial objects acquire false-colored borders, especially at high magnification levels.

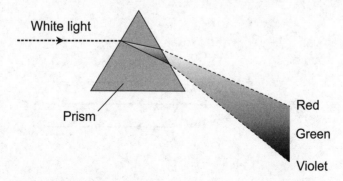

FIGURE 14-5 · Dispersion causes a glass prism to "split up" white light into its constituent colors.

Lenses and Mirrors

Specially shaped pieces of glass can make an object appear larger or smaller than its actual size. People have used lenses for centuries to help correct deficiencies in human vision. Lenses work because they refract light more or less depending on where, and at what angle, the light strikes their surfaces. Curved mirrors produce similar effects when they reflect rays of visible light.

The Convex Lens

You can buy a *convex lens* in almost any novelty store or department store. In a good hobby store, you should be able to find a "magnifying glass" measuring 10 or 15 cm in diameter. The term *convex* arises from the fact that one or both faces of the glass "bulge outward" at the center. We can also call a convex lens a *converging lens*. It brings parallel light rays to a sharp *focal point* or *focus* (Fig. 14-6A) when those rays arrive parallel to the axis of the lens. A convex lens can also *collimate* (make parallel) the light from a point source as shown in Fig. 14-6B.

The properties of a convex lens depend on the lens diameter, and also on the difference in thickness between the edges and the center. As the diameter increases, so does the *light-gathering power*. As the difference in thickness

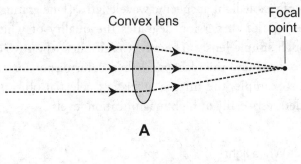

A

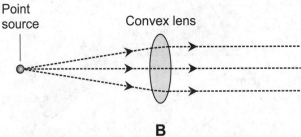

B

FIGURE 14-6 · At A, a convex lens focuses parallel light rays to a point. At B, the same lens collimates light from a point source at the focus.

between the center and the edges increases, the distance between the lens and the point at which it brings parallel light rays to a focus decreases.

We call the effective area of a lens, measured in a plane perpendicular to its axis, the *light-gathering area*. We call the distance between the center of the lens and the focal point (as shown in Fig. 14-6A or 14-6B) the *focal length*. If you look through a convex lens at a close-up object such as a coin, the features all appear magnified. The light rays from an object at a great distance from a convex lens converge to form a *real image* at the focal point.

The surfaces of most convex lenses have spherical contours. If you could find a large ball having exactly the correct diameter, the curve of the lens face would fit precisely inside the ball. Some convex lenses have the same radius of curvature on each face; others have different radii of curvature on their two faces. Some converging lenses have one curved face and one flat face; we call them *plano-convex lenses*.

The Concave Lens

You might not find a *concave lens* in a department store, but you can order one from a specialty catalog or Web site. The term *concave* refers to the fact that one or both faces of the glass "cave inward" at the center. Some opticians or optometrists call this type of lens a *diverging lens* because it spreads incoming parallel light rays outward as shown in Fig. 14-7A. A concave lens can also collimate incident converging rays (Fig. 14-7B) if the incoming rays converge at a certain angle.

The properties of a concave lens depend on the lens diameter and the extent to which the surface(s) depart from flat. As the difference in thickness between the edges and the center of the glass increases, the lens causes parallel rays of light to diverge more. If you look through a concave lens at an object such as a coin, the object appears reduced in size. The same thing happens with a scene such as a landscape; the scene might also appear blurred.

The surfaces of concave lenses, like those of their convex counterparts, have spherical contours. Some concave lenses have the same radius of curvature on each face; others have different radii of curvature on their two faces. Some diverging lenses have one flat face; we call them *plano-concave lenses*.

The Convex Mirror

A *convex mirror* reflects light rays to produce an effect similar to that of a concave lens, except that the rays reverse direction. Incident rays, when parallel, tend to spread out (Fig. 14-8A) after they reflect from the convex surface. Converging incident rays, if the angle of convergence is just right, reflect from a convex mirror in parallel alignment (Fig. 14-8B).

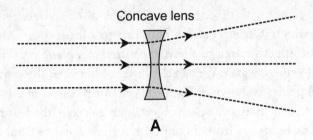

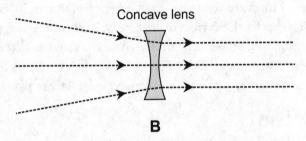

FIGURE 14-7 · At A, a concave lens spreads parallel light rays. At B, the same lens collimates converging light rays.

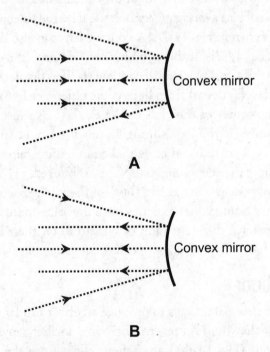

FIGURE 14-8 · At A, a convex mirror spreads parallel incident light rays. At B, the same mirror collimates converging incident light rays.

When you look at the reflection of a scene in a convex mirror, the objects all appear reduced in size, and therefore more distant. The field of vision enlarges. External automotive rear-view mirrors sometimes have slightly convex surfaces, allowing the driver to see vehicles lurking in the "blind spot." The extent to which a convex lens spreads light rays depends on the radius of curvature. As the radius of curvature decreases, the extent to which parallel incident rays diverge after reflection increases.

The Concave Mirror

A *concave mirror* reflects light rays in a manner similar to the way a convex lens refracts them, except that the rays reverse direction. When incident rays arrive parallel to each other and to the axis of the mirror, they reflect to converge at a focal point, as shown in Fig. 14-9A. When we place a point source of light at the focus, the concave mirror reflects the rays so that they emerge parallel, as shown in Fig. 14-9B.

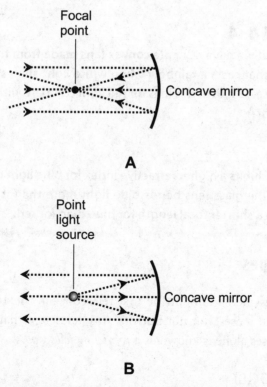

FIGURE 14-9 · At A, a concave mirror focuses parallel light rays to a point. At B, the same mirror collimates light from a point source at the focus.

The properties of a concave mirror depend on the size of the reflecting surface, and also on the radius of curvature. As the mirror area increases, the light-gathering power increases in direct proportion. As the radius of curvature decreases, so does the focal length. If you look at your reflection in a large concave mirror, you'll see the same effect that you would observe if you placed a large *convex* lens up against a flat mirror.

Concave mirrors can have spherical surfaces, but the finest mirrors have surfaces that follow the contour of an idealized three-dimensional figure called a *paraboloid*. A paraboloid results from the rotation of a *parabola*, such as that having the equation $y = x^2$ in rectangular coordinates, around its axis. When a concave mirror's radius of curvature greatly exceeds the radius of the mirror itself, a casual observer won't notice the difference between a *spherical mirror* and a *paraboloidal mirror* (more commonly called a *parabolic mirror*). But when we use a concave mirror as the objective in a precision optical instrument such as a telescope at high magnification, we will easily notice the difference, even if the mirror has a large radius of curvature, and therefore a long focal length.

PROBLEM 14-4

Imagine that we have a simple convex lens made from the same material as a prism that casts a rainbow when white light shines through it. How does the focal length of this lens for red light compare with the focal length for blue light?

SOLUTION

The glass exhibits a higher refractive index for blue light than for red light. Therefore, the glass lens bends blue light more than it bends red light, resulting in a shorter focal length for blue than for red.

Refracting Telescopes

Astronomers designed and built the first telescopes in the 1600s. These telescopes employed lenses, but not mirrors. Any telescope that enlarges distant images with lenses alone is known as a *refracting telescope*.

Galilean Refractor

Galileo Galilei became famous (and, among Church officials, infamous) during the 1600s for noticing craters on the earth's moon, distant moons orbiting

Jupiter, and other celestial phenomena that defied explanation at the time. He saw these things using a telescope constructed with a convex-lens *objective* and a concave-lens *eyepiece*. His first telescope magnified the apparent diameters of distant objects by a factor of only a few times. Some of his later telescopes magnified up to 30 times. The so-called *Galilean refractor* (Fig. 14-10A) produces an *erect image*, that is, a right-side up view of distant objects. In addition to appearing right-side-up, images also "look true" in the left-to-right sense. The *magnification factor*, defined as the number of times the angular diameters of distant objects appear to increase, depends on the focal length of the objective, and also on the distance between the objective and the eyepiece.

You can find Galilean refractors today, mainly as novelties intended for terrestrial viewing. Galileo's original refractors had objective lenses that measured 2 to 3 cm (about 1 in) across. Most Galilean telescopes that you'll find in hobby stores have similar small diameters. Some of these telescopes, also known as *spy glasses*, have sliding, concentric tubes, providing variable magnification. When you push the inner cylinder all the way into the outer one, you get the lowest possible magnification factor. When you pull the inner cylinder all the way out, you get the most magnification. The image remains fairly clear over the entire magnification-adjustment range, so you don't have to "fiddle around" with any sort of *focus control*.

Keplerian Refractor

The astronomer *Johannes Kepler*, who got more friendly treatment from the Church establishment than Galileo did, refined Galileo's telescope design. Kepler's refracting telescope had a convex-lens objective with a long focal length and a smaller, convex-lens eyepiece with a short focal length. Unlike the Galilean telescope, the so-called *Keplerian refractor* (Fig. 14-10B) produces an *inverted image* that appears upside-down and backward compared to the actual scene. To obtain a clear image, the distance between the objective and the eyepiece must precisely equal the sum of the focal lengths of the two lenses; the user must adjust a focus control to get the distance exactly right. The magnification factor equals the ratio of the focal length of the objective to the focal length of the eyepiece.

Most astronomers prefer the Keplerian telescope over the Galilean type, primarily because the Keplerian design provides a larger *apparent field of view*. That's the angular size of the "sky disk" that an observer sees through the eyepiece. Galilean telescopes in general have apparent fields of view so narrow that looking through them produces a sensation of "tunnel vision." We can adjust the magnification factor of a Keplerian telescope by using multiple eyepieces having various focal lengths. As we reduce the focal length of the eyepiece, the

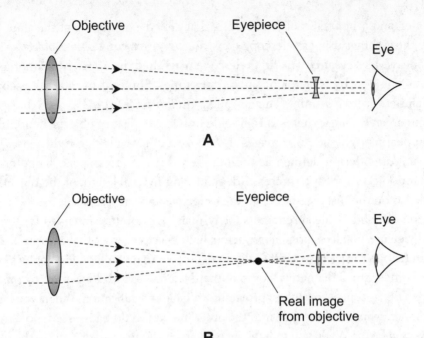

FIGURE 14-10 · The Galilean refractor (A) uses a convex objective and a concave eyepiece. The Keplerian refractor (B) has a convex objective and a convex eyepiece.

magnification factor, informally known as "power," increases, assuming the focal length of the objective lens remains constant.

The largest refracting telescope in the world inhabits a dome-shaped building at the *Yerkes Observatory* in Wisconsin. Its objective lens has a diameter of 40 in (slightly more than 1 m). Amateur astronomers all over the world use and enjoy Keplerian refractors because of their relatively low cost, mechanical stability, and "user-friendliness."

Limitations of Refractors

Certain problems can occur with poorly designed refracting telescopes. These problems fall mainly into three categories: *spherical aberration, chromatic aberration*, and *lens sag*.

Spherical aberration results from the fact that spherical convex lenses don't bring parallel light rays to a perfect focus. A refracting telescope with a spherical objective focuses a ray passing through its edge a little differently than a ray passing closer to the center. When we use the telescope to look at a distant star, the focused light rays don't come to a perfect point. Instead, they converge to a short line along the lens axis. This effect causes blurring in the images of celestial objects having relatively large angular diameters, such as nebulae and galaxies. We can correct the problem by grinding the objective lens so that its faces have paraboloidal, rather than spherical, contours. As you might expect, this process costs money, so paraboloidal convex lenses tend to be expensive.

Chromatic aberration occurs because the glass in a simple lens refracts the shortest wavelengths of light slightly more than the longest wavelengths. A plain glass convex lens has a shorter focal length for violet light than for blue light, shorter for blue than for yellow, and shorter for yellow than for red. This color dispersion produces rainbow-colored halos around star images, and also along sharply defined edges in objects with large angular diameter. We can almost (but not completely) get rid of chromatic aberration by employing *compound lenses* in telescopes. A compound lens has multiple, layered sections made from different types of glass, glued together with transparent adhesive. We call them *achromatic lenses*. They're "standard issue" in refracting telescopes these days, even at the lay consumer level.

Lens sag can pose a problem in large refracting telescopes. When we have a refracting telescope whose objective lens radius exceeds approximately 200 mm, the lens has such large mass that its own weight pulls it out of shape. Glass isn't perfectly rigid, as you've noticed if you've ever watched the reflection of the landscape from a large window on a windy day.

Still Struggling

You might ask, "How can we get rid of lens sag in a large refracting telescope?" The short answer is, "We can't." Astronomers haven't found a good way to get rid of lens sag entirely in large refractors, except to locate the telescope in outer space so that gravitation can't "warp" the lens.

Reflecting Telescopes

We can largely overcome the troubles that plague refracting telescopes, particularly lens sag, by using mirrors instead of lenses as the objectives. We can manufacture a *first-surface mirror*, with the silvering on the outside so the light never passes through any glass, to bring light rays to a focus that doesn't vary with the wavelength. We can support mirrors from behind, allowing us to employ them in diameters much larger than lenses without having to worry about sag.

Newtonian Reflector

The physicist, astronomer, and mathematician *Isaac Newton* designed a *reflecting telescope* that did not suffer from chromatic aberration. We still find his design in common use today. The *Newtonian reflector* employs a concave objective mirror, mounted at one end of a long tube. The other end of the tube remains open to admit incoming light. A small, flat mirror rests at a 45° angle near the open end of the tube to reflect the focused light through an opening in the side of the tube containing the eyepiece. Figure 14-11A illustrates the basic design.

In a Newtonian reflector, the flat mirror obstructs some of the incoming light, slightly reducing the effective surface area of the objective mirror. For example, suppose that the objective mirror measures 10 cm in radius. The total

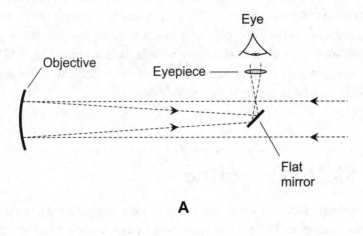

A

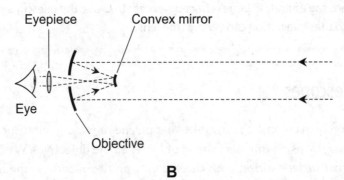

B

FIGURE 14-11 • The Newtonian reflector (A) has an eyepiece set into the side of the tube. The Cassegrain reflector (B) has an eyepiece in the center of the objective mirror.

surface area of this mirror equals approximately 314 centimeters squared (cm²). If we use an eyepiece mirror measuring 3 cm², it has a total surface area of 9 cm², only about 3 percent of the total surface area of the objective. That's not enough for a casual observer to notice, but it will make a difference if we intend to use the telescope to take time-exposure photographs.

Newtonian reflectors have limitations. Some people find it unnatural to "look sideways" at objects. If the telescope has a long tube, observers must climb a step ladder to view objects at high elevations in the sky. We can overcome these annoyances by getting the light to the eyepiece by means of a different geometry.

Cassegrain Reflector

Figure 14-11B shows the basic design of a *Cassegrain reflector*. (Some people call it a *Schmidt-Cassegrain reflector*.) The convex eyepiece mirror sits closer to the objective than the flat eyepiece mirror inside a Newtonian reflector. The convex contour of the eyepiece mirror increases the effective focal length of the objective. Light reflects from the convex mirror and passes through a hole in the center of the objective to the eyepiece.

We can build a Cassegrain reflector with a physically short tube and an objective mirror having a smaller radius of curvature than that of a Newtonian telescope of the same diameter. As a result, a Cassegrain reflector has less mass and bulk than an equivalent Newtonian reflector. Cassegrain reflectors with heavy-duty mountings offer excellent physical stability.

Amateur and professional astronomers use Cassegrain reflectors at low magnification to view or photograph large regions of the sky. A Cassegrain reflector, even a small one, can function as an excellent telephoto lens for viewing and photographing wildlife! For example, from a distance of several hundred meters you can get a close-up image of a deer without alarming the animal. You can photograph a fruit fly on the side of a house from 20 or 30 m away and get an image so clear that an enlarged print will reveal the hairs on its legs. (I did that a few years ago with my own telescope and a standard 35-mm film camera.)

Telescope Specifications

We must take several parameters into account when we want to determine the effectiveness of a telescope for specific applications. These parameters include the *magnification*, the *resolving power*, the *light-gathering area*, and the *absolute field of view*.

Magnification

We define the magnification, also called "power" and symbolized ×, as the extent to which a telescope makes objects look "closer." (Actually, telescopes increase the observed sizes of distant objects, but they do not look closer in terms of perspective.) The magnification tells us the factor by which the apparent angular radius or diameter of an object increases. For example, a 20× telescope makes the moon, whose disk has an angular diameter of approximately 0.5° as observed with the unaided eye, appear about 10° in diameter. A 180× telescope makes a crater on the moon with an angular diameter of one *minute of arc* (1/60 of a degree) appear to have an angular diameter of 3°.

We can calculate a telescope's magnification in terms of the focal lengths of the objective and the eyepiece. If f_o represents the effective focal length of the objective and f_e represents the focal length of the eyepiece (in the same units as f_o), then the magnification factor m equals the ratio of f_o to f_e; that is,

$$m = f_o / f_e$$

For a given eyepiece, as the effective focal length of the objective increases, the magnification of the telescope also increases. For a given objective, as the effective focal length of the eyepiece increases, the magnification of the telescope decreases.

Resolving Power

We define the resolving power of a telescope, also called the *resolution*, as the instrument's ability to separate two objects that do not occupy the same point in the sky. Astronomers express resolving power in an angular sense, usually in seconds of arc (units of 1/3600 of a degree). The smaller the number, the better the resolving power.

We can measure a telescope's resolving power by scanning the heavens for known pairs of stars that appear close to each other in the angular sense. Astronomical data charts can tell us which pairs of stars best serve this purpose. We can also examine the moon, and use a detailed map of the lunar surface to ascertain how much detail a particular telescope can render. Of course, we must base all of these determinations on the assumption that we have good eyesight! Some people must wear corrective lenses (either contacts or glasses) to enjoy the full resolving power of a fine telescope.

Resolving power increases with magnification up to a certain point. The greatest image resolution that a telescope can theoretically provide varies in direct proportion to the diameter or radius of the objective. However, in practice, atmospheric turbulence prevents us from getting unlimited resolving power from any surface-based telescope.

Still Struggling

If you've ever wondered why astronomers put telescopes in space, you can figure out the answer now! Scientists built earth-orbiting optical instruments such as the *Hubble telescope* to get the instruments above the image-degrading atmosphere. Some astronomers and cosmologists want humans to return to the moon and set up a permanent base there. The moon has essentially no atmosphere. Imagine a 100-m Schmidt-Cassegrain reflector on the Sea of Tranquility!

Light-Gathering Area

Astronomers define the light-gathering area of a telescope in centimeters squared (cm²) or meters squared (m²), telling us the effective surface area of the objective lens or mirror as light rays arrive along its axis. Some people express a telescope's light-gathering area in inches squared (in²).

As the term suggests, a telescope's light-gathering area specification tells us how much light a telescope can "pull in," a direct indicator of the "sensitivity" of the instrument. A telescope with large light-gathering area allows an observer to see dimmer objects than a telescope with smaller light-gathering area.

For a refracting telescope, given an objective radius of r, we can calculate the light-gathering area A with the formula

$$A = \pi r^2$$

where π equals approximately 3.14159. If we express r in centimeters, then we obtain A in centimeters squared. If we express r in meters, then we get A in meters squared. If we express r in inches, then we get A in inches squared.

For a reflecting telescope, given an objective radius of r, we can calculate the light-gathering area A according to the formula

$$A = \pi r^2 - B$$

where B represents the area obstructed or taken away by the secondary mirror assembly. If we express r in centimeters and B in centimeters squared, then we obtain A in centimeters squared. If we express r in meters and B in meters squared, then we obtain A in meters squared. If we express r in inches and B in inches squared, then we obtain A in inches squared.

Absolute Field of View

When we look through the eyepiece of a telescope, we see a circular "sky disk." Actually, we can see anything within a cone-shaped region whose apex coincides with the location of the telescope, as shown in Fig. 14-12. We define the absolute field of view as the angular diameter q of this cone. We can express q in degrees, minutes, and/or seconds of arc. We can also specify the angular radius instead of the angular diameter. In Fig. 14-12 the angular radius would equal $q/2$.

The absolute field of view depends on several factors. The magnification of the telescope has an effect; when we hold all other factors constant, the absolute field of view varies inversely with the magnification. If we double the magnification, for example, then we cut q in half. If we reduce the magnification to 1/4 of its previous value, then we increase q by a factor of 4.

The *viewing angle* (apparent field of view) of a telescope's eyepiece affects the absolute field of view of the entire instrument. Some types of eyepieces exhibit a wide field, such as 60° or even 90°. Others have narrower apparent fields, in some cases less than 30°. If we hold all other factors constant, then a telescope's absolute field of view varies in direct proportion to the eyepiece's apparent field of view.

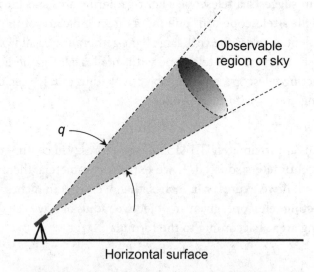

Observable
region of sky

q

Horizontal surface

FIGURE 14-12 · We can express a telescope's absolute field of view, q, in angular degrees, minutes, and/or seconds of arc.

The ratio of a telescope's objective diameter to its focal length also affects the absolute field of view. In general, as this ratio increases, so does the maximum absolute field of view that we can obtain with the telescope. Long, narrow telescopes tend to have small (narrow) maximum obtainable values of q. Short, fat telescopes offer large (wide) maximum obtainable values of q.

 PROBLEM 14-5

How much more light can a refracting telescope with a 15.0-cm-diameter objective gather, compared with a refracting telescope having a 6.00-cm-diameter objective? Express the answer as a percentage.

 SOLUTION

Light-gathering area varies directly in proportion to the square of the objective's radius (and also, therefore, directly in proportion to the diameter). The ratio of the larger telescope's light-gathering area to the smaller telescope's light-gathering area therefore equals the square of the ratio of their objectives' diameters. If we call the ratio k, we have

$$k = 15.0/6.00$$
$$= 2.50$$

Squaring, we obtain

$$k^2 = 2.50^2$$
$$= 6.25$$

The larger telescope gathers 6.25 times, or 625 percent, as much light as the smaller one.

 PROBLEM 14-6

Suppose that a telescope has a magnification factor of 100× with an eyepiece of 20.0 mm focal length. What is the focal length of the objective?

 SOLUTION

We can use the formula given previously in the section "Magnification." The value of f_o constitutes our unknown. We know that $f_e = 20.0$ mm and $m = 100$. We start with the formula

$$m = f_o/f_e$$

When we plug in the known values, we get

$$100 = f_o / 20.0$$

Solving for f_o, we obtain

$$f_o = 100 \times 20.0$$
$$= 2000 \text{ mm}$$

We can justify this figure to only three significant digits. Let's say that $f_o = 2.00$ m.

PROBLEM 14-7

Suppose that the absolute field of view provided by the telescope in the above example spans 20 arc minutes. If we replace the 20-mm eyepiece with a 10-mm eyepiece that provides the same viewing angle as the 20-mm eyepiece, what happens to the absolute field of view provided by the telescope?

✔SOLUTION

The 10-mm eyepiece provides twice the magnification of the 20-mm eyepiece. Therefore, the absolute field of view that we get with the 10-mm eyepiece equals half the absolute field of view that we get with the 20-mm eyepiece, or 10 arc minutes.

The Compound Microscope

Optical microscopes magnify the images of objects too small to resolve with the unaided eye. Microscopes, in contrast to telescopes, work at close range. The simplest microscopes consist of single convex lenses that provide magnification factors up to 10× or 20×. In the laboratory, scientists use a multiple-lens instrument called a compound microscope, which offers much greater magnification than a single lens.

Basic Principle

A basic compound microscope employs two lenses. The objective has a short focal length, in some cases 1 mm or less. It "hovers over" the specimen or sample that we want to examine. The objective produces a real image at a certain distance above itself, where the light rays come to a focus. The distance (let's call it s) between the objective and the image always exceeds the focal length of the objective, sometimes by a considerable distance.

The eyepiece has a longer focal length than the objective—exactly the opposite state of affairs from a telescope! However, just as it does in a telescope, the eyepiece magnifies the real image cast by the objective. In a typical microscope, we can illuminate a translucent observed specimen by directing the light from a bright lamp upward through the specimen. Some microscopes allow for light to shine downward on opaque specimens. Figure 14-13 portrays the geometry of a compound microscope, showing how we can illuminate the specimen. We bring the viewed sample into focus by adjusting the distance between the objective and the sample (not the distance between the objective and the eyepiece, as we do in a telescope).

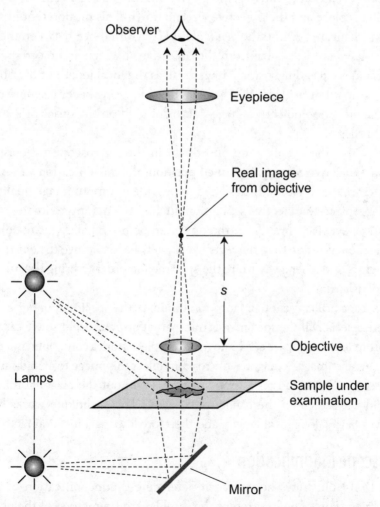

FIGURE 14-13 · Illumination and focusing in a compound optical microscope.

Laboratory-grade compound microscopes have two or more objectives. We can select one of them at a time by rotating a wheel with all of the objectives attached. This ingenious mechanical feature provides several different levels of magnification for a given eyepiece. In general, as the focal length of the objective decreases, the magnification of the microscope increases. Some compound microscopes can magnify images up to approximately 2500×. A hobby-grade compound microscope can provide decent image quality at magnifications up to 1000× or so.

Focusing

We focus a compound microscope by moving the entire assembly, including both the eyepiece and the objective, up and down. This motion requires a precision mechanism, because the *depth of field* (the difference between the shortest and the greatest distance from the objective at which an object appears in good focus) is exceedingly small. In general, as the focal length of the objective decreases, depth of field also decreases, and the focusing becomes more critical. High-magnification objectives have depths of field on the order of 2 μm (2×10^{-6} m) or less.

If we move the eyepiece up and down in the microscope tube assembly while the objective remains in a fixed position, the magnification varies. However, microscopes are usually designed to provide optimum image quality for a specific eyepiece-to-objective separation, such as 16 cm (approximately 6.3 in). Moving the eyepiece away from the optimum point will allow us to adjust the magnification. If we do that, however, we'll likely have to move the entire assembly closer to, or farther away from, the specimen in order to bring the image into sharp focus again.

If we use a brilliant enough lamp to illuminate the specimen under examination, and especially if the specimen is transparent or translucent so we can illuminate it from underneath, we can remove the eyepiece from a compound microscope and project the image. A diagonal mirror can reflect this image to an old-fashioned "movie screen" mounted on a tripod or wall. (Without the diagonal mirror, we must put the screen on the ceiling of our lab!) This technique works best for objectives having long focal length, and therefore low magnification factors.

Microscope Magnification

Figure 14-14 illustrates how we can determine the magnification of a compound microscope. Let f_o represent the focal length (in meters) of the objective

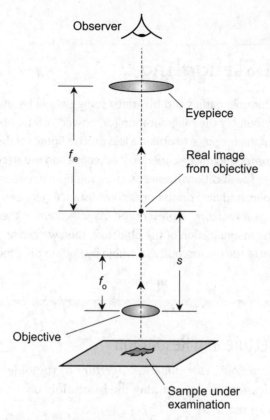

FIGURE 14-14 · Calculation of the magnification factor in a compound microscope.

lens, and let f_e represent the focal length (in meters) of the eyepiece. Assume that the objective and the eyepiece lie along a common axis, and that we adjust the distance between their centers for proper focus. Let s represent the distance (in meters) from the objective to the real image it forms of the object under examination. The *microscopic magnification* (a dimensionless quantity, denoted m in this context) is given by

$$m = [(s - f_o)/f_o] \, [(f_e + 0.25)/f_e]$$

The quantity 0.25 represents the average *near point* of the human eye: the closest distance (in meters) over which a "normal eye" can focus on an object.

Still Struggling

The foregoing formula baffles and frustrates some people because of its complexity. We can multiply the magnification (or "power") of the objective by the magnification of the eyepiece to obtain a less precise figure for the overall magnification of a compound microscope. Most objective and eyepiece manufacturers provide these specifications, based on an air medium between the objective and the specimen, and also on the optimum distance between the objective and the eyepiece. If we let m_e represent the magnification of the eyepiece and m_o represent the magnification of the objective, then we can approximate the magnification m of the microscope as a whole by finding the product

$$m \approx m_e m_o$$

Numerical Aperture and Resolution

In an optical microscope, the *numerical aperture* of the objective provides an important specification for determining the resolution, or the amount of detail the microscope can render.

Refer to Fig. 14-15. Consider a line L that passes through a point P in the specimen, and that also lies precisely along the objective's axis. Let K represent a line passing through P and intersecting the outer edge of the objective lens. (We assume that this outer edge forms a perfect circle). Let q represent the measure of the angle between lines L and K. Let M represent the medium between the objective and the sample under examination. Let r_m represent the refractive index of M. Under these conditions, we can calculate the numerical aperture of the objective, A_o, using the formula

$$A_o = r_m \sin q$$

In general, as A_o increases, the resolution improves. We can maximize the numerical aperture of a microscope objective having a given focal length in three ways:

- Maximize the diameter of the objective.
- Maximize the value of r_m.
- Minimize the wavelength of the illuminating light.

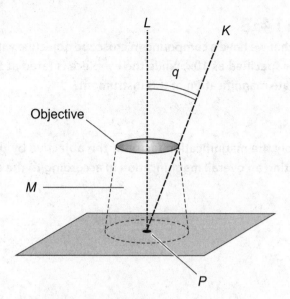

FIGURE 14-15 · Determination of the numerical aperture for a microscope objective.

Large-diameter objectives having short focal length, thereby providing high magnification, have proven difficult to manufacture. Therefore, when scientists want to improve the resolution of a microscope, they can use blue light, which has a relatively short wavelength. Alternatively (or in addition), the medium M between the objective and the specimen can be changed to a substance that has a high index of refraction, such as clear oil. This action reduces the wavelength of the illuminating beam that strikes the objective, because it slows down the speed of light in the medium M. (Remember the relation between the speed of an EM disturbance, the wavelength, and the frequency!) As a side-effect of this tactic, we will notice a reduction in the effective magnification of the objective lens (even though the resolution of the whole instrument improves!), but we can compensate for the loss of magnification by increasing the distance between the objective and the eyepiece.

In a compound microscope, the use of a monochromatic light source, rather than white light, for illuminating the specimen offers another advantage. Chromatic aberration affects the light passing through a microscope in the same way as it affects the light passing through a telescope. If the illuminating light exists at a single wavelength, chromatic aberration does not occur. In addition, the use of various colors of monochromatic light (red, orange, yellow, green, blue) can accentuate structural or anatomical features in a specimen that don't show up clearly in white light.

PROBLEM 14-8

Suppose that we have a compound microscope objective with a magnification factor specified as 10×, while the eyepiece is rated at 5×. What is the approximate magnification of this instrument?

SOLUTION

We multiply the magnification factor of the objective by that of the eyepiece, getting an overall magnification m according to the approximation formula

$$m \approx (5 \times 10)\times$$
$$= 50\times$$

QUIZ

Refer to the text in this chapter if necessary. A good score is eight correct. Answers are at the back of the book.

1. Suppose that a certain transparent liquid has a refractive index of 1.80 for visible green light. How fast does green light propagate through this liquid?

    A. 3.00×10^8 m/s
    B. 2.48×10^8 m/s
    C. 1.67×10^8 m/s
    D. We need more information to calculate the speed.

2. Which of the following lens types can collimate the rays of light originating from a nearby point source?

    A. Plano-convex
    B. Diverging
    C. Concave
    D. All of the above

3. In which of the following situations can total internal reflection occur?

    A. We aim a visible-light laser beam toward a flat boundary between air and water, with the source located in the air.
    B. We aim a visible-light laser beam toward a flat boundary between air and glass, with the source located in the air.
    C. We aim a visible-light laser beam toward a flat boundary between water and air, with the source located in the water.
    D. All of the above

4. What is the critical angle for light rays inside a sample of clear, solid material whose refractive index is 1.75, surrounded by fresh water whose refractive index is 1.33?

    A. 40.5°
    B. 49.5°
    C. 72.3°
    D. No critical angle exists in this situation.

5. Figure 14-16 illustrates a situation in which we place the objective lens of a compound microscope near a sample for observation, using an oily medium with a refractive index of 1.50. Given the information in this drawing, what is the numerical aperture of the objective?

    A. 3.5
    B. 1.6
    C. 0.63
    D. 0.42

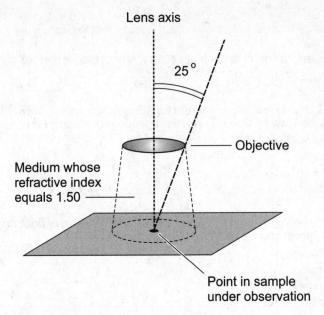

Lens axis

25°

Objective

Medium whose
refractive index
equals 1.50

Point in sample
under observation

FIGURE 14-16 · Illustration for Quiz Question 5.

6. **The resolving power of a refracting telescope depends primarily on the**
   A. diameter of the eyepiece.
   B. focal length of the eyepiece.
   C. focal length of the objective.
   D. diameter of the objective.

7. **What will happen if we try to use a first-surface concave paraboloidal mirror having a focal length of 80 cm as the objective in a Newtonian reflector?**
   A. It will work just fine.
   B. It will work, but we'll have trouble with color dispersion.
   C. It will work, but we'll have trouble with spherical aberration.
   D. It won't work at all.

8. **Suppose that a Newtonian reflecting telescope has an objective mirror with a focal length of 1.50 m and an eyepiece with a focal length of 30 mm. What's the magnification of the whole instrument?**
   A. 450×
   B. 50×
   C. 45×
   D. We need more information to calculate it.

9. Imagine that we immerse a flat pane of clear, solid material with a refractive index of 1.05 in a pool of fresh water whose refractive index is 1.33. A ray of light traveling in the water strikes the solid pane at an angle of 66° relative to the normal. When the ray leaves the solid pane and reenters the water, what angle will the ray subtend relative to the normal?

    A. 66°
    B. 52°
    C. 24°
    D. The premise is wrong! The ray will not enter the clear sample at all, but will instead reflect back into the water.

10. What is the approximate overall magnification of microscope with an objective rated at 60× and an eyepiece rated at 15×?

    A. 900×
    B. 75×
    C. 4.0×
    D. We need more information to calculate the magnification.

chapter 15

Relativity Theory

Albert Einstein's relativity theory comprises two main parts: the *special theory* and the *general theory*. The special theory involves relative motion at constant velocity, and the general theory involves acceleration and gravitation. Before we delve into these theories, let's find out what follows from the hypothesis that the speed of light in free space appears the same from all nonaccelerating reference frames.

CHAPTER OBJECTIVES

In this chapter, you will

• Synchronize clocks separated by great distances.

• Explore the effects of time dilation.

• See how extreme speeds can affect the mass of an object.

• Watch a space ship "shrink" as it approaches the speed of light.

• Compare the effects of acceleration and gravitation.

• Envision "warped" space.

Telling Time

When Einstein became interested in light, space, and time, he pondered the results of experiments intended to discover how the earth moves relative to the "cosmic jelly" that carries electromagnetic (EM) waves such as visible light. Einstein came to believe that no such "cosmic jelly" exists, at least not in material form. He couldn't explain why light and other EM waves can propagate through space that evidently contains nothing whatsoever, but he accepted the observed fact that they do.

The Luminiferous Ether

In the 1800s, physicists determined that light has wavelike properties, and in some ways resembles sound. But light travels much faster than sound. Also, light can travel through a vacuum, while sound cannot. Sound waves require a material medium such as air, water, or metal to propagate. Most scientists believed that light waves must also require some sort of medium, something that "does the waving"—but what? What sort of "cosmic jelly" could exist everywhere, even in a vacuum? Scientists gave the mysterious medium a name: *luminiferous ether*, or simply *ether*.

If the ether exists, some scientists wondered, how could it pass through all ordinary matter, even the entire planet earth? How could the ether get inside an evacuated bottle, allowing light rays to pass through the vacuum? Scientists began to concoct ideas for experiments that might detect the presence of the luminiferous ether. One particularly ingenious experiment involved an attempt to find out if the ether "blows" against the earth as it travels through the universe. Physicists reasoned that an "ether wind" must cause the speed of light to vary depending on the orientation of the rays, just as a gale affects the speed of sound in the atmosphere.

In 1887, two physicists named *Albert Michelson* and *Edward Morley* performed an experiment designed to find out how fast the "ether wind" blows, and from what direction it comes. The so-called *Michelson-Morley experiment* showed that light rays travel at precisely the same speed regardless of their orientation. A light ray coming from, say, a star in the constellation Gemini during October (as the earth's orbit around the sun carries us toward Gemini) has exactly the same speed as a light ray coming from the same star in April, 6 months later (as the earth's orbit carries us away from Gemini). This result cast doubt on the ether theory. If the ether exists, then according to the results obtained by Michelson and Morley, it must travel along with the earth, and therefore remain stationary to the earth near the surface, just like the atmosphere! Einstein could not accept that idea; he might have suspected that these people had fallen into the trap of "molding reality to fit a theory." Einstein took the results of the Michelson-Morley experiment literally. He extrapolated to conclude that the experiment would have the same outcome for observers on the moon, on Mars, or on a spaceship.

The Ultimate Constant

Laying the foundation for his fledgling theory, Einstein formulated a *postulate*—an *axiom*—in the same manner as theoretical mathematicians propose axioms to "plant the seeds" for knowledge systems such as geometry, set theory, and number theory. Here's the axiom in a single sentence:

- In a vacuum, all EM waves propagate at the same constant speed as seen from any nonaccelerating point of view.

Armed with this axiom (with which the results of the Michelson-Morley experiment were consistent), Einstein set out to deduce what would logically follow. As we know today, he got plenty of fascinating and far-reaching results! Einstein did all his work by using a combination of mathematics and daydreaming that he called "mind journeys."

No Absolute Time?

One of the first results of Einstein's speed-of-light axiom suggested that no single absolute time standard exists for all points in space. In recent decades, engineers have built atomic clocks, and scientists claim that these clocks can keep time accurate to within billionths of a second (where a billionth equals a thousand-millionth or 10^{-9}). But this specification has meaning only when we find ourselves right next to such a clock. If we move a little distance away from the clock, the light (or any other signal that we know of) takes some time to get to us, so the clock "looks slow."

The speed of EM-field propagation, the fastest speed known, equals approximately 3.00×10^8 m/s (1.86×10^5 mi/s). A beam of light therefore travels about 300 m (984 ft) in 1.00×10^{-6} s (1.00 μs). If you move a little more than three times the length of an American football field away from a super-accurate atomic clock, the clock will appear in error by 1.00 μs. If you go to the other side of the world, where a radio signal from that clock must travel 20,000 km (12,500 mi) to reach you, the time reading will appear off by about 0.067 s.

If you go to the moon, which revolves around the earth at a distance of roughly 4.0×10^5 km (2.5×10^5 mi), the clock will appear off by approximately 1.3 s.

If scientists ever discover an energy field that can travel through space instantaneously regardless of the distance, then the conundrum of absolute time might meet a resolution. But in practical scenarios, nothing can travel faster than the speed of light in a vacuum. (Some recent experiments suggest that certain effects can propagate faster than the speed of light in a vacuum over short distances, but no one has demonstrated this on a large scale yet, much less used such effects to transmit any information such as data from an atomic clock.)

The Clock Conundrum

Imagine eight clocks in outer space, arranged at the vertices of a gigantic cube. Suppose that each edge of the cube has a length of one *light minute* or approximately 1.8×10^7 km (1.1×10^7 mi), as shown in Fig. 15-1. We want to synchronize the clocks so that they all agree within the limit of visibility.

Each edge measures
1 light minute
in length

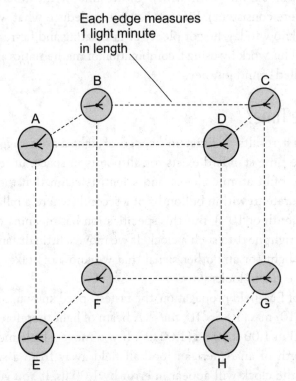

FIGURE 15-1 • A hypothetical set of eight clocks, arranged at the vertices of a cube that measures 1 light minute on each edge. How can we synchronize these clocks?

Suppose that we have a telescope so powerful that we can read the clocks directly by sight from any point in and near the cube, despite its vast expanse. The "data" that tells us what the clocks say travels to us at the speed of light. Now imagine that we climb into a spaceship and maneuver ourselves to the center of the cube, equidistant from all eight clocks, and synchronize the clocks using remote-control wireless equipment. The process takes a little while, because our command signals take the better part of a minute to reach the clocks from our central location, and the signals coming back from the clocks take just as long to get to us so we can see what they say. But soon enough, we see that clocks A through H all tell the same time to within a fraction of a second.

Satisfied with our work, we take our spaceship to some point outside of the cube, and then look back at the clocks. We see that the clocks no longer appear synchronized. We take our ship back to the center of the cube, thinking that something has gone wrong with the system, and we prepare to take corrective action. But when we get to the center of the cube, we find no problem to resolve; the clocks all appear to agree once more.

Can you guess what has taken place during our little "mind journey"? The clock readings *as we see them* depend on how far their signals must travel to reach us. For an observer at the center of the cube, the signals from all eight clocks, A through H, arrive from exactly the same distance. But this state of affairs does not prevail for an observer located anywhere else in space. If we go to another location, we can resynchronize the clocks so that they all agree as we see them, but the new "settings" will not guarantee that they clocks will all agree as we see them if we move yet again.

Still Struggling

No particular location in space offers more validity, in any absolute sense, than any other point in space when it comes to synchronizing the clocks as we see them. If the cube remains stationary relative to some favored reference point such as New York City, we can synchronize the clocks, for convenience, from that reference point. But if the cube moves relative to our frame of reference, we cannot keep the clock readings, *as we observe them*, in sync for any significant period of time.

PROBLEM 15-1

Imagine an atomic clock on the moon (clock M) that broadcasts its time signals using a powerful radio transmitter. Someone sets this clock to precisely agree with another atomic clock in your home town on earth (clock E), also equipped with a radio transmitter. If you travel to the moon, how will the readings of the clocks compare, as determined by listening to the radio signals?

✔SOLUTION

Radio signals travel through space at about 3.00×10^5 km/s. The moon orbits approximately 4.0×10^5 km, or 1.3 light-seconds, away from the earth. When you travel to the moon, the reading of clock M will appear shifted approximately 1.3 s ahead in time, because the time lag for its signals to reach you will no longer exist. The reading of clock E will appear to shift about 1.3 s behind in time, because a time lag will exist where none previously did. When you get to the moon, clock M will therefore appear to run approximately 2.6 s ahead of clock E.

Time Dilation

Isaac Newton hypothesized that time must "flow" at an absolute, unvarying rate, regardless of the observer's location or speed. In that sense, he believed that the *progression of time* represents a fundamental constant in the universe. Einstein hypothesized, based on experimental results, that the *speed of light* remains constant, and left time to fend for itself. Einstein's axiom gave rise to the notion of *time dilation*, thereby proving Newton wrong.

A Laser Clock

Imagine a spaceship equipped with a laser/sensor on one wall and a mirror on the opposite wall, as shown in Fig. 15-2. Suppose that we orient the laser/sensor and the mirror so that the light ray from the laser must travel at a right angle to the axis of the ship, and (once we get the ship moving) perpendicular to the ship's direction of motion. We adjust the laser and mirror to a separation distance of 3.00 m. Because light travels at approximately 3.00×10^8 m/s through air at "breathable pressure," it takes 1.00×10^{-8} s, or 10.0 nanoseconds (10.0 ns), for the light ray to get across the ship from the laser to the mirror, and another 10.0 ns for the ray to return to the sensor. The ray therefore requires 20.0 ns to make one round trip from the laser/sensor to the mirror and back again.

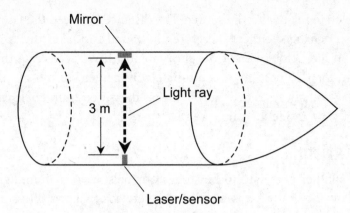

FIGURE 15-2 · A spaceship equipped with a laser clock. An observer in the ship always sees the situation this way.

Imagine that our laser emits pulses of extremely brief duration—far shorter than the time required for the beam to get across the ship. We might even suppose that only one photon comes out of the laser with each pulse! We decide to construct a time standard using a sophisticated oscilloscope, so we can observe the pulses going out and coming back, and measure the time lag between them. This apparatus constitutes a special sort of clock, because its timekeeping ability derives from the speed of light. Remember, Einstein proposed that the speed of light always appears constant, as long as we don't accelerate.

Clock Stationary

Imagine that we fire up the ship's rocket engines and get the ship moving. We accelerate with the eventual goal of reaching nearly the speed of light. After a long, tedious effort, we manage to attain a sizable fraction of the speed of light. Then we shut off the engines so that the ship coasts through space at a constant velocity. You ask, "Relative to *what* are we moving?" That, as we shall see, constitutes a crucial question! For now, let's measure our ship's speed with respect to the earth.

We use our "super clock" to determine the time it takes for a single laser pulse to propagate across the ship and back again. We ride along with the laser, the mirror, and all the dubious comforts of a small spacecraft. We find that the time lag has not changed from its value when the ship was not moving relative to earth. The oscilloscope still shows a delay of 20.0 ns. This result follows directly from Einstein's axiom. The speed of light does not change, because we don't accelerate relative to the source. The distance between the laser and the mirror hasn't changed either. Therefore, the round trip takes the same length of time as it did before we got the ship moving. If we accelerate the ship to 60 percent, then 70 percent, and ultimately 99 percent of the speed of light, the time lag will always remain at 20.0 ns as measured from a *reference frame*, or point of view, inside the ship.

Another Axiom

Let's add another postulate to Einstein's originals: In a vacuum, light beams always follow the shortest possible distance between two points (assuming that no reflecting or refracting media such as lenses or mirrors intervene, of course). Normally, such paths appear to us as straight lines. You ask, "How can the shortest path between two points in a vacuum be anything other than a straight line?" That's another good question; we'll deal with it later in this chapter. For now, let's merely assert that light beams appear to follow straight lines through free space as long as the observer does not accelerate relative to the light source.

Clock in Motion

Imagine that we have completed our space travels and have settled back safely on the earth. Someone else takes off in the ship, ready to conduct more experiments. We equip ourselves with a telescope that allows us to see inside the ship as it whizzes by at a significant fraction of the speed of light. We can see the laser and the mirror clearly. We can even observe the paths of the laser pulses themselves, because the new occupants of the space vessel happen to love cigars. Their activities constantly fill the cabin with smoke, making light rays show up vividly!

Figure 15-3 shows what we see. The laser beam still travels in straight lines, and it still travels at 3.00×10^8 m/s relative to us. These facts hold true because of Einstein's axiom concerning the speed of light, and also because of our own axiom to the effect that light rays always appear to follow straight lines if we do not accelerate relative to the source. However, the rays must travel farther than 3.00 m to get across the ship. The ship moves so fast that, by the time a pulse of light reaches the mirror from the laser, the ship has moved a significant distance forward. The same thing happens as the pulse returns to the sensor from the mirror. As a result, it will seem to us, as we watch the ship from earth, to take more than 20.0 ns for the laser beam to go across the ship and back.

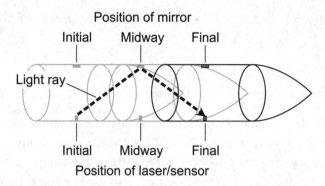

FIGURE 15-3 · An external observer sees this scene as the laser-clock-equipped spaceship whizzes by at a sizable fraction of the speed of light.

We base our time concept on the light pulses; the laser/mirror apparatus constitutes the ultimate time-keeping system. Because the pulses take more time to propagate back and forth across the ship as it whizzes by at high speed (compared to when we rode in it), we conclude that the rate of time "flow" inside the ship has slowed. But inside the ship, as far as the cigar-smoking astronauts can tell, time "flows" at normal speed. As the ship's speed increases, so does the time-rate discrepancy between our point of view and the occupants' point of view. That's relativistic time dilation.

As the speed of the ship approaches the speed of light, the so-called *time-dilation factor* can attain large values. In theory, it can increase without limit. We can visualize this effect by imagining Fig. 15-3 stretched out horizontally, so the light rays have to travel almost parallel to the direction of motion, as seen from the "stationary" reference frame. Instead of traversing a distance of 3.00 m to get from one wall of the ship to the other, a pulse might have to go 4, 5, 10, 100 m, or more.

Formula for Time Dilation

A mathematical relationship exists between the speed of the spaceship in the foregoing "mind experiment" and the extent to which time dilates. Let t_{ship} represent the number of seconds that appear to elapse on the moving ship, as precisely 1 s elapses as measured by a clock next to us as we sit in our earth-based observatory. Let u represent the speed of the ship, as a fraction of the speed of light. Then

$$t_{\text{ship}} = (1 - u^2)^{1/2}$$

The time dilation factor (call it k) equals the reciprocal of this quantity, so that

$$k = 1/[(1 - u^2)^{1/2}]$$
$$= (1 - u^2)^{-1/2}$$

In these formulas, the number 1 represents a mathematically exact value, so we can consider it accurate to as many significant digits as we want.

Let's calculate the time dilation for a space ship traveling at 1.50×10^8 m/s. In that case, $u = 0.500$. If 1.00 s passes on earth, then an earthbound observer sees a shipboard elapsed time interval of

$$t_{ship} = (1.00 - 0.500^2)^{1/2}$$
$$= (1.00 - 0.250)^{1/2}$$
$$= 0.750^{1/2}$$
$$= 0.866 \text{ s}$$

That is, 0.866 s will seem to pass on the ship, as 1.00 s passes as we measure it while watching the ship from the earth. The time dilation factor equals 1.00/0.866, or approximately 1.15.

Now let's calculate the time dilation for a ship traveling at 2.97×10^8 m/s. That's 99 percent of the speed of light, so we have $u = 0.990$. If 1.00 s passes on the earth, then we, as earthbound observers, will see a shipboard elapsed time interval of

$$t_{ship} = (1.00 - 0.990^2)^{1/2}$$
$$= (1.00 - 0.98)^{1/2}$$
$$= 0.0200^{1/2}$$
$$= 0.141 \text{ s}$$

In this scenario, only 0.141 s will seem to pass on the ship, as 1.00 s passes on the earth. The time dilation factor k equals 1.00/0.141, or approximately 7.09. From the point of view of an observer on earth, time "flows" more than seven times more slowly on a ship moving at 99 percent of the speed of light than it "flows" on the earth.

Still Struggling

As you can imagine, time dilation could have profound ramifications for future astronauts. According to the special theory of relativity, if we could get into a spaceship and travel fast enough and far enough, we could propel ourselves into the earth's future. We might travel to a distant star, return to the earth after a few months by our reckoning, and find ourselves in earth's year 5000 A.D.

PROBLEM 15-2

Why don't we notice relativistic time dilation on short trips by car, train, or airplane? When we arrive at our destination, clocks still appear synchronized (except for time-zone differences in some cases). Why doesn't the traveling clock fall behind the earth-based clocks?

SOLUTION

They do fall behind, but by an amount far too small to notice. The time dilation factor does not become appreciable unless a vessel travels at a significant fraction of the speed of light. We can't even detect, let alone measure, the effect of time dilation at common everyday speeds unless we employ atomic clocks, capable of measuring time intervals on the order of 10^{-9} s or less, to measure the time in both reference frames.

PROBLEM 15-3

How fast would we have to travel to produce a time dilation factor of $k = 2.00$?

SOLUTION

Let's use the formula for the time dilation factor k, and let u represent the unknown speed as a fraction of the speed of light. Then we have

$$k = (1 - u^2)^{-1/2}$$

Substituting 2.00 for k, we obtain

$$2.00 = (1 - u^2)^{-1/2}$$

When we take the reciprocal of both sides, we get

$$0.500 = (1 - u^2)^{1/2}$$

We can square both sides, obtaining

$$0.250 = 1 - u^2$$

Subtracting 1 from each side yields

$$-0.750 = -u^2$$

which is equivalent to

$$u^2 = 0.750$$

This equation solves as

$$u = (0.750)^{1/2}$$
$$= 0.866$$

We must travel at 0.866 times the speed of light, or 2.60×10^8 m/s, to produce a time-dilation factor of 2.00.

Spatial Distortion

So-called *relativistic speeds*—that is, speeds high enough to cause significant time dilation—cause objects to appear foreshortened in the direction of their motion. As with time dilation, *relativistic spatial distortion* occurs only from the point of view of an observer watching an object traveling at a sizable fraction of the speed of light.

Point of View: Length

If we ride inside a spaceship, regardless of its speed, everything appears normal as long as our ship does not accelerate. We can cruise along at 99.9 percent of the speed of light relative to the earth, but if we remain inside a spaceship, the ship never moves from our point of view. Time, space, and mass appear "normal" to the passengers on a relativistic space journey. However, as an observer on the earth watches the spaceship sail by, its length decreases as its speed increases. (Its lateral diameter does not change.) The extent to which the axial foreshortening occurs is the same as the extent to which time slows down.

Let L represent the apparent length of the moving ship, as a fraction of its length when stands still relative to an external observer. Let u represent the speed of the ship, as a fraction of the speed of light. Then

$$L = (1 - u^2)^{1/2}$$

Figure 15-4 illustrates the spatial distortion for a spaceship traveling at various forward speeds with respect to an external observer. The foreshortening takes place entirely in the direction of motion, causing apparent physical distortion of the ship and everything inside, including the passengers. As the speed of the ship approaches the speed of light, its observed length approaches zero.

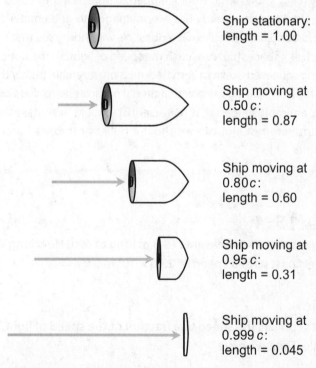

Ship stationary:
length = 1.00

Ship moving at 0.50 c:
length = 0.87

Ship moving at 0.80 c:
length = 0.60

Ship moving at 0.95 c:
length = 0.31

Ship moving at 0.999 c:
length = 0.045

FIGURE 15-4 · As an object moves at extreme and increasing speeds relative to an observer, that object appears to get shorter along the axis of its motion.

Suppositions and Cautions

Based on the foregoing results, we can speculate about the shapes of photons, the particles that compose visible light and all other EM radiation. Photons travel at the speed of light. Does that mean that if we could look at a photon, we'd see a flat disk hurtling sidelong through space? No one has ever seen a photon, so nobody can tell us from experience. We might imagine that photons comprise two-dimensional "stuff" of some sort, and as such, that they must have zero volume. However, such "extrapolation to infinity" can get us into trouble, as any mathematician who has attempted to define division by zero can tell you.

Still Struggling

Scientists know what happens to objects as they *approach* the speed of light, but no one really knows what would happen to objects if they could *attain* the speed of light, because nobody has ever managed to get a material thing to travel that fast and then watch what it does! We will shortly see that no physical object (such as a space ship) can reach the speed of light, so the notion of a real object being squeezed down to zero thickness must remain purely theoretical. As for photons, comparing them with material particles like bullets or baseballs represents an intuitive leap that few scientists would likely take. We cannot bring a photon to rest, nor can we shoot a bullet or throw a baseball at the speed of light.

 PROBLEM 15-4

Suppose a spaceship measures 19.5 m long at rest. How long will it appear if it travels past us at a speed of 2.40×10^8 m/s?

 **SOLUTION**

First, let's convert the speed to a fraction of the speed of light, and call this fraction u. Then we have

$$u = (2.40 \times 10^8)/(3.00 \times 10^8)$$
$$= 0.800$$

We can now use the formula for spatial distortion to find L, the fraction of our ship's at-rest length. The formula is

$$L = (1 - u^2)^{1/2}$$

When we plug in $u = 0.800$, we get

$$L = (1 - 0.800^2)^{1/2}$$
$$= (1 - 0.640)^{1/2}$$
$$= 0.360^{1/2}$$
$$= 0.600$$

Finally, we must multiply the at-rest length of the vessel, 19.5 m, by 0.600, getting 11.7 m as the apparent length of the ship as it zooms past us at 2.40×10^8 m/s.

Mass Distortion

When material objects travel at relativistic speeds, their mass increases as witnessed (and as experienced) by external observers. This mass increase occurs to the same extent as the decrease in length and the dilation of time.

Point of View: Mass

If we travel inside a spaceship, regardless of its speed, the masses of all the objects in the ship with us appear normal, as long as our ship does not accelerate. As an observer on the earth sees the situation, the mass of the ship, and the masses of all the atoms inside it, increase as the vessel moves faster.

Let m represent the relative mass of the moving ship (the multiple of its mass when it does not move with respect to an external observer). Let u represent the speed of the ship, as a fraction of the speed of light. Then

$$m = 1/(1 - u^2)^{1/2}$$
$$= (1 - u^2)^{-1/2}$$

Let's work out an example. Suppose that a spaceship has a *rest mass* (the mass when stationary) of 10 metric tons. When it speeds by at half the speed of light, its mass increases to between 11 and 12 metric tons. At 80 percent of the speed of light, it attains a mass of roughly 17 metric tons. At 95 percent of the speed of light, the ship masses approximately 32 metric tons. At 99.9 percent of the speed of light, the ship's mass increases to more than 220 metric tons. As the speed of the ship approaches the speed of light, its mass increases without limit.

Speed Limits Itself

If we could get a material object to travel at the speed of light, would its mass reach infinity? If an object has zero rest mass, will it attain some finite mass at the speed of light? As u approaches 1 (or 100 percent) in the above formula, the value of m, the mass-increase factor, gets larger without limit. It seems reasonable to suppose that m could become infinite, until we realize that if we set $u = 1$, we end up dividing by zero. The equation "blows up" because division by zero remains undefined in mathematics.

As the mass of a speeding spaceship increases, it needs more and more rocket thrust to get it moving even faster. As a spaceship approaches the speed of light, its mass becomes gigantic, so it gets harder and harder to give it any more speed. Using a mathematical technique known as *integral calculus*, physicists have shown that no finite amount of energy can propel any material object to the speed of light.

High-Speed Particles

You've heard expressions such as *electron rest mass*, which refers to the mass of an electron when it does not move relative to an observer. If we observe an electron whizzing by at relativistic speed, it acquires a mass greater than its rest mass, and it therefore has momentum and kinetic energy greater than the values implied by the formulas used in classical physics. When electrons move at high enough speed, they attain some of the properties of far more massive particles. We have a name for high-speed electrons that act this way: *beta particles*.

Physicists take advantage of the relativistic effects on the masses of protons, helium nuclei, neutrons, and other subatomic particles. When these particles encounter the powerful electric and magnetic fields in a *particle accelerator* (such as the *Large Hadron Collider* near Geneva, Switzerland), the particles get moving so fast that their masses increase because of relativistic effects. When such "relativistically enhanced" particles strike atoms of matter, the nuclei of the target atoms sometimes break apart, liberating energy in the form of infrared (IR), visible light, ultraviolet (UV), x rays, and γ rays, as well as a potpourri of exotic particles.

If astronauts ever travel long distances through space in ships moving at speeds near the speed of light, relativistic mass increase will become a practical concern. Particles whizzing by outside will attain increased mass in a real, and dangerous, way. It's scary enough to think about what will happen when a 1-kg meteoroid strikes a space vessel at any speed. But that 1-kg stone will have mass more than 22 kg at 99.9 percent of the speed of light! In addition, every atom outside the ship will strike the vessel's hull at relativistic speed, producing deadly radiation.

Experimental Confirmation

Scientists have measured relativistic time dilation and relativistic mass increase under controlled conditions, and the results concur with Einstein's formulas stated earlier. These effects constitute more than academic "mind journeys." They really happen.

To measure time dilation, experimenters placed a super-accurate atomic clock on board an aircraft, and then sent the craft up to fly around for a while at several hundred kilometers per hour. The experimenters kept a second atomic clock at the place where the aircraft took off and landed. Although the aircraft's speed never amounted to more than a tiny fraction of the speed of light, and although the resulting time dilation remained exceedingly small, the experimenters got a meaningful result. When the aircraft arrived back at the point of its original departure, the clocks, which the experimenters had synchronized (when placed next to each other) before the trip began, were checked. The clock that had traveled on the aircraft registered a time reading slightly earlier than the clock that had rested on the earth's surface.

Still Struggling

If we want to measure the extent of relativistic mass increase at high speed, we can take advantage of a particle accelerator. We can determine the mass of a moving particle based on its known rest mass and the kinetic energy it possesses as it moves. When we grind out all of the arithmetic, we'll inevitably find that Einstein's formulas prove correct.

PROBLEM 15-5

Suppose that a small meteoroid, having a mass of 300 milligrams (300 mg), strikes the hull of a space vessel moving at 99.9 percent of the speed of light. What's the apparent mass of the meteoroid?

SOLUTION

We can use the formula for relativistic mass increase, considering $u = 0.999$, calculating the mass-increase factor m as

$$m = (1 - 0.999^2)^{-1/2}$$
$$= (1 - 0.998)^{-1/2}$$
$$= 0.002^{-1/2}$$
$$= 1/(0.002)^{1/2}$$
$$= 22.36$$

When the meteoroid strikes the vessel, it masses 300×22.36 mg, or 6.71 grams (g).

General Relativity

No absolute standard for location exists in the universe. No absolute standard exists for velocity, either. All nonaccelerating reference frames have equal "validity." But Einstein nevertheless noticed something different about *accelerating* reference frames. This difference gave rise to the general theory of relativity.

"Acceleration Force"

Imagine that we sit inside a spaceship traveling through deep space. Suppose that we cover up all the portholes so that we can't see anything outside. We disable the radar and navigational equipment so we can't observe the surrounding environment in any way whatsoever. Even in this blind state, we can tell whether or not the ship accelerates!

If the ship's velocity remains constant (no acceleration), everything inside the ship hovers in a "weightless" state (Fig. 15-5A). When we fire the ship's rockets so that it gains speed in the forward direction, all the objects inside the vessel experience force that pulls them toward the rear (Fig. 15-5B). If we fire the ship's retro thrusters to slow the ship down, everything in the ship

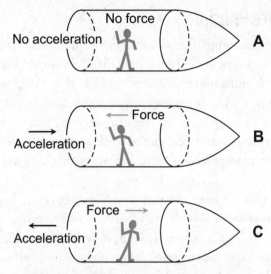

FIGURE 15-5 · When a vessel in deep space does not accelerate, the objects inside never experience force (A). When the ship accelerates, the objects inside always experience force (B and C).

experiences force directed toward the front (Fig. 15-5C). If we activate the directional thrusters to change the ship's direction without changing its speed, everything inside experiences a sideways force.

Regardless of the direction of the ship's acceleration vector, a simple rule holds true: The strength of the "acceleration force" on any given object inside the ship varies in direct proportion to the magnitude of the ship's acceleration vector. We learned this principle in Chapter 2; it involves nothing more than basic classical mechanics. If m represents the mass of an object in our ship (in kilograms) and a represents the ship's acceleration magnitude (in meters per second per second), then we can calculate the force F (in newtons) using the formula

$$F = ma$$

Of course, "acceleration force" occurs whether or not we can see through the ship's portholes, whether or not we use the radar, and whether or not we have the navigational equipment online. We can blind our eyes, but we can't block the force. Einstein used this reasoning to conclude that the interplanetary and interstellar travelers of the future will always have a way to determine their ship's acceleration vector, no matter where in the universe they go. Evidently, there exists an absolute cosmic reference frame for acceleration, making it fundamentally different from location or velocity.

The Equivalence Principle

Imagine that our spaceship, instead of accelerating in deep space, rests on the surface of some distant planet. Our ship might sit tail-downward, in which case the force of gravity pulls on the objects inside as if the ship were accelerating forward. The ship might sit nose-downward, so that gravity "pulls" on the objects inside as if the ship were decelerating forward (accelerating backward). The ship might rest at some odd angle, so that gravity "pulls" on the objects inside as if the ship were accelerating sideways or changing its direction of travel.

Now imagine that we keep the ship's windows covered, keep the radar off, and keep the navigational aids on standby. Also imagine that we've somehow forgotten if we're still in space or if we have landed on a planet. (Maybe we took a long nap and our captain took some unexpected action!) How can we, confined inside the vessel and unable to directly observe any aspect of the outside environment, tell whether the forces on every object inside the ship occur because of gravitation or because of acceleration? Einstein's answer: *We can't.* "Acceleration force" and "gravitational force" manifest themselves in exactly the same way.

As a result of a "mind journey" similar to the one we've just gone through, Einstein created a new axiom for his theories: the *equivalence principle*. According to this axiom, "acceleration force" and "gravitational force" constitute the same phenomenon. Einstein reasoned that these two forces act in identical ways on everything, from people's perceptions to atoms, and from light rays to the "fabric" of space-time. Whenever we encounter gravitation, we can think of it as acceleration. Whenever we experience acceleration, we can think of it as gravitation. The equivalence principle became the cornerstone of Einstein's theory of general relativity.

Spatial Curvature

Once again, let's imagine that we ride in a spaceship, hurtling through deep space. We fire the ship's rockets at their full capacity, causing our vessel to accelerate forward at an extreme rate. Suppose that we've installed a laser apparatus, similar to that described earlier in this chapter, in our ship. Instead of a mirror on the wall opposite the laser, we place a screen. Before the acceleration begins, we align the laser so that its beam strikes the center of the screen as shown in Fig. 15-6. What will happen when we get the rocket engines going and the ship accelerates furiously forward?

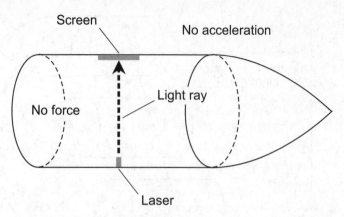

FIGURE 15-6 · As seen from inside a nonaccelerating deep-space vessel, a laser beam travels in a straight line from wall to wall.

In a real-life scenario, the spot from the laser will not move on the screen enough for us to notice, because any reasonable (non-life-threatening) rate of acceleration will not give rise to enough force to visibly deflect the beam. But let's "suspend our disbelief" and imagine that we can accelerate the vessel at any rate, no matter how great, without having our bodies crushed against the ship's rear wall. If we accelerate fast enough, the ship pulls away from the laser beam as the beam travels across the ship. As we witness the situation from inside the ship, we see the light beam follow a curved path (Fig. 15-7). A stationary observer on the outside sees the beam propagate in a straight line, but the accelerating vessel pulls out ahead of the beam as time passes (Fig. 15-8).

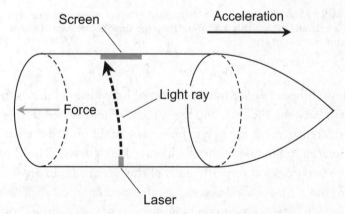

FIGURE 15-7 · As seen from inside a deep-space vessel accelerating forward at an extreme rate, a laser beam follows a curved path from wall to wall.

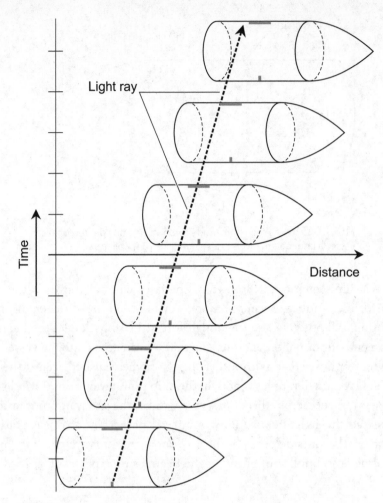

FIGURE 15-8 · An observer in a "stationary" external reference frame sees an accelerating space ship "pull away" from the straight-line path of a laser beam.

Regardless of the reference frame, the ray of light always follows the shortest possible path between the laser and the screen. When viewed from any nonaccelerating reference frame, light rays appear as straight lines. But when observed from accelerating reference frames, light rays look curved. The shortest path between the two points at opposite ends of the laser beam in Fig. 15-7 is not a straight line, but a curve called a *geodesic*—the mathematical expression for the shortest distance between two points in *non-Euclidean* ("warped" or "curved") space. An apparently "straight" path is actually *longer* than the geodesic as seen from inside the accelerating vessel.

According to Einstein's principle of equivalence, gravitation always associates with the same *spatial curvature* as acceleration does. These phenomena constitute more than mere figments of cosmologists' imaginations. The simple notion of *Euclidean* ("flat") space cannot explain the manifestations of powerful gravitation or extreme acceleration.

For spatial curvature to manifest itself as vividly as Figs. 15-7 and 15-8 illustrate, a space vessel must accelerate at an incredible rate. We normally express and measure acceleration magnitude in meters per second per second, or *meters per second squared* (m/s^2). Astronauts and aerospace engineers often express and measure acceleration in units called *gravities* (symbolized g), where 1 gravity (1 g) represents the acceleration that produces the same force as the gravitational field of the earth at the surface, approximately 9.8 m/s^2. (Don't confuse the abbreviation for gravity or gravities with the abbreviation for grams! Pay attention to the context if you see a unit symbolized g.) Figures 15-7 and 15-8 show situations for many thousands of g's. If you weigh 150 pounds on the earth, you would weigh thousands of tons in a ship experiencing acceleration or gravitation intense enough to "warp" space that much.

Time Dilation Caused by Acceleration or Gravitation

The spatial curvature caused by intense acceleration or gravitation produces an effective slowing-down of time. The light ray must travel farther to get from the laser to the screen in the situation of Fig. 15-7 than in the situation of Fig. 15-6. The acceleration-induced light-beam curvature increases the time dilation even more that would be the case if the ship were not accelerating. From the vantage point of a passenger in the spaceship, the curved path shown in Fig. 15-7 represents the shortest possible path the light ray can take (that is, the geodesic) across the vessel between the point at which it leaves the laser and the point at which it strikes the screen.

We might turn the laser device slightly, pointing it a little bit toward the front of the ship; this maneuver will cause the beam to arrive at the center of the screen as shown in Fig. 15-9. But the laser will still follow a curved path across the vessel, and the length of that path will still exceed the length we would observe under conditions of zero acceleration. The laser represents the most accurate possible timepiece because it functions on the basis of speed of light, an absolute constant. Acceleration produces time dilation, not only as seen by observers looking at the ship from the outside, but for passengers within the vessel itself. In this respect, acceleration and gravitation constitute more powerful "time dilators" than relative motion.

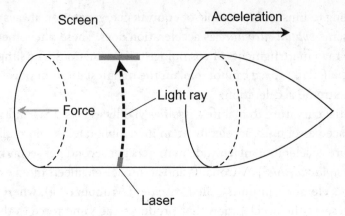

FIGURE 15-9 · Even if we orient the laser so that its beam hits the center of the screen, the path of the ray curves when the spaceship accelerates at a high rate.

If we could experience unlimited acceleration or gravitation and remain alive, we would perceive time as slowing down inside the vessel under conditions such as those shown in Fig. 15-7 or 15-9. Clocks would run more slowly, even from reference frames inside the ship. In addition, everything inside the ship would appear optically distorted. If we had powerful enough rockets, we might cause the laser beam to miss the opposite wall altogether (Fig. 15-10), "warping" the scene in a way that would make any Hollywood science-fiction producer envious.

Observational Confirmation

When Einstein developed his general theory of relativity, the scientific world expressed skepticism. The formulas looked right, but how could experimenters demonstrate their validity? In their attempts to prove (or disprove) Einstein's notions concerning the "curvature of space," astronomers chased *solar eclipses* around the world during the early part of the 20th century.

During the total phases of several solar eclipses, when the moon dimmed the sun' light, astronomers used powerful telescopes and special photographic equipment to observe the light rays from distant stars as those rays passed close to the sun. If Einstein's theory were correct, the powerful gravitation would bend the light rays. The path curvature would show up as slight changes in the apparent positions of distant stars as the sun passed between them and the earth.

It took some time, and a lot of failure and frustration for experimenters, to come to the final verdict: Einstein's theory held true. The apparent positions of

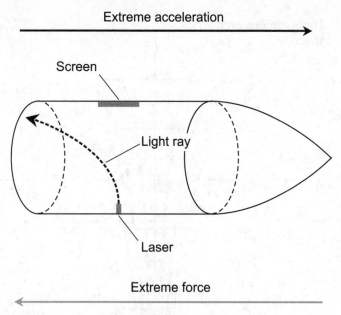

Extreme acceleration

Screen

Light ray

Laser

Extreme force

FIGURE 15-10 · If acceleration becomes great enough, space warps to a tremendous extent.

distant stars did indeed change as the sun grazed their light paths. Moreover, the effect always took place to the same extent as Einstein's general relativity formulas said that it should.

More recently, astronomers have observed the light from a distant celestial object called a *quasi-stellar source* (or *quasar*) as it passes behind a dark source of gravitation called a *black hole*. On its way to us, the light from the quasar follows multiple curved paths around the black hole, producing multiple images of the quasar in the form of a "cross" with the black hole at the center.

Some astronomers, cosmologists, and mathematicians compare the curvature of space in the presence of a strong gravitational field to a funnel (Fig. 15-11), except that the surface has three dimensions rather than two. The shortest distance in three-dimensional space between any two points near the gravitational source always constitutes a curve in four-dimensional space. On a large scale, countless millions of these "funnels" might produce general "warping" of space-time.

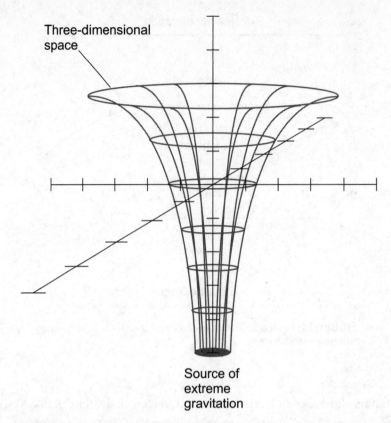

Three-dimensional
space

Source of
extreme
gravitation

FIGURE 15-11 · Spatial curvature in the vicinity of an object that produces
an intense gravitational field.

Still Struggling

What sort of geometric figure might the cosmos comprise? A "hypersphere"?
A "hyperparaboloid"? A "hypercone"? Speculation abounds. Evidence runs short.
The shape of the universe remains a mystery—for now.

QUIZ

Refer to the text in this chapter if necessary. A good score is eight correct. Answers are at the back of the book.

1. In a spaceship that accelerates at an extreme rate, which of the following phenomena can occur?
   A. Some light beams follow curved paths.
   B. The vessel exceeds the speed of light.
   C. Gravitational force attains infinite magnitude.
   D. Time flows backward.

2. As the foundation for his general theory of relativity, Einstein postulated that
   A. light beams never travel in straight lines.
   B. time "flows" at an absolute constant rate.
   C. acceleration and gravitation produce the same effects.
   D. all accelerating reference frames are equivalent.

3. Suppose you have a spherical ball with a mass of 2.4 kg and a radius of 100 mm at rest. If you fire the ball out of a powerful cannon so that it travels at 200,000 km/s, what will its mass become, as witnessed by an observer stationary relative to the cannon?
   A. 2.4 kg
   B. 3.2 kg
   C. 3.6 kg
   D. 4.3 kg

4. What will be the ball's *longitudinal radius* (the radius in the direction of its travel) as witnessed by an observer stationary relative to the cannon, after the ball emerges from the cannon as described in Question 3?
   A. 55.6 mm
   B. 74.5 mm
   C. 100 mm
   D. 134 mm

5. What will be the ball's *lateral radius* (the radius perpendicular to the direction of its travel) as witnessed by an observer stationary relative to the cannon, after the ball emerges from the cannon as described in Question 3?
   A. 55.6 mm
   B. 74.5 mm
   C. 100 mm
   D. 134 mm

6. In a ship traveling through space in a straight line at a constant speed of 0.999c, which of the following effects would passengers on the vessel *not* observe?

A. The relativistic mass effect would increase the danger of radiation from sub-atomic particles striking the front of the vessel.

B. The relativistic mass effect would increase the danger posed by small meteor-oids in space.

C. All clocks inside the ship would appear to run at normal speed.

D. Lateral light beams would follow curved paths.

7. Imagine that you see a spaceship whiz by in a straight-line path at a constant speed of 100,000 km/s. If 60.00 s elapse according to a clock stationary relative to you, how much time appears to elapse on board the ship?

A. 56.57 s

B. 60.00 s

C. 63.64 s

D. 67.50 s

8. If a spaceship traveling at constant relativistic speed takes an extreme and sudden turn toward the left, then in theory, a laser beam directed from the rear of the ship toward the front will momentarily

A. veer toward the right-hand wall.

B. veer toward the left-hand wall.

C. increase speed in a straight line.

D. decrease speed in a straight line.

9. According to Isaac Newton's hypothesis concerning time, space, and the speed of light,

A. the speed of light appears constant from all nonaccelerating points of view.

B. the speed of light appears constant from all points of view.

C. time "flows" at a rate that depends on the speed of the observer.

D. time "flows" at the same rate from all points of view.

10. Albert Einstein hypothesized that the luminiferous ether

A. moves relative to the earth, as the earth travels through space.

B. remains stationary relative to the earth.

C. consists of a substance that can pass through ordinary matter.

D. does not exist at all.

Test: Part III

Don't look back at any of the text while taking this test. The correct answer choices appear at the back of the book. Consider having a friend check your score the first time you take this test, without telling you which questions you got right and which ones you missed. That way, you won't subconsciously memorize the answers in case you want to take the test again later.

1. Consider two sinusoidal acoustic waves in the air. Call them waves X and Y. Wave X has a frequency of 400 Hz. Wave Y has a period of 0.500 ms. What can we say about these two waves?

    A. Wave X oscillates at the fifth harmonic of wave Y.
    B. The two waves have the same frequency.
    C. The two waves propagate at different speeds.
    D. The two waves have different shapes.
    E. The two waves beat to produce another wave at 1.60 kHz.

2. Consider two acoustic disturbances in a concert hall, called waves C and P. Suppose that C and P have identical frequency, but C comes from a clarinet while P comes from a piano so that the sounds differ in timbre (or "tone"). Which of the following statements A, B, C, or D, if any, is *false*?

    A. C and P have identical wavelengths in the air.
    B. C and P have identical waveforms.
    C. C and P have identical periods.
    D. C and P propagate at identical speeds in the air.
    E. All of the above statements are true.

3. After Michelson and Morley found that light rays from distant stars approach the earth at the same speed regardless of the star's position in the sky, Einstein hypothesized that

    A. the speed of light constitutes an absolute constant as observed from all nonaccelerating reference frames.
    B. time "flows" smoothly, and at an absolutely constant rate, as observed from all nonaccelerating reference frames.
    C. the luminiferous either travels along with the earth, "masking" the actual differences in the speeds at which light rays reach the earth from various directions.
    D. the earth constitutes a reference frame with absolute zero velocity relative to the universe as a whole.
    E. Michelson's and Morley's equipment did not have sufficient sensitivity to detect differences in the speeds of the light from stars in various directions.

4. Gamma rays have more penetrating power than

    A. x rays.
    B. UV rays.
    C. visible-light rays.
    D. IR rays.
    E. All of the above

5. Imagine that a spaceship has a mass of 7500 kg at rest. We get it moving so fast that its mass increases to 9375 kg. We start a stopwatch and let it run for 60.0 s, all the while watching the second hand of a huge clock mounted on the outside of the ship. By how much time does the ship's second hand advance in 60.0 s by our reckoning?

    A. 75.0 s
    B. 60.0 s

   C. 48.0 s

   D. 38.4 s

   E. We need more information to answer this question.

6. Einstein theorized that photons in outer space don't always follow straight-line paths. Astronomers verified this aspect of the general theory of relativity by observing the behavior of light rays as they passed through strong

   A. magnetic fields.

   B. electric fields.

   C. streams of beta particles.

   D. gravitational fields.

   E. solar winds.

7. Imagine that a small, dark, extremely massive object in our galaxy partially obstructs a bright, distant galaxy. The dark object distorts the image of the distant galaxy. This phenomenon gives us a visual example of relativistic

   A. time dilation.

   B. spatial curvature.

   C. mass distortion.

   D. color dispersion.

   E. chromatic aberration.

8. Suppose that a sample of transparent material has a refractive index of 1.25 for visible blue light. How fast does blue light propagate through it?

   A. 3.75×10^8 m/s

   B. 3.00×10^8 m/s

   C. 2.40×10^8 m/s

   D. 2.00×10^8 m/s

   E. We need more information to calculate it.

9. How can we make a two-lens telescope produce an image that appears right-side up and also true in the left-to-right sense (that is, "not backward")?

   A. We can't.

   B. We can use a concave objective and a convex eyepiece.

   C. We can use a concave objective and a concave eyepiece.

   D. We can use a convex objective and a convex eyepiece.

   E. We can use a convex objective and a concave eyepiece.

10. Figure Test III-1 shows one full cycle of a sine wave, broken down into degrees of phase and relating to a circular revolution. If the wave has a frequency of 50 kHz, how much time elapses between points X and Y?

   A. 1.0×10^{-6} s

   B. 2.0×10^{-6} s

   C. 5.0×10^{-6} s

   D. 4.0×10^{-5} s

   E. 2.0×10^{-4} s

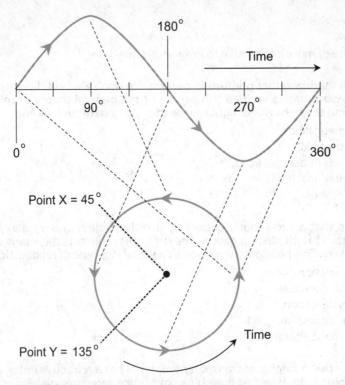

FIGURE TEST III-1 · Illustration for Part III Test Question 10.

11. **Which of the following sources commonly emits gamma rays?**

    A. A sample of uranium
    B. A bright light bulb
    C. A computer monitor
    D. A hydrogen fuel cell
    E. All of the above

12. **Which of the following EM-wave frequencies falls in the visible-light range?**

    A. 50 GHz
    B. 500 GHz
    C. 5 THz
    D. 50 THz
    E. 500 THz

13. **If we want to directly observe primary cosmic particles, we must**

    A. employ a powerful telescope with a first-surface mirror.
    B. use a spectrophotometer at the focus of a telescope.
    C. travel to an altitude well above the earth's lower atmosphere.
    D. employ a radiation counter with a thick metal shield to filter out other radiation.
    E. employ a sensitive EM receiver equipped with an efficient mixer.

14. We can describe a waveform whose amplitude rises and falls instantaneously, but whose peaks appear flat and constant, as

    A. triangular.
    B. sinusoidal.
    C. ramped.
    D. sawtoothed.
    E. rectangular.

15. Suppose that you adjust your car radio to receive a station in the AM broadcast band late at night, and you discover that the station transmits from a town more than 2000 km away. This phenomenon most likely occurs because the station's EM waves

    A. interact with the earth's ionosphere.
    B. travel along the earth's surface like an electrical impulse.
    C. reflect from the moon.
    D. propagate underground in a straight line through the earth's crust.
    E. reflect from the aurora.

16. Imagine that we build a Keplerian refracting telescope with two convex lenses, one at each end of a long, hollow tube. One lens has a focal length of 800 mm and serves as the objective, while the other lens has a focal length of 16 mm and serves as the eyepiece. When we look at a distant object through this telescope, the object's image appears

    A. upside-down, and also backward in the left-to-right sense.
    B. upside-down, but true in the left-to-right sense.
    C. right-side-up, but backward in the left-to-right sense.
    D. stretched-out horizontally, right-side up, and true in the left-to-right sense.
    E. stretched-out vertically, right-side-up, and true in the left-to-right sense.

17. What is the magnification factor of the Keplerian refractor described in Question 16?

    A. 400×
    B. 200×
    C. 100×
    D. 50×
    E. We need more information to calculate it.

18. Figure Test III-2 shows rays of visible light crossing a flat boundary from a medium called X into a medium called Y. Which of the following statements A, B, C, or D, if any, can we make with certainty?

    A. Medium Y has more mass per unit volume than medium X.
    B. Medium Y has a higher index of refraction than medium X.
    C. Medium Y propagates light waves at a higher speed than medium X.
    D. For some incidence angles (not shown), a ray originating in medium X will reflect from the boundary back into medium X.
    E. We can't make any of the above statements A, B, C, or D with certainty.

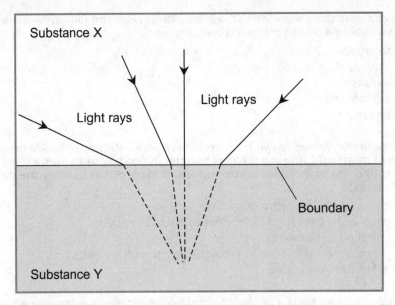

FIGURE TEST III-2 · Illustration for Part III Test Questions 18 and 19.

19. Suppose that, in the situation shown by Fig. Test III-2, all of the light rays in medium X measure precisely 500 nm in length. Which of the following statements A, B, C, or D, if any, can we make with certainty?

    A. The waves in medium Y measure less than 500 nm in length.
    B. The waves in medium Y measure more than 500 nm in length.
    C. The waves in medium Y measure precisely 500 nm in length.
    D. We waves in medium Y have lower frequency than the waves in medium X.
    E. We need more information to know how the wavelength or frequency in medium Y compares with the wavelength or frequency in medium X.

20. All EM waves have four distinct properties (among others): wavelength, frequency, propagation speed, and intensity. Which of these parameters can vary independently any of the other three?

    A. Wavelength
    B. Frequency
    C. Propagation speed
    D. Intensity
    E. None of them

21. The convex mirror in a Cassegrain reflecting telescope

    A. minimizes the effects of chromatic aberration in the eyepiece.
    B. minimizes the effects of lens sag in the objective.
    C. increases the effective focal length of the objective.
    D. reduces color dispersion, enhancing the image clarity.
    E. All of the above

22. **Which of the following statements *never* holds true in *any* circumstance?**

    A. A concave mirror can collimate the light rays from a point source.
    B. A concave lens can spread parallel light rays out so that they diverge.
    C. A convex mirror can reflect light rays to the eyepiece in a telescope.
    D. We can combine two convex lenses to make a telescope.
    E. A concave lens can function as the objective in a telescope.

23. **Which of the following conditions can cause a ray of light *not* to follow a geodesic between two points?**

    A. Relative motion
    B. A gravitational field
    C. Acceleration
    D. Refraction
    E. All of the above

24. **If we want to decrease the magnification factor in a compound microscope, we can**

    A. increase the focal length of the objective.
    B. reduce the magnification of the eyepiece.
    C. reduce the distance between the eyepiece and the objective.
    D. increase the refractive index of the medium between the objective and the sample under observation.
    E. Do any of the above

25. **The F layer in the earth's ionosphere**

    A. protects life on the surface by filtering out UV radiation from the sun.
    B. can return EM waves to the surface at certain frequencies.
    C. blocks all solar EM radiation at wavelengths shorter than 500 nm.
    D. generates gamma rays that the lower atmosphere blocks.
    E. prevents excessive ozone from forming near the earth's surface.

26. **Figure Test III-3 shows the basic principle of operation for**

    A. an oscilloscope.
    B. a radiation counter.
    C. a spectrophotometer.
    D. a dispersion analyzer.
    E. a wavelength spreader.

27. **In Fig. Test III-3, the component marked "X" is a**

    A. diffraction grating.
    B. signal mixer.
    C. concave lens.
    D. convex lens.
    E. thin sheet of paper.

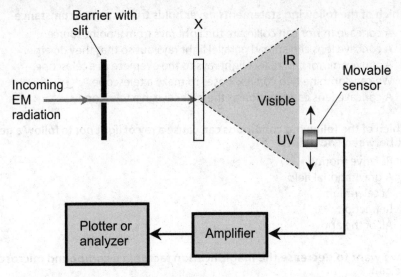

FIGURE TEST III-3 · Illustration for Part III Test Questions 26 and 27.

28. **According to the theory of special relativity, a long-distance, high-speed space traveler can**

    A. travel into the past.
    B. travel into the future.
    C. prevent human aging.
    D. exceed the speed of light.
    E. All of the above

29. **When parallel rays of light from the sun pass through a concave glass lens and emerge from the other side of the glass, the rays**

    A. remain parallel.
    B. converge.
    C. increase in wavelength.
    D. increase in frequency.
    E. diverge.

30. **Low-frequency sound waves can propagate around obstacles as a result of**

    A. dispersion.
    B. heterodyning.
    C. diffraction.
    D. mixing.
    E. beating.

31. **If a periodic wave has a frequency of 250 Hz, what's the period?**

    A. 2.50×10^5 ms
    B. 0.00400 ms

C. 0.250 ms

D. 4.00 ms

E. It depends on the medium through which the wave propagates.

32. **Suppose that a wireless signal, as it propagates through a vacuum, comprises photons whose wavelengths all measure 17 m. If we double all of their wavelengths to 34 m, the energy contained in each photon**

A. does not change.

B. doubles.

C. increases by a factor of 4.

D. becomes half as great.

E. becomes 1/4 as great.

33. **Relativistic mass distortion occurs**

A. at relative speeds amounting to a significant fraction of the speed of light.

B. only perpendicular to the axis of relative motion.

C. at relative speeds faster than the speed of light.

D. at relative speeds great enough to reverse the flow of time.

E. only for objects that have theoretically zero rest mass.

34. **Fill in the blank to make the following sentence true: "When a spaceship moves at a speed approaching the speed of light relative to an observer, the observer will see the ship appear to _____ compared to its condition at rest."**

A. shrink in the lateral dimension (side-to-side)

B. shrink in the longitudinal dimension (front-to-back)

C. expand in the lateral dimension

D. expand in the longitudinal dimension

E. keep the same size and shape

35. **Which of the following types of radiation can ionize atoms in living tissue?**

A. High-intensity acoustic waves

B. High-intensity infrared rays

C. High-intensity radio waves

D. High-intensity gamma rays

E. All of the above

36. **At any particular point in free space, an electromagnetic (EM) field propagates in a direction**

A. perpendicular to both the electric (E) flux lines and the magnetic (M) flux lines.

B. parallel to the M flux lines but perpendicular to the E flux lines.

C. parallel to the E flux lines but perpendicular to the M flux lines.

D. parallel to both the E flux lines and the M flux lines.

E. that depends on how fast the charge carriers accelerate.

37. Consider a sample of radioactive material that emits 80 alpha particles per minute as indicated by a radiation counter. We set the sample aside for 30 days. After that time has passed, we test the sample again. Our counter shows only 10 alpha particles per minute. What's the radioactive half-life of this sample for alpha particles?

   A. 20 days
   B. 15 days
   C. 10 days
   D. 5 days
   E. We need more information to calculate it.

38. When we talk about the propagation of EM fields through space and mention the term *tropospheric bending*, we refer to a phenomenon in which

   A. gamma rays from distant sources in outer space refract as they pass through the earth's lower atmosphere.
   B. IR waves from the sun tend to increase in wavelength as they pass through the earth's lower atmosphere.
   C. electromagnetic waves originating from a source on the earth's surface refract as they pass through ionized gases in the upper atmosphere.
   D. visible-light rays from distant stars refract as they pass through the earth's upper atmosphere.
   E. radio waves at certain frequencies refract as they propagate through the earth's lower atmosphere.

39. What type of energy emission does the device shown in Fig. Test III-4 primarily produce?

   A. X rays
   B. Visible laser light
   C. IR rays
   D. RF microwaves
   E. Alpha particles

40. Suppose that the advertisement for a wireless device claims that it operates at 15 GHz. How long is an EM wavelength in free space at this frequency?

   A. 4.5 mm
   B. 20 mm
   C. 45 mm
   D. 20 cm
   E. 45 cm

41. Which of the following statements is false?

   A. Infrared rays have shorter wavelengths than radio waves.
   B. The visible-light range constitutes a small part of the EM spectrum.
   C. Gamma rays can ionize living tissue.
   D. Radio waves have higher frequencies than visible-light waves.
   E. The energy contained in a photon depends on its wavelength.

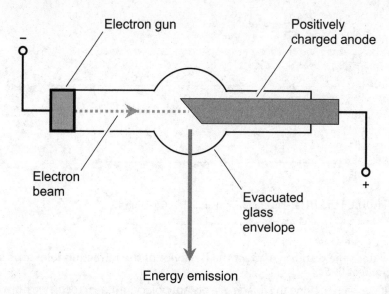

FIGURE TEST III-4 · Illustration for Part III Test Question 39.

42. **When we use monochromatic light to illuminate the specimen in a compound microscope, we gain certain advantages, including**

    A. increased eyepiece magnification.
    B. the elimination of chromatic aberration.
    C. enhanced wavelength dispersion.
    D. increased aperture diffraction.
    E. increased objective focal length.

43. **Scientists sometimes use the Ångström unit to express**

    A. the wavelength of an EM field.
    B. the frequency of visible-light waves.
    C. the energy content of a photon.
    D. the penetrating power of ionizing radiation.
    E. the intensity of ionizing radiation.

44. **Scientists have proven that high levels of ELF radiation cause**

    A. atoms of all substances to fluoresce.
    B. blindness in humans and animals.
    C. common materials to become radioactive.
    D. nonferromagnetic materials to ionize.
    E. None of the above

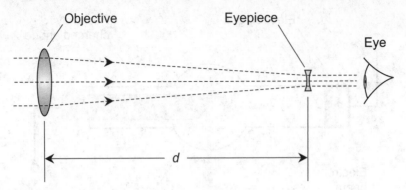

FIGURE TEST III-5 · Illustration for Part III Test Question 45.

45. **How does the distance *d* affect the behavior of the refracting telescope shown in Fig. Test III-5?**

   A. If we want to bring the image of a distant object into clear focus, we must adjust *d* to a specific distance that depends on the refractive qualities of the lenses.

   B. If we want to get the image of a distant object to appear right-side-up, we must adjust *d* to a specific distance that depends on the refractive qualities of the lenses.

   C. The distance *d* affects both the focusing and the magnification of the telescope. The image appears upside-down for all values of *d*.

   D. The distance *d* affects the magnification of the telescope, but not the focusing. The image appears right-side-up for all values of *d*.

   E. The distance *d* affects the magnification of the telescope, but not the focusing. The image appears upside-down for all values of *d*.

46. **Which of the following statements constitutes the foundation of Einstein's general theory of relativity?**

   A. The speed of light depends on the speed of the observer.

   B. Light rays can never follow curved paths through space.

   C. Acceleration and gravitation produce equivalent effects.

   D. The shortest distance between two points is always a straight line.

   E. Three-dimensional space is Euclidean ("flat") everywhere in the universe.

47. **Time dilation can result from extreme**

   A. movement of the ether.

   B. gamma radiation.

   C. solar activity.

   D. relative motion.

   E. ionizing radiation.

48. Suppose that a spaceship zooms past the earth so fast that clocks on board seem, from our point of view on the earth's surface, to run at only 0.100 percent of their normal speed. If the ship encounters a meteoroid massing 100 g at rest, how much mass will that meteoroid have, in effect, as the moving vessel collides with it?

    A. 707 kg
    B. 100 kg
    C. 1.41 kg
    D. 7.07×10^6 kg
    E. We need more information to calculate it.

49. Suppose that two pure sine-wave signals having frequencies of 440 Hz and 1.770000 MHz combine in a circuit designed to mix them together. In addition to the original signals, we should theoretically expect to observe

    A. only an output signal at 1.769560 MHz.
    B. only an output signal at 1.770440 MHz.
    C. output signals at both 1.769560 MHz and 1.770440 MHz.
    D. broadband electrical noise.
    E. no other signals or noise.

50. A Newtonian reflecting telescope

    A. has an eyepiece opening in the center of the objective lens.
    B. has a concave objective mirror and a flat secondary mirror.
    C. suffers from lens sag if the diameter of the objective is too small.
    D. has a convex objective mirror and a concave secondary mirror.
    E. requires the use of a concave lens as the eyepiece.

Final Exam

Don't look back at any of the text while taking this exam. The correct answer choices appear at the back of the book. Consider having a friend check your score the first time you take this exam, without telling you which questions you got right and which ones you missed. That way, you won't subconsciously memorize the answers in case you want to take the exam again later.

1. Which of the following characteristics A, B, or C, if any, limits the performance of a Keplerian refracting telescope?

   A. Chromatic aberration in cheap lenses
   B. Lens sag in large objectives
   C. Spherical aberration in cheap objectives
   D. All of the above factors A, B, and C
   E. None of the above factors A, B, or C

2. Which of the following events A, B, or C, if any, causes a complex-impedance vector in the resistance-reactance half-plane to grow shorter?

   A. The reactance remains constant while the resistance decreases.
   B. The resistance remains constant while the inductive reactance gets closer to zero.
   C. The resistance remains constant while the capacitive reactance gets closer to zero.
   D. Any of the three events A, B, or C
   E. None of the three events A, B, or C

3. With respect to a solid, the term *malleability* refers to the ease with which we can

   A. liquefy it.
   B. solidify it.
   C. evaporate it.
   D. pound it flat.
   E. harden it.

4. Imagine that you stand on top of a towering cliff on a planet where the gravitational acceleration equals 4.50 m/s². You drop a stone of mass 1.57 kg from the top of the cliff, letting the stone fall straight down. After 1.66 s have passed, and assuming that atmospheric resistance has no effect, what instantaneous downward speed will the stone have attained?

   A. 36.9 cm/s
   B. 57.9 cm/s
   C. 7.47 m/s
   D. 11.7 m/s
   E. We need more information to answer this question.

5. Consider three identical capacitors connected in parallel, each having a value of 900 pF. What's the total capacitance of the combination, assuming that no mutual capacitance exists among the components?

   A. 100 pF
   B. 300 pF
   C. 900 pF
   D. 2700 pF
   E. It depends on the frequency.

6. **Suppose that a transparent substance exhibits a refractive index of 1.25 for green light. How fast does green light travel through this medium? Consider the speed of light in a vacuum as 3.00×10^8 m/s.**

    A. 3.75×10^8 m/s

    B. 3.00×10^8 m/s

    C. 2.68×10^8 m/s

    D. 2.40×10^8 m/s

    E. 1.92×10^8 m/s

7. **Figure Exam-1 illustrates a device that we can use to indirectly measure the**

    A. weight of an object in a gravitational field.

    B. electrostatic charge on an object.

    C. interaction of an object's magnetic field with the earth's magnetic field.

    D. gravitational acceleration on an object.

    E. mass of an object in a "weightless" environment.

8. **Imagine that we ride in a spaceship traveling at half the speed of light relative to the earth. From our point of view, to what extent do objects inside the ship, traveling right along with us, appear foreshortened along axes parallel to the direction of the ship's motion, compared with their apparent lengths if the ship weren't moving?**

    A. They don't appear foreshortened at all.

    B. They appear slightly foreshortened, but more than half as long as they would if the ship weren't moving.

    C. They appear half as long as they would if the ship weren't moving.

    D. They appear less than half as long as they would if the ship weren't moving.

    E. They appear completely flattened out, with zero length.

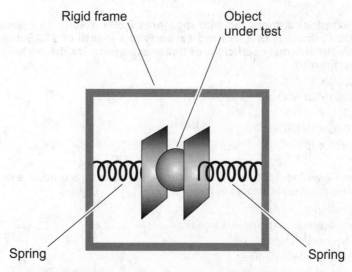

FIGURE EXAM-1 · Illustration for Final Exam Question 7.

9. When operating in a state of saturation, a conventional PNP bipolar transistor loses its ability to efficiently amplify alternating-current signals, although it can still work as

   A. a variable capacitor.
   B. an electronic switch.
   C. a variable inductor.
   D. a photovoltaic device.
   E. a light-emitting or IR-emitting device.

10. Geomagnetic *inclination* can cause

   A. like magnetic poles to uncharacteristically attract.
   B. opposite magnetic poles to uncharacteristically repel.
   C. one end of a magnetic compass needle to dip slightly.
   D. a lodestone to orient in an east-west direction.
   E. a magnetic compass needle to point away from true north.

11. When we design a circuit using an NPN bipolar transistor, we'll usually provide the collector with

   A. a positive direct-current (DC) voltage relative to the emitter.
   B. the same DC voltage as the emitter.
   C. a negative DC voltage relative to the emitter.
   D. a source of alternating-current (AC) voltage.
   E. a source of incident visible light, infrared (IR), or ultraviolet (UV).

12. Suppose that *s* represents the difference in linear dimension (in meters) produced by a temperature change of *T* (in degrees Celsius) for an object whose original or "starting" linear dimension (in meters) equals *d*. We can calculate the thermal coefficient of linear expansion *a* using the equation

$$a = s/(dT)$$

   Now consider a metal rod that measures 3.000 m long at a temperature of +10.00°C. Imagine that this rod expands to a length of 3.045 m at +40.00°C. What's the thermal coefficient of linear expansion for the material that composes this rod?

   A. 2.000×10^{-3}/°C
   B. 1.500×10^{-4}/°C
   C. 5.000×10^{-4}/°C
   D. 3.000×10^{-5}/°C
   E. 5.000×10^{-6}/°C

13. For an IR, visible-light, or UV wave propagating in a vacuum, the period in seconds equals the reciprocal of the

   A. wavelength in meters.
   B. amplitude in watts peak-to-peak.
   C. frequency in hertz.
   D. phase angle in degrees.
   E. None of the above

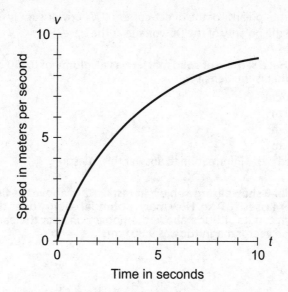

FIGURE EXAM-2 · Illustration for Final Exam Question 14.

14. **Figure Exam-2 graphically portrays the instantaneous speed of an object traveling in a straight-line path as a function of time. What can we say about the acceleration magnitude as a function of time?**

    A. The object starts out accelerating rapidly. As time passes, its instantaneous acceleration magnitude decreases.
    B. The object starts out accelerating slowly. As time passes, its instantaneous acceleration magnitude increases.
    C. The object accelerates at a constant, slow rate.
    D. The object accelerates at a constant, rapid rate.
    E. The object starts out accelerating forward at a decreasing rate, then comes to a complete halt, and then begins accelerating backward at an increasing rate.

15. **What's the wavelength of a ray of electromagnetic energy at a frequency of 7.5 GHz, propagating through the earth's lower atmosphere at 3.0×10^8 m/s?**

    A. 4.0 cm
    B. 2.5 cm
    C. 2.3 mm
    D. 0.40 mm
    E. We need more information to calculate it.

16. **We can usually replace an N-channel JFET with a P-channel JFET having the appropriate specifications and obtain a new circuit that will perform the same functions as the old one did, provided that we**

    A. reverse the phase of the alternating-current power-supply voltage.
    B. reverse the phase of the input signal.
    C. interchange the source and drain electrodes.

D. reverse the polarity of the direct-current (DC) power-supply or battery voltage.

E. reverse the polarity of the DC voltage at the base.

17. Suppose that a sample of solid matter has a volume of 0.150 cm³ and a mass of 300 mg. What is the density?

A. 0.500 g/cm³

B. 2.00 g/cm³

C. 5.00 g/cm³

D. 45.0 g/cm³

E. We need more information to answer this question.

18. Figure Exam-3 shows a large object at rest, 2.42 m above the floor. Suppose that the object masses 10.0 kg. How much potential energy does the mass have as a consequence of its altitude above the floor? Consider the earth's gravitational acceleration vector magnitude as 9.807 m/s².

A. 0.246 J

B. 2.37 J

C. 2.46 J

D. 23.3 J

E. 237 J

19. Imagine the situation of Fig. Exam-3, except that we relocate to a distant Planet X where the gravitational acceleration magnitude equals exactly half its value on earth. How does the potential energy of the mass on Planet X compare with its potential energy on earth?

A. It's the same on Planet X as on earth.

B. It's 0.707 times as great on Planet X as on earth.

C. It's half as great on Planet X as on earth.

D. It's 1/4 as great on Planet X as on earth.

E. It's 1/8 as great on Planet X as on earth.

20. In an alternating-current electrical circuit, the complex number $30 + j40$ represents

A. 30 ohms of resistance and 40 farads of capacitance.

B. 30 ohms of resistance and 40 ohms of capacitive reactance.

C. 30 ohms of resistance and 40 henrys of inductance.

D. 30 ohms of resistance and 40 ohms of inductive reactance.

E. 50 ohms of resistance.

21. Imagine that we ride in a spaceship that travels so fast that the time dilation factor equals 4.00. If we encounter a "stationary" meteoroid whose rest mass equals 16.0 g, how much mass will it have, in effect, when it strikes the hull of our vessel?

A. 256 g

B. 64.0 g

C. 16.0 g

D. 4.00 g

E. 1.00 g

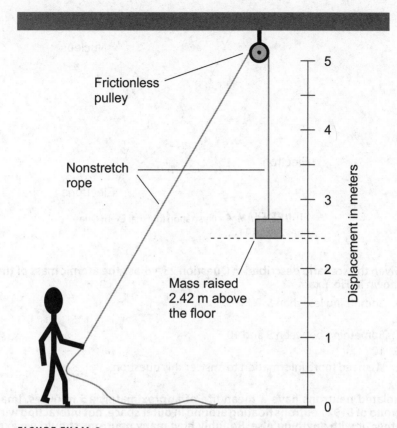

FIGURE EXAM-3 · Illustration for Final Exam Questions 18 and 19.

22. **Consider seven identical lamps connected in series with a 24-V battery. If we insert a new lamp to get a circuit with eight identical lamps connected in series with the same battery, the voltage across any single one of the original lamps will**

    A. remain the same.
    B. increase slightly.
    C. increase a lot.
    D. decrease slightly.
    E. decrease a lot.

23. **Figure Exam-4 shows a simplistic view of an atom. Suppose that the atom as a whole lacks electrical charge. What's its atomic number?**

    A. Something less than 5
    B. 5
    C. Something between 5 and 10
    D. 10
    E. We need more information to answer this question.

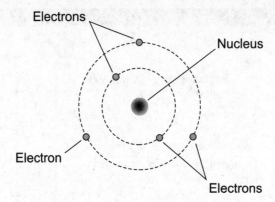

FIGURE EXAM-4 · Illustration for Final Exam Questions 23 and 24.

24. **Given the scenario described in Question 23, what's the atomic mass of the atom shown in Fig. Exam-4?**

    A. Something less than 5
    B. 5
    C. Something between 5 and 10
    D. 10
    E. We need more information to answer this question.

25. **Isolated neutrons have a mean life of approximately 15 minutes. Imagine a group of 8192 neutrons floating around in outer space, not interacting with each other or with anything else. Roughly how many neutrons should we expect to find remaining in this hypothetical swarm after one full earth day has passed?**

    A. 8192
    B. 512
    C. 64
    D. 16
    E. None, or in an exceptional case, maybe one

26. **As we travel in a spaceship and measure the speed of visible light rays around us, assuming that we don't accelerate and we experience no gravitational force, the speed of any particular photon depends on**

    A. our absolute location in the universe.
    B. our speed relative to fixed objects.
    C. the mass of the vessel in which we travel.
    D. the direction from which the photon comes.
    E. None of the above

27. **When we forward-bias a semiconductor diode with a steady direct-current (DC) voltage exceeding the forward breakover voltage, the device**

    A. conducts current continuously.
    B. conducts current some of the time.

C. conducts current poorly or not at all.
D. exhibits capacitive reactance.
E. exhibits inductive reactance.

28. **Which of the following pairs of radio-spectrum wave propagation modes produce effects so similar that we might confuse one for the other?**
    A. Surface-wave propagation and tropospheric bending
    B. Ionospheric F-layer propagation and meteor-scatter propagation
    C. Tropospheric scatter and auroral propagation
    D. Tropospheric bending and sporadic-E propagation
    E. Ionospheric D-layer propagation and line-of-sight propagation

29. **Which of the following events A, B, C, or D, if any, will *not* cause a change in the momentum vector of a moving object?**
    A. The object's mass increases while the velocity vector remains constant.
    B. The object's straight-line speed decreases while the mass remains constant.
    C. The object becomes more dense while the velocity vector and the mass remain constant.
    D. The object veers to the right or left, while the mass and the speed remain constant.
    E. Any of the above events will cause a change in the momentum of a moving object.

30. **Figure Exam-5 shows a sealed container filled with incompressible liquid, along with two pistons in pipes. Suppose that the areas of the pistons' faces exist in the exact ratio $A = 2B$. If we push down on the left-hand piston with force $F = 16$ N, how much upward force G will we observe at the right-hand piston?**
    A. 4.0 N
    B. 8.0 N
    C. 16 N
    D. 32 N
    E. 64 N

31. **How distant is a standing-wave loop from the nearest loop on either side, in terms of degrees of phase?**
    A. 45°
    B. 90°
    C. 180°
    D. 270°
    E. 360°

32. **In cylindrical coordinates, we can uniquely define the location of a point in space as**
    A. three direction angles.
    B. one displacement component and two direction angles.
    C. two displacement components and one direction angle.
    D. three displacement components.
    E. Any of the above

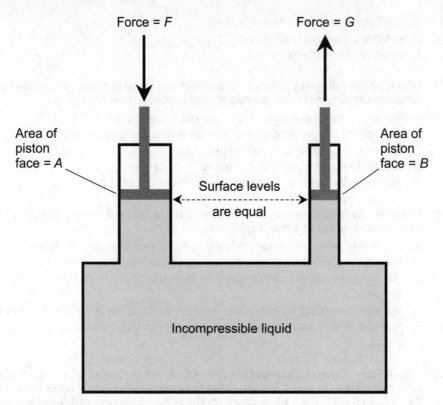

FIGURE EXAM-5 · Illustration for Final Exam Question 30.

33. **The extent (in terms of displacement) to which we can deform a perfectly elastic ideal solid sample, either by stretching or compression, varies in**

   A. direct proportion to the external force that we apply.
   B. direct proportion to the square of the external force that we apply.
   C. direct proportion to the square root of the external force that we apply.
   D. inverse proportion to the external force that we apply.
   E. inverse proportion to the square root of the external force that we apply.

34. **Suppose that we wind a coil of wire around a solenoidal metal core. We find that by connecting a variable direct-current (DC) source to the coil, we can drive the magnetic flux density in the core up to 3000 gauss (G) but no higher. When we shut down the current source, the flux density inside the core drops to 150 G. What's the retentivity of this metal?**

   A. 2.50 percent
   B. 5.00 percent
   C. 10.0 percent
   D. 22.4 percent
   E. We need more information to calculate it.

35. **What's the permeability of the metal described in Question 34?**
    A. 400
    B. 20.0
    C. 4.47
    D. 2.00
    E. We need more information to calculate it.

36. **We can express the force that a gravitational field exerts on a known mass in terms of**
    A. kilogram-meter-seconds.
    B. kilogram-meters per second.
    C. kilogram-seconds.
    D. kilogram-meters per second squared.
    E. kilograms per second squared.

37. **The meter-kilogram-second (mks) system of units is**
    A. preferred by scientists.
    B. widely used in the United States but not in the rest of the world.
    C. cumbersome because it doesn't rely on powers of 10.
    D. also called the English system.
    E. outmoded.

38. **Figure Exam-6 illustrates the behavior of a sample of an unknown "substance X" as the temperature rises. What property of substance X does this graph portray?**
    A. Thermal coefficient of linear expansion
    B. Heat of fusion
    C. Specific heat
    D. Heat of vaporization
    E. Thermal heat density

39. **Suppose that an alternating-current sine wave with a constant period presents itself at 2000 rad/s. What's the _frequency?_**
    A. 318.3 Hz
    B. 636.6 Hz
    C. 6.283 kHz
    D. 12.57 kHz
    E. To calculate the frequency, we must know the nature of the wave disturbance (acoustic, visible light, or whatever) and the characteristics of the medium through which it travels.

40. **Suppose that an alternating-current sine wave with a constant period presents itself at 2000 rad/s. What's the _wavelength?_**
    A. 942.4 km
    B. 471.2 km
    C. 47.71 km

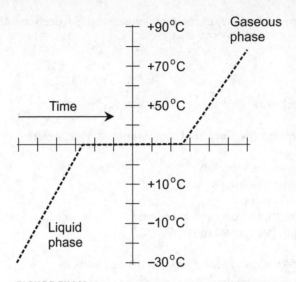

FIGURE EXAM-6 · Illustration for Final Exam Question 38.

    D. 23.86 km
    E. To calculate the wavelength, we must know the nature of the wave disturbance
       (acoustic, visible light, or whatever) and the characteristics of the medium
       through which it travels.

41. **In terms of force per unit area, the pressure at any particular depth below the
surface of a body of incompressible fluid varies in direct proportion to the**

    A. weight density of the liquid.
    B. viscosity of the liquid.
    C. temperature of the liquid.
    D. atomic mass of the liquid.
    E. All of the above

42. **Geomagnetic *declination* can cause**

    A. a magnetic compass needle to point away from true north.
    B. like magnetic poles to uncharacteristically attract.
    C. one end of a magnetic compass needle to dip slightly.
    D. a lodestone to orient in an east-west direction.
    E. opposite magnetic poles to uncharacteristically repel.

43. **In a battery-powered electrical circuit, the conventional current**

    A. increases as the resistance increases, if we hold all other factors constant.
    B. decreases as the conductance increases, if we hold all other factors constant.
    C. flows from the positive battery pole to the negative battery pole.
    D. can never actually equal zero.
    E. always dissipates in the form of heat or visible light.

44. Suppose that we have a 6.4-g sample of a certain pure substance. We transfer 9.6 cal of energy to the sample, and its temperature rises uniformly by 3.0°C. None of the material changes state during this process. What's its specific heat?

   A. 3.2 cal/g/°C
   B. 3.0 cal/g/°C
   C. 1.5 cal/g/°C
   D. 0.5 cal/g/°C
   E. We need more information to answer this question.

45. Suppose that we connect a pair of 20-mH inductors in parallel, and call the resulting component P. Then we connect a pair of 40-mH inductors in parallel and call the resulting component Q. Finally we connect P in series with Q. Assuming that no mutual inductance exists among any of the four individual inductors, what's the net inductance of the series combination of P and Q?

   A. 10 mH
   B. 20 mH
   C. 30 mH
   D. 40 mH
   E. We need more information to calculate it.

46. The magnitude of the momentum vector for a moving object varies in direct proportion to the object's

   A. mass.
   B. altitude.
   C. acceleration.
   D. displacement.
   E. density.

47. Suppose that the direct-current (DC) source shown in Fig. Exam-7 produces 45.000 V and the potentiometer has a resistance of 450.00 ohms. How much current flows through meter A?

   A. 4.5000 A
   B. 10.000 A
   C. 20.250 A
   D. 100.00 mA
   E. 0.22222 mA

48. How can we get a spaceship to travel at the speed of light, strictly according to the principles of Einstein's special theory of relativity?

   A. We can use a device that makes space non-Euclidean.
   B. We can use a matter/antimatter rocket engine.
   C. We can use an antigravity propulsion system.
   D. We can use a device that reverses the flow of time.
   E. We can't.

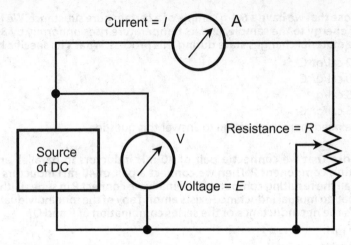

FIGURE EXAM-7 · Illustration for Final Exam Question 47.

49. **What optical phenomenon lends support to the wave theory of light, but defies explanation based on the particle theory of light?**

    A. Reflection of light rays from a mirror.
    B. Refraction of light rays at the surface of a lake.
    C. The tendency of light rays to scatter when they reflect from a rough surface.
    D. The tendency of light rays to travel in straight lines through a vacuum.
    E. All of the above

50. **Imagine that a certain tiny particle has a diameter of 5.00×10^{-5} centimeters (cm). We can also call this displacement**

    A. 500 nanometers (nm).
    B. 500 millimeters (mm).
    C. 500 picometers (pm).
    D. 500 femtometers (fm).
    E. 500 attometers (am).

51. **Suppose that we have 13 identical lamps connected in series with a 24-V battery. If we insert a new lamp to get a circuit with 14 identical lamps in series with the same battery, the current through any single one of the original lamps will**

    A. remain the same.
    B. decrease slightly.
    C. decrease a lot.
    D. increase slightly.
    E. increase a lot.

52. **An electromagnetic field having a frequency of 500 Hz falls into the**

    A. x-ray spectrum.
    B. gamma-ray spectrum.
    C. microwave spectrum.

D. radio-wave spectrum.

E. extremely low frequency (ELF) spectrum.

53. **Imagine that you stand on the surface of a planet where the gravitational acceleration equals 4.00 m/s². You lift a 5.00-kg mass straight upward over a distance of 7.50 m. How much work have you done?**

A. 0.167 J

B. 6.00 J

C. 9.38 J

D. 75.0 J

E. 150 J

54. **What's the approximate peak-to-peak voltage of the wave shown in Fig. Exam-8?**

A. 1.1 V pk-pk

B. 1.3 V pk-pk

C. 1.8 V pk-pk

D. 2.4 V pk-pk

E. 3.7 V pk-pk

55. **The number of moles of atoms or molecules per meter cubed (mol/m³) in a sample of a liquid, where 1 mol represents approximately 6.02 × 10²³ atoms or molecules, tells us the liquid's**

A. mass density.

B. weight density.

C. particle density.

D. nuclear density.

E. chemical density.

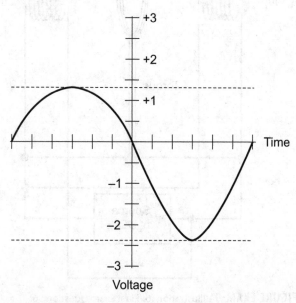

FIGURE EXAM-7 · Illustration for Final Exam Question 54.

56. **Figure Exam-9 is a simplified functional diagram of an electric motor designed to operate from a source of direct current (DC). What function does component X serve?**

   A. It ensures that the torque on the rigid wire coil keeps it rotating.
   B. It prevents excessive current from flowing in the rigid wire coil.
   C. It optimizes the interaction between the two electromagnets.
   D. It keeps the rigid wire coil from rotating too fast.
   E. It brings the rigid wire coil to a smooth stop after powering-down.

57. **Suppose that an isotope of the metallic element molybdenum contains 42 protons and 54 neutrons in the nucleus of a single atom. What, approximately, is the atomic mass?**

   A. 96 AMU
   B. 54 AMU
   C. 42 AMU
   D. 12 AMU
   E. We need more information to answer this question.

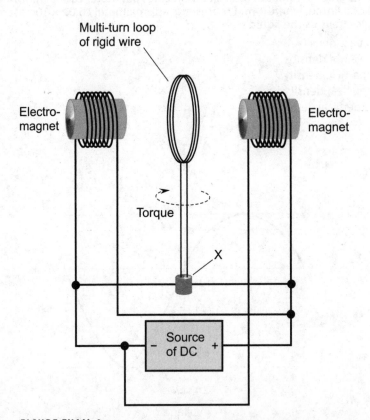

FIGURE EXAM-9 · Illustration for Final Exam Question 56.

58. **The magnetosphere has practically no effect on the paths of primary cosmic particles as they approach our planet. Why?**

    A. The particles have long wavelengths, which remain immune to the effects of magnetic and electric fields.
    B. Because of their extreme speed, the particles have so much momentum that the magnetosphere can't significantly deflect them.
    C. The particles have low frequencies, so they can penetrate the ionosphere and magnetosphere with ease.
    D. The particles lack electric charge, so neither an electric field nor a magnetic field has any effect on them.
    E. All of the above

59. **Which of the following effects or phenomena takes the form of a vector?**

    A. Temperature
    B. Speed
    C. Mass
    D. Force
    E. All of the above

60. **We can express thermal energy in terms of**

    A. British thermal units per hour (Btu/h).
    B. calories per gram (cal/g).
    C. kelvins per gram (K/g).
    D. kelvins per kilocalorie (K/kcal).
    E. kilocalories (kcal).

61. **Imagine that a rocket engine imposes a force vector $F = 4000$ N on small spacecraft of mass $m = 200$ kg, causing the ship's speed to increase in the direction of Betelgeuse, a star in the constellation Orion. What's this spacecraft's acceleration vector a?**

    A. 5.00 cm/s^2 away from Betelgeuse
    B. 5.00 cm/s^2 toward Betelgeuse
    C. 20.0 m/s^2 away from Betelgeuse
    D. 20.0 m/s^2 toward Betelgeuse
    E. We need more information to answer this question.

62. **An alternating-current wave with a frequency of 20 MHz has a period of**

    A. 0.5 ms.
    B. 20 μs.
    C. 500 ns.
    D. 50 ns.
    E. 20 ns.

63. Suppose that you connect a lamp across a 12-V battery. As the lamp shines, it exhibits 24 ohms of resistance. Assuming that the circuit contains no resistance other than that in the lamp, how much power does the lamp consume?

   A. 0.50 W
   B. 2.0 W
   C. 6.0 W
   D. 288 W
   E. More information is needed to calculate it.

64. Imagine that a ray of light passes from a vacuum into a clear substance X having a refractive index of 1.50, striking perpendicular to a plane tangent to the surface. Suppose that we replace this clear substance with another clear material Y having a refractive index of 2.00, and position it so that the same ray of light once again strikes perpendicular to a plane tangent to the surface. How does the angle of refraction change when we replace substance X with substance Y in this fashion?

   A. It doesn't change at all.
   B. It increases a little.
   C. It increases a lot.
   D. It decreases a little.
   E. The angle of refraction becomes irrelevant, because total internal reflection occurs.

65. At which of the following electromagnetic wavelengths would we most likely observe tropospheric bending?

   A. 2 km
   B. 200 m
   C. 20 m
   D. 2 m
   E. All of the above

66. Imagine an ideal system of moving masses, in which no friction or other real-world imperfection occurs. If any two of the objects collide, the total system momentum will remain constant

   A. unless the total mass of the system changes.
   B. only if a certain compensating force appears from the outside.
   C. only if a certain compensating impulse appears from the outside.
   D. except in a weightless environment.
   E. no matter what else happens to the system.

67. Consider three perfect sine waves X, Y, and Z, none of which has a direct-current (DC) component, and all of which have identical frequency. Suppose that wave X *leads* wave Y by 120°, while wave Y *leads* wave Z by precisely 1/6 of a cycle. How do waves X and Z relate?

   A. Waves X and Z coincide in phase.
   B. Wave X leads wave Z by 60°.
   C. Wave Z leads wave X by 60°.
   D. Wave X leads wave X by 120°.
   E. Waves X and Z oppose in phase.

68. Consider three perfect sine waves X, Y, and Z, none of which has a DC component, and all of which have identical frequency. Suppose that wave X *leads* wave Y by 120°, while wave Y *lags* wave Z by precisely 1/6 cycle. How do waves X and Z relate?

    A. Waves X and Z coincide in phase.
    B. Wave X leads wave Z by 60°.
    C. Wave Z leads wave X by 60°.
    D. Wave X leads wave X by 120°.
    E. Waves X and Z oppose in phase.

69. Figure Exam-10 shows four graphs of relative reactance versus relative capacitance. In each case, the capacitance increases as we move toward the right along the horizontal axis; the reactance increases negatively as we move downward along the vertical axis. Which one, if any, of the four graphs correctly portrays the general relation for reactance versus capacitance at a constant alternating-current frequency?

    A. Graph A
    B. Graph B

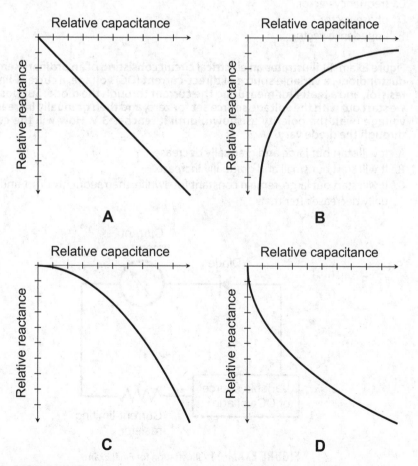

FIGURE EXAM-10 · Illustration for Final Exam Question 69.

C. Graph C
D. Graph D
E. None of them

70. In which of the following processes can the atomic number of an atom change,
     thereby converting one element into another?

A. Ionization
B. Irradiation
C. Isotope transformation
D. Nuclear fusion
E. Evaporation

71. In an alternating-current electrical circuit, the resistance, the capacitive reactance,
     and the inductive reactance combine as vectors to give us a complex-number

A. conductance vector.
B. inductance/capacitance vector.
C. frequency vector.
D. phase vector.
E. impedance vector.

72. Figure Exam-11 illustrates an electrical circuit consisting of an ordinary semicon-
     ductor diode, a variable source of direct-current (DC) voltage, a current-limiting
     resistor, and a meter for measuring the current through the diode. Suppose that
     we start out with the voltage source set for zero, and then gradually increase the
     voltage (with the polarity as shown) until it reaches 3 V. How will the current
     through the diode vary?

A. It will start out large and gradually decrease.
B. It will start out small and gradually increase.
C. It will start out large, remain constant for awhile, then suddenly drop, and grad-
    ually decrease after that.

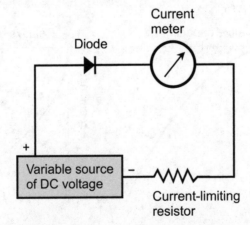

FIGURE EXAM-11 · Illustration for Final Exam
Question 72.

D. It will start out at zero, suddenly begin to flow at a certain voltage, and gradually increase after that.

E. It will remains constant regardless of the voltage provided by the DC source.

73. **If we double the period of an electromagnetic wave propagating through a vacuum, the free-space wavelength**

    A. quadruples.
    B. doubles.
    C. remains the same.
    D. gets half as long.
    E. gets 1/4 as long.

74. **Cerenkov radiation occurs when subatomic particles pass through a substance at a speed greater than**

    A. the speed of visible light in a vacuum.
    B. the speed of visible light in that substance.
    C. the speed of sound in dry air at sea level.
    D. the speed of sound in that substance
    E. the speed of those same particles in a vacuum.

75. **Suppose that a sample of gas has a volume of 1.5 m³. It masses 6.0 kg. What is its particle density?**

    A. 4.0 mol/m³
    B. 2.0 mol/m³
    C. 0.50 mol/m³
    D. 0.25 mol/m³
    E. We need more information to answer this question.

76. **Suppose that two objects, both moving at constant velocity, collide in a closed system, bouncing off of each other after the impact. How does the total system momentum after the collision compare with the total system momentum before the collision?**

    A. It depends on the relative masses of the objects.
    B. It depends on the relative velocities of the objects.
    C. It stays the same.
    D. It increases.
    E. It decreases.

77. **If a Newtonian reflecting telescope has an objective mirror with a focal length of 1500 mm and we use a 30-mm eyepiece with it, how much magnification should we expect to obtain?**

    A. 7.1×
    B. 17×
    C. 50×
    D. 270×
    E. We must know the radius of the objective to calculate the magnification.

78. While driving your car along a straight, level highway on a windless night, you see an animal crossing the road. You press on the brake pedal but keep the car traveling in a straight line. The car suddenly slows down. While the speed decreases, the acceleration vector

   A. points backward.
   B. points forward.
   C. points toward the pavement.
   D. points straight up.
   E. vanishes altogether.

79. Figure Exam-12 illustrates an example of acoustic wave

   A. diffraction.
   B. dispersion.
   C. diffusion.
   D. refraction.
   E. heterodyning.

80. Suppose that we cut a wedge-shaped section out of a circular pie. When we take the wedge out of the pie, we measure the lengths of its straight and curved edges, and we find that they all equal the pie's radius. In this situation, the angle

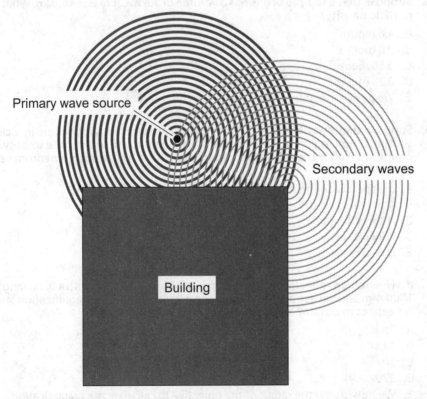

Primary wave source

Secondary waves

Building

FIGURE EXAM-12 · Illustration for Final Exam Question 79.

at the wedge's apex (which lay at the center of the pie before we removed the wedge) has a measure of one

A. angular degree.
B. steradian.
C. radian.
D. radial increment.
E. conical degree.

81. Suppose that a ball of mass 80.0 g travels at a constant speed of 30.0 m/s in a straight line. What's the magnitude of its momentum vector?

A. 0.240 kg · m/s
B. 0.375 kg · m/s
C. 2.40 kg · m/s
D. 267 kg · m/s
E. 375 kg · m/s

82. Consider a solid object that travels in a straight line through interplanetary space. The object will *inevitably* accelerate if it experiences a change in

A. mass.
B. speed.
C. weight.
D. density.
E. Any of the above

83. Einstein's equivalence principle tells us that

A. the apparent speed of light depends on the motion of the observer.
B. time flows at the same rate everywhere in space.
C. gravitation and acceleration produce the same effects.
D. acceleration and gravitation cause space to become Euclidean ("flat").
E. All of the above

84. Figure Exam-13 illustrates a device for detecting electromagnetic energy. For what type of radiation would we most likely find this technology employed?

A. X rays
B. Visible light

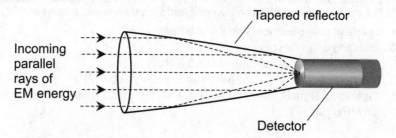

FIGURE EXAM-13 · Illustration for Final Exam Question 84.

   C. Infrared (IR)

   D. Radio waves

   E. Extremely low frequency (ELF) radiation

85. **In a metal-oxide-semiconductor field-effect transistor (MOSFET), how does the forward breakover voltage specification affect the performance of the device?**

   A. It limits the current gain that we can obtain with the device operating at a constant frequency.

   B. It limits the gate bias voltage that we can apply to get linear amplification without driving the device into saturation.

   C. It limits the maximum frequency at which the device can amplify, assuming constant drain voltage.

   D. It keeps the device from oscillating when we don't want that to happen, as long as the frequency remains below a certain critical level.

   E. The premise is wrong; a MOSFET doesn't exhibit the forward breakover phenomenon at all, because it doesn't contain a P-N junction.

86. **According to some pre-Einstein scientists, the speed of a light ray through space does not depend on the ray's orientation (the result demonstrated by Michelson and Morley in 1887), because the luminiferous ether must**

   A. pass through matter as easily as it does through space.

   B. travel along with the earth near the surface, just as the atmosphere does.

   C. not exist at all.

   D. constitute the absolute standard for motion in the universe.

   E. consist of real matter whose particles remain unidentified to this day.

87. **After Michelson and Morley published the results of their experiments, Albert Einstein concluded that the luminiferous ether must**

   A. pass through matter as easily as it does through space.

   B. travel along with the earth near the surface, just as the atmosphere does.

   C. not exist at all.

   D. constitute the absolute standard for motion in the universe.

   E. consist of real matter whose particles remain unidentified to this day.

88. **Suppose that two signals in a wireless transmitter, one at 4.5 MHz and the other at 2.1 MHz, combine in a circuit deliberately designed to produce heterodyning. At the output, along with the two original signals, we will observe**

   A. one (and only one) new signal at 6.6 MHz.

   B. one (and only one) new signal at 2.4 MHz.

   C. one (and only one) new signal at 3.3 MHz.

   D. two new signals, one at 2.4 MHz and the other at 6.6 MHz.

   E. three new signals, one at 2.4 MHz, another at 3.3 MHz, and yet another at 6.6 MHz.

89. Imagine a long, thin, solenoidal wire coil containing hundreds of turns and having an air core. Suppose that we place a battery across the coil, thereby driving direct current (DC) through the wire. Under these circumstances, the magnetic flux *inside the coil* takes a form that we can best describe as
    A. straight lines perpendicular to the coil axis.
    B. straight lines parallel to the coil axis.
    C. circles in planes perpendicular to the coil axis.
    D. circles in planes containing the coil axis.
    E. a helix that follows the contour of the coil itself.

90. Auroral communication can occur in the very high frequency (VHF) part of the radio spectrum over distances up to about 2000 km (1200 mi). The maximum possible communications range depends on all of the following factors *except one*. Which one?
    A. The waveform of the transmitted signal
    B. The orientations of the ionized trails produced by the meteors
    C. The altitudes of the ionized trails produced by the meteors
    D. The location of the transmitting station
    E. The location of the receiving station

91. Suppose that two samples of matter differ in temperature by 20.0 K. What's this difference in degrees Fahrenheit?
    A. 36.0°F
    B. 20.0°F
    C. 11.1°F
    D. 10.4°F
    E. We need more information to answer this question.

92. Figure Exam-14 illustrates a compound microscope. What will happen if we replace the lower convex lens (the one closer to the sample under examination) with a lens having the same radius but a longer focal length?
    A. The microscope will attain enhanced light-gathering power.
    B. The microscope's magnification factor will decrease.
    C. We will find it more difficult to bring the observed image into good focus.
    D. We will observe all three phenomena A, B, and C.
    E. We won't observe any of the phenomena A, B, or C.

93. In which of the following situations would you expect to observe the *greatest* amount of diffraction?
    A. Visible-light waves around a wooden ship's mast.
    B. Ocean swells around the piling of a pier.
    C. Radio waves at 500 GHz around a steel storage shed.
    D. Sound waves at 15 kHz around the end of a brick wall.
    E. X rays traveling through a vacuum.

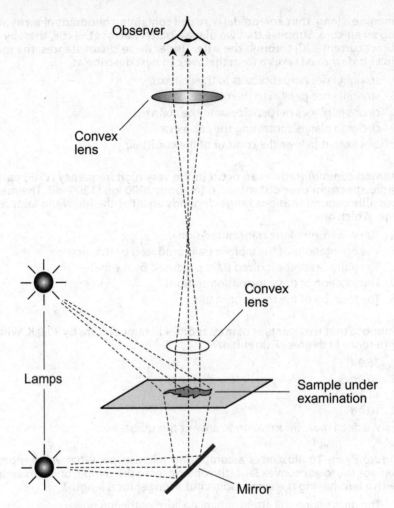

Observer

Convex
lens

Convex
lens

Lamps

Sample under
examination

Mirror

FIGURE EXAM-14 · Illustration for Final Exam Question 92.

94. **Figure Exam-15 illustrates a compound microscope with three adjustable distances *x*, *y*, and *z*. Which one, or combination, of these distances should we adjust in order to bring the sample into clear focus without altering the microscope's magnification?**

    A. Distance *x* only
    B. Distance *y* only
    C. Distance *z* only
    D. Distances *x* and *y* only
    E. All three distances *x*, *y*, and *z*

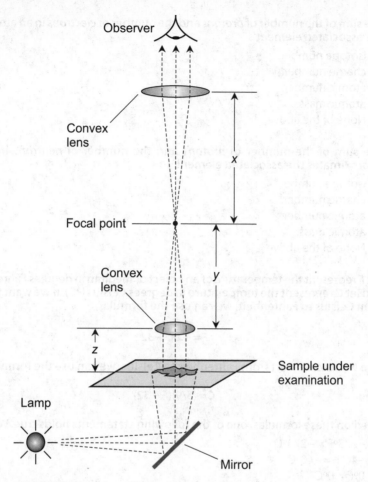

Observer

Convex
lens

x

Focal point

Convex
lens

y

z

Sample under
examination

Lamp

Mirror

FIGURE EXAM-15 · Illustration for Final Exam Question 94.

95. **How much light can a refracting telescope with a 20-in-diameter objective gather, compared with a refracting telescope having a 5-in-diameter objective?**

    A. 64 times as much
    B. 16 times as much
    C. 4 times as much
    D. Twice as much
    E. We need more information to calculate this ratio.

96. **In a direct-current (DC) electrical circuit, watts represent the equivalent of**

    A. volts times amperes.
    B. volts per ohm.
    C. amperes times ohms.
    D. ohms per ampere.
    E. ohms squared per ampere squared.

97. The sum of the number of *protons* and the number of *electrons* in an atom defines the associated element's
    A. isotope number.
    B. charge number.
    C. atomic number.
    D. atomic mass.
    E. None of the above

98. The sum of the number of *protons* and the number of *neutrons* in an atom approximates the associated element's
    A. isotope number.
    B. charge number.
    C. atomic number.
    D. atomic mass.
    E. None of the above

99. Let *F* represent the temperature of an object or medium in degrees Fahrenheit (°F), and let *C* represent the temperature in degrees Celsius (°C). If we want to convert from Celsius to Fahrenheit, we can use the formula

$$F = 1.8C + 32$$

If we want to convert from Fahrenheit to Celsius, we can use the formula

$$C = (5/9)(F - 32)$$

Based on these formulas, one of the following statements holds true. Which one?
    A. $-273°F = -273°C$
    B. $-40°F = -40°C$
    C. $0°F = 0°C$
    D. $+32°F = +32°C$
    E. $+212°F = +212°C$

100. In an alternating-current (AC) electromagnet, the magnetic field polarity
    A. remains constant.
    B. varies in intensity, but not in direction.
    C. periodically reverses direction.
    D. revolves in a circle.
    E. has no relevance, because AC electromagnets never work in the real world.

Answers to Quizzes, Tests, and Final Exam

Chapter 0	Chapter 2	Chapter 4	Chapter 6
1. C	1. C	1. C	1. C
2. A	2. D	2. A	2. A
3. D	3. C	3. B	3. B
4. A	4. A	4. B	4. B
5. B	5. C	5. A	5. D
6. B	6. D	6. A	6. D
7. B	7. C	7. B	7. C
8. D	8. D	8. C	8. B
9. C	9. A	9. D	9. C
10. B	10. B	10. D	10. A

Chapter 1	Chapter 3	Chapter 5	Test: Part I
1. D	1. A	1. B	1. C
2. C	2. C	2. A	2. D
3. A	3. D	3. B	3. E
4. D	4. B	4. A	4. A
5. B	5. B	5. C	5. A
6. C	6. A	6. C	6. C
7. D	7. B	7. C	7. B
8. A	8. C	8. A	8. B
9. C	9. D	9. C	9. C
10. C	10. D	10. D	10. E

11. A
12. A
13. C
14. E
15. D
16. B
17. C
18. D
19. D
20. E
21. D
22. A
23. B
24. C
25. A
26. E
27. B
28. B
29. B
30. E
31. C
32. B
33. E
34. E
35. B
36. D
37. B
38. C
39. D
40. A
41. A
42. B
43. D
44. E
45. A
46. C
47. D

48. B
49. E
50. A

Chapter 7
1. D
2. B
3. C
4. C
5. D
6. A
7. C
8. A
9. D
10. B

Chapter 8
1. B
2. C
3. A
4. D
5. D
6. D
7. D
8. B
9. B
10. A

Chapter 9
1. C
2. A
3. D
4. C
5. B
6. A
7. A
8. B
9. D
10. B

Chapter 10
1. B
2. D
3. B
4. B
5. C
6. C
7. A
8. A
9. A
10. B

Chapter 11
1. C
2. D
3. A
4. D
5. B
6. A
7. B
8. D
9. C
10. B

Test: Part II
1. D
2. D
3. E
4. B
5. A
6. D
7. C
8. B
9. C
10. B
11. A
12. E
13. E

14. C
15. D
16. D
17. A
18. C
19. C
20. A
21. B
22. C
23. D
24. E
25. D
26. A
27. D
28. B
29. E
30. C
31. C
32. A
33. D
34. B
35. E
36. C
37. D
38. E
39. D
40. E
41. C
42. D
43. A
44. E
45. E
46. B
47. C
48. C
49. A
50. A

Chapter 12
1. B
2. D
3. A
4. D
5. D
6. A
7. B
8. C
9. D
10. A

Chapter 13
1. C
2. D
3. A
4. D
5. A
6. C
7. B
8. B
9. B
10. A

Chapter 14
1. C
2. A
3. C
4. B
5. C
6. D
7. A
8. B
9. D
10. A

Chapter 15
1. A
2. C

3. B
4. B
5. C
6. D
7. A
8. A
9. D
10. D

Test: Part III
1. E
2. B
3. A
4. E
5. C
6. D
7. B
8. C
9. E
10. C
11. A
12. E
13. C
14. E
15. A
16. A
17. D
18. B
19. A
20. D
21. C
22. E
23. D
24. E
25. B
26. C
27. A
28. B

29. E
30. C
31. D
32. D
33. A
34. B
35. D
36. A
37. C
38. E
39. A
40. B
41. D
42. B
43. A
44. E
45. D
46. C
47. D
48. B
49. C
50. B

Final Exam
1. D
2. D
3. D
4. C
5. D
6. D
7. E
8. A
9. B
10. C
11. A
12. C
13. C
14. A

15. A
16. D
17. B
18. E
19. C
20. D
21. B
22. D
23. B
24. E
25. E
26. E
27. A
28. D
29. C
30. B
31. C
32. C
33. A
34. B
35. E
36. D
37. A
38. D
39. A
40. E
41. A
42. A
43. C
44. D
45. C
46. A
47. D
48. E
49. B
50. A
51. B
52. E

53. E
54. E
55. C
56. A
57. A
58. B
59. D
60. E
61. D
62. D
63. C
64. A
65. D
66. A
67. E
68. B
69. B
70. D
71. E
72. D
73. B
74. B
75. E
76. C
77. C
78. A
79. A
80. C
81. C
82. B
83. C
84. A
85. E
86. B
87. C
88. D
89. B
90. A

91. A
92. B
93. B
94. C
95. B
96. A
97. E
98. D
99. B
100. C

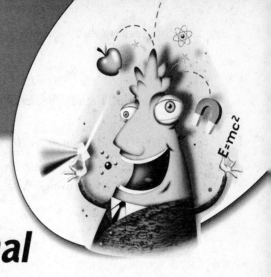

Suggested Additional Reading

Epstein, Lewis Carroll, *Thinking Physics: Understandable Practical Reality*. Insight Press, 2009.

Gautreau, Ronald and Savin, William, *Schaum's Outline of Modern Physics*. McGraw-Hill Text, 1999.

Giancoli, Douglas, *Physics: Principles with Applications*, 6th ed. Benjamin Cummings, 2004.

Gibilisco, Stan, *Teach Yourself Electricity and Electronics*, 4th ed. McGraw-Hill, 2006.

Gibilisco, Stan, *Advanced Physics Demystified*. McGraw-Hill, 2007.

Gibilisco, Stan, *Algebra Know-It-All*. McGraw-Hill, 2008.

Gibilisco, Stan, *Astronomy Demystified*. McGraw-Hill, 2002.

Gibilisco, Stan, *Mastering Technical Mathematics*, 3rd ed. McGraw-Hill, 2007.

Gibilisco, Stan, *Optics Demystified*. McGraw-Hill, 2009.

Gibilisco, Stan, *Pre-Calculus Know-It-All*. McGraw-Hill, 2009.

Gibilisco, Stan, *Technical Math Demystified*. McGraw-Hill, 2006.

Halpern, Alvin, *3000 Solved Problems in Physics*. McGraw-Hill Professional Publishing, 1988.

Halpern, Alvin, *Beginning Physics II: Waves, Electromagnetism, Optics and Modern Physics*. McGraw-Hill, 1998.

Hecht, Eugene, *Optics*, 4th ed. Addison Wesley, 2001.

Holzner, Steve, *Physics for Dummies*. For Dummies, 2004.

Kuhn, Karl F., *Basic Physics: A Self-Teaching Guide*, 2nd ed. John Wiley & Sons, Inc., 1996.

Orzel, Chad, *How to Teach Physics to Your Dog*. Scribner, 2009.

Seaborn, James B., *Understanding the Universe: An Introduction to Physics and Astrophysics*. Springer Verlag, 1997.

Index

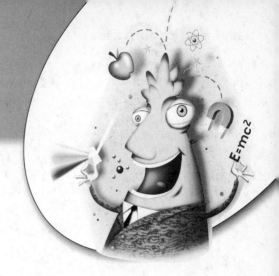

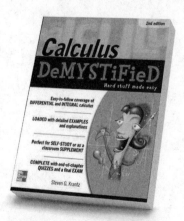